Ordering No.: SKIT

Supportability Toolkit

AF704595

Prepared by:

Reliability Analysis Center
201 Mill Street
Rome, NY 13440-6916

Under Contract to:

Army Contracting Agency, ITEC4
ATTN: SFCA-IT-B
2461 Eisenhower Avenue
Alexandria, VA 22331

Reliability Analysis Center

RAC is a DoD Information Analysis Center sponsored by the Defense Technical Information Center and operated by Alion Science and Technology, Rome, NY

ISBN 0-9741979-3-9

REPORT DOCUMENTATION PAGE

Form Approved
OMB No. 0704-0188

1. REPORT DATE (DD-MM-YYYY)	2. REPORT TYPE	3. DATES COVERED (From - To)
15-02-2005	Technical	03/1/2004 - 9/1/2004

4. TITLE AND SUBTITLE

Supportability Toolkit

5a. CONTRACT NUMBER: SP0700-97-D-4006

5b. GRANT NUMBER: None

5c. PROGRAM ELEMENT NUMBER: None

6. AUTHOR(S)

Professor Benjamin S. Blanchard and John W. Langford

5d. PROJECT NUMBER: None

5e. TASK NUMBER: None

5f. WORK UNIT NUMBER: None

7. PERFORMING ORGANIZATION NAME(S) AND ADDRESS(ES)

Reliability Analysis Center
201 Mill Street
Rome, NY 13440-6916

8. PERFORMING ORGANIZATION REPORT NUMBER

SKIT

9. SPONSORING/MONITORING AGENCY NAME(S) AND ADDRESS(ES)

Defense Technical Information Center (DTIC-AI)
8725 John J. Kingman Road, Suite 0944
Ft. Belvoir, VA 22060-6218

10. SPONSOR/MONITOR'S ACRONYM(S)

DTIC-AI

11. SPONSOR/MONITOR'S REPORT NUMBER(S)

SP0700-97-D-4006

12. DISTRIBUTION/AVAILABILITY STATEMENT

Approved for public release; distribution unlimited.

13. SUPPLEMENTARY NOTES

Hard copies available from the Reliability Analysis Center, 201 Mill Street, Rome, NY 13440-6916.
(Price: $50.00 U.S., $60.00 Non-U.S.).

14. ABSTRACT

The Supportability Toolkit was developed as a useful guide for government, government contractors, and commercial industry. It provides common sense and up-to-date information for planning and carrying out those activities needed to develop supportable systems and products and to support those products and systems in operation. It discusses the fundamental concept of supportability and related disciplines. It provides guidance on identifying and implementing the activities necessary to address supportability throughout the life cycle of an item. In today's environment, systems must be robust, reliable of high quality, supportable, cost effective, and responsive to the needs of the customer. This toolkit is dedicated to helping meet the objective of design for supportability.

15. SUBJECT TERMS

Supportability	Cost Effective	Reliability	Integrated Logistics Support
Maintainability	Quality	Safety	Logistics Support Analysis
System Effectiveness	Supportability Analysis	System Engineering	Supply Chain Management

16. SECURITY CLASSIFICATION OF: UNCLASSIFIED			17. LIMITATION OF ABSTRACT	18. NUMBER OF PAGES	19a. NAME OF RESPONSIBLE PERSON
a. REPORT	b. ABSTRACT	c. THIS PAGE			Ned Criscimagna
Unclassified	Unclassified	Unclassified	Unlimited	420	19b. TELEPHONE NUMBER (Include area code) 301-918-1526

Standard Form 298 (Rev. 8/98)
Prescribed by ANSI Std. Z39.18

The Reliability Analysis Center (RAC) is a Department of Defense Information Analysis Center sponsored by the Defense Technical Information Center, managed by the Air Force Research Laboratory (formerly Rome Laboratory), and operated by Alion Science and Technology Corporation (formerly IIT Research Institute). RAC is chartered to collect, analyze and disseminate reliability, maintainability and quality information pertaining to systems and products, as well as the components used in them. The RAC addresses both military and commercial perspectives and includes such reliability related topics as testability, Total Quality Management and lifetime extension.

The data contained in the RAC databases is collected on a continuous basis from a broad range of sources, including testing laboratories, device and equipment manufacturers, government laboratories and equipment users (government and industry). Automatic distribution lists, voluntary data submittals and field failure reporting systems supplement an intensive data solicitation program. Users of RAC are encouraged to submit their reliability, maintainability and quality data to enhance these data collection efforts.

RAC publishes documents for its users in a variety of formats and subject areas. While most are intended to meet the needs of reliability practitioners, many are also targeted to managers and designers. RAC also offers reliability consulting, training, and responses to technical and bibliographic inquiries.

REQUESTS FOR TECHNICAL ASSISTANCE AND INFORMATION ON AVAILABLE RAC SERVICES AND PUBLICATIONS MAY BE DIRECTED TO:

Reliability Analysis Center
201 Mill Street
Rome, NY 13440-6916

General Information: (888) RAC-USER
(888) 722-8737
Product Ordering: (800) 526-4802
Technical Inquiries: (315) 337-9933
TeleFax: (315) 337-9932
E-Mail: rac@alionscience.com
Internet: http://rac.alionscience.com

ALL OTHER REQUESTS SHOULD BE DIRECTED TO:

Air Force Research Laboratory
AFRL – Information Directorate
Attn: R. Hyle
525 Brooks Road
Rome, NY 13441-4505

Telephone: (315) 330-4857
DSN: 587-4857
TeleFax: (315) 330-7647
E-Mail: Richard.Hyle@rl.af.mil

PREFACE

The *Supportability Toolkit* is the fourth in a series that started with the *Reliability Toolkit: Commercial Practices Edition*. The others in the series are the *Maintainability Toolkit* and the *Quality Toolkit*. All feature a style and format that make them excellent reference documents for engineers, logisticians, and technical managers.

I gratefully acknowledge the work of Professor Benjamin S. Blanchard and Mr. John W. Langford in authoring this toolkit. Their experience, expertise, and professionalism made the development of the toolkit a learning experience for me. More importantly, it resulted in a genuinely useful and practical guide for those working in supportability and related fields. On behalf of the RAC, I thank them for their work.

Developing a document of this magnitude requires the contributions of "behind-the-scene" individuals. Peter Tabbagh worked many hours helping to convert analog figures and tables to digital format and proofing the technical content of the book. Pamela Kinstle, as she has done for all our products, prepared the final manuscripts, formatted the document, often having to redo figures, and did her typically thorough copy edit. Any errors that may remain are solely my responsibility.

We dedicate this *Supportability Toolkit* to the many dedicated people who work to ensure that our consumer, industrial, and government products and systems are inherently supportable and supported throughout their life cycle. In the words of the authors, "... a greater depth of material coverage could have been provided in any one of the topics discussed within. However, it is hoped that what has been included provides a good overview of the subject, and will be helpful in the implementation of your day-to-day activities pertaining to the *design for supportability*."

Ned H. Criscimagna

This page intentionally left blank

TABLE OF CONTENTS

TABLE OF CONTENTS (CONT'D)

TABLE OF CONTENTS (CONT'D)

TABLE OF CONTENTS (CONT'D)

TABLE OF CONTENTS (CONT'D)

List of Figures

List of Figures (Cont'd)

Page

LIST OF FIGURES (CONT'D)

List of Figures (Cont'd)

List of Tables

Section 1: Introduction to Supportability

Insight

In today's environment, there is an ever-increasing need to develop and produce *systems* that are robust in nature, reliable and of high quality, supportable, cost-effective, and responsive to the needs of the customer in a satisfactory manner. Of particular significance is the need to provide for the effective and efficient support of a given system throughout its life cycle, including all of the related activities that are required to realize this objective. It is essential that a comprehensive and well-integrated logistics and maintenance support infrastructure be developed, that it be considered as a critical element of the system in question, and that it be addressed from the beginning as new systems are developed and/or re-engineered systems are introduced. This infrastructure should include all logistics and system support activities, and the requirements for such must be an inherent consideration in the system design and development process. This objective is basically accomplished through the *design for system supportability.*

The purpose of this Toolkit is to introduce the reader to the subject of *supportability.* Supportability is a "design characteristic" that must be addressed from the beginning, while the ultimate system logistics and maintenance support infrastructure is the result of design. If one is to influence this very critical element of a system, in terms of producing a cost-effective product output, then it becomes essential that the *design for supportability* be a major thrust throughout the system development process. This Toolkit is dedicated to this very important objective.

As a start, it is important to gain some familiarity with a few general principles and concepts. More specifically, it is intended to deal with such questions as: *What is supportability? How can it be measured? Why is supportability important? How can the principles and concepts of supportability be applied? When can these principles be applied?* The objective of this section is to respond to these basic questions through the introduction of a few key terms and definitions, an abbreviated discussion of some closely-related concepts, and ending with an overall management approach dedicated to ensuring that supportability is an inherent consideration in all activities throughout the system life cycle.

While the principles and concepts associated with supportability, and much of the literature on the subject, have been primarily applied to large-scale military defense systems, these same principles and concepts can be equally applicable in the design and development of any type of system, whether a communications system, a chemical processing plant, an electric power distribution system, a health care system, a production system, a transportation system, or any comparable type of functional entity. Thus, the reader should review the material through this Toolkit from a "generic" approach, versus being applicable to only defense systems.

Topic 1.1: The Current Environment

A major prerequisite to determining the logistics and system support requirements, and the specific supportability requirements in design, for any system is dependent on a good understanding of the overall environment; i.e., the geographical location where the system is likely to be deployed and utilized, the nature and culture of the operating agency (organization), the availability of appropriate technologies, the system procurement and acquisition process, the corporate or political structure, and so on. Additionally, it should be recognized that we are operating in a highly dynamic world, and the need for agility and flexibility predominates.

Although individual perceptions as to today's challenges will differ depending on personal experiences and observations, there are a number of trends that appear to be significant. Some of these trends are noted as follows:

1. *Constantly changing requirements.* The requirements for new systems are constantly changing owing to the dynamic conditions worldwide, changes in mission thrusts and priorities, and to the introduction of new technologies on an evolving basis. Thus, there is a need for an *open-architecture* approach in the design of systems, and for a highly flexible (agile) logistics and maintenance support capability.

2. *More emphasis on systems.* There is greater emphasis on total *systems* versus the *components* of systems. The system needs to be addressed in total, and throughout its entire life cycle, to ensure that the necessary functions are being accomplished in an effective and efficient manner. Thus, the logistics and maintenance support infrastructure must be considered as a major element of the system, must be in place and reliable, and must be readily available to support the prime mission-related elements of the system when needed. The requirements pertaining to the *design for supportability* must address both the prime mission-related elements of the system and the elements of the logistics and maintenance support infrastructure on a concurrent basis.

3. *Increasing system complexities.* It appears that the structures of many systems are becoming more complex as new technologies are introduced and evolve. The system (to include the logistics and maintenance support infrastructure) must be designed such that changes can be incorporated quickly, efficiently, and without significantly affecting the overall configuration of the system. Given these constraints, the logistics and maintenance support infrastructure must be designed to support the added complexities.

4. *Extended system life cycles -- shorter technology life cycles.* The life cycles of many current systems are being extended for one reason or another while, at the same time, the life cycles of most technologies are becoming relatively shorter (due to obsolescence). As for the situation mentioned in items 1 and 3, it is necessary to design systems such that a new technology can be

incorporated easily and efficiently. At the same time, the logistics and maintenance support infrastructure must be designed to be responsive, and the duration (life cycle) of the support capability is likely to be longer due to the extended system life cycles.

5. *Greater utilization of commercial off-the-shelf (COTS) products.* With current defense system goals pertaining to lower initial costs and shorter and more efficient procurement and acquisition cycles, there has been more emphasis on the utilization of best commercial practices, standard processes, and commercial off-the-shelf (COTS) equipment and software. As a result, there is a greater need for a good definition of requirements at the outset, and there is a greater emphasis on the design of systems (and their major subsystems) versus the design of components. Further, much of the required logistics and maintenance support activity has shifted from a major producer to one or more suppliers. This shift, in turn, has increased the complexity of the overall logistics and maintenance support network, with more organizations participating, as well as some added challenges in determining the detailed support requirements for many of the COTS items being utilized in various military system design configurations.

6. *Increasing globalization.* The world is shrinking (as they say) and there is more trading with and dependency on different nationalities (and manufacturers) throughout the world. This has been facilitated thanks to the introduction of rapid and improved communication practices, the availability of quicker and more efficient packaging and transportation methods, the application of electronic commerce (EC) methods for expediting the accomplishment of procurement and related processes, and so on. At the same time, the logistics and support configuration must be responsive to the many challenges pertaining to rapid and reliable communications, short transportation times and safe/secure transportation routes, and quick turnaround times in support of maintenance requirements.

7. *More outsourcing.* There is more outsourcing and the procurement of COTS items (equipment, software, processes, services) from external sources of supply than ever before. Thus, there are more suppliers associated with almost any given program. Consequently, there must be greater emphasis on the early definition and allocation of system-level requirements, the development of a good and complete set of specifications, and a closely coordinated and integrated level of activity throughout the system development and acquisition process. At the same time, a closely and well-integrated logistics and maintenance support capability must be developed and operational when required. This can best be accomplished through the effective implementation of the system engineering process and the proper specification of the *design for supportability* requirements from the beginning.

8. *Greater international competition.* Along with the noted trends toward increasing globalization and more outsourcing, there is more international competition than ever before, owing not only to improvements in

communications and transportation methods but to the greater utilization of COTS items and the establishment of effective partnerships worldwide. A major goal is, of course, to deliver in a short time frame a product and/or service that is reliable and high quality, cost-effective, and with complete customer satisfaction in mind. This goal requires a highly responsive, effective, and efficient logistics and maintenance support infrastructure.

9. *Higher overall life-cycle costs.* In general, experience has indicated that the life-cycle costs of many current systems are increasing. Whereas much emphasis always has been placed on minimizing the costs associated with the procurement and acquisition of systems, relatively little attention until recent years has been dedicated to the costs of system operation and support. In designing systems, one needs to view all decisions in the context of *total* cost to properly assess the risks associated with the decision(s) in question. As the logistics and maintenance support infrastructure is a major element of the system, and often represents a high-cost contributor later on, the various alternative approaches in the design of such must be justified on the basis of total life-cycle cost. Thus, the design for supportability must consider not only the "technical" characteristics of design but the "economical" aspects as well. It should be emphasized that these trends are all interrelated and need to be addressed as an integrated set when determining the requirements for systems and for the logistics and maintenance support infrastructure necessary to support those systems. Further, an awareness of these issues associated with the environment is essential in the *design for supportability.*

Although some of the foregoing and related trends have evolved over time, the tendency is to ignore the changes that have taken place and continue with a business-as-usual approach by implementing past practices, many of which tend to inhibit innovation and growth. Because the operating environment has undergone a major transition in recent years, the requirements for logistics and system support have also undergone significant changes, and it is anticipated that such changes will continue to evolve. In any event, it is the recognition of these changes that forms the basis in determining the requirements in the *design for system supportability*, and for much of the material discussed throughout this Toolkit.

Topic 1.2: Definition of Supportability

According the MIL-HDBK-502, supportability may be defined as:

> *The degree to which system design characteristics and planned logistics resources meet system peacetime and wartime requirements. Supportability is the capability of a total system design to support operations and readiness needs throughout the system's service life at an affordable cost. It provides a means of assessing the suitability of a total system design for a set of operational needs within the intended operations and support environment (including cost constraints). Supportability characteristics may be specified in terms of many different performance measures.*[1]

Referring to a published DoD Guide, supportability can be described as:

> *The inherent quality of a system – including design, technical support data, and maintenance procedures – to facilitate detection, isolation, and timely repair/replacement of system anomalies. This includes factors such as diagnostics, prognostics, real-time maintenance data collection, "design for support" and "support the design" aspects, corrosion protection and mitigation, reduced logistics footprint, and other factors that contribute to optimum environment for developing and sustaining a stable, operating system.*[2]

Supportability is a *design characteristic*, with the appropriate attributes built-into (inherent within) the design such that the ultimate system configuration can be both effectively and efficiently supported throughout its planned life cycle. Such characteristics, to be incorporated in both the design of the prime mission-related elements of the system and the design of the elements of the logistics and maintenance support infrastructure, may include such features as standardization and commonality, functional packaging and modularity, interchangeability, the utilization of open-system standards, an effective self-test and diagnostic capability, direct accessibility, the application of reliable components, good labeling, and so on.

Supportability can be *specified.* Referring to MIL-HDBK-502, *supportability factors are integral elements of program performance specifications. However, support requirements are not to be stated as distinct logistics elements, but instead as performance requirements that relate to a system's operational effectiveness,*

---

1 MIL-HDBK-502, "Department of Defense Handbook on Acquisition Logistics," Department of Defense, Washington, DC, May 1997, pages 4-14.

2 "Designing and Assessing Supportability in DOD Weapon Systems: A Guide to Increased Reliability and Reduced Logistics Footprint," Prepared by the Office of Secretary of Defense, October 24, 2003, page 14.

operational suitability, and life-cycle cost reduction.[3] *The objective is to mesh supportability requirements with system operational requirements from the beginning.*

Supportability can be measured. Such factors as operational availability (A_o), system operational effectiveness (SOE), mission reliability, mean-time-between-maintenance (MTBM), maintenance downtime (MDT), logistics effectiveness (LE), information system response time, maintenance turnaround time (TAT), facility utilization, overall equipment effectiveness (OEE), life-cycle cost (LCC), and so on, constitute different measures of some aspect of supportability. The appropriate measures, as applied to any given system, must (of course) be "tailored" to the specific mission(s) of that system. Further, the measures that are applicable must be properly specified as requirements from the beginning, and then the system must be evaluated later in terms of these requirements to ensure ultimate compliance.

[3] MIL-HDBK-502, "Department of Defense Handbook on Acquisition Logistics," Department of Defense, Washington, DC, May 1997, Sections 5.1 and 6.

Topic 1.3: Definition of Logistics

Historically, the principles and concepts of logistics stem from specific facets of activity within both the commercial and the defense sectors. In the *commercial* sector, where the emphasis is on the business-oriented activities associated primarily with the distribution of relatively small products (i.e., consumables), logistics can be defined as:

> That part of the supply chain process that plans, implements, and controls the efficient, effective, and reverse flow and storage of goods, services, and related information between the point of origin and the point of consumption in order to meet customers' requirements.[4]

In this context, logistics activities include (1) the identification and management of suppliers, procurement and order processing, and the physical supply of materials and/or services from the sources of supply to the manufacturer or producer; (2) the materials handling and inventory management of materials/services during and throughout the manufacturing process; and (3) the subsequent transportation, warehousing, and physical distribution of products from the manufacturer to the consumer (i.e., customer). Referring to Figure 1.3-1, there is a *forward flow* of activities in the supply chain, evolving from the supplier(s) to the ultimate user of the product. This forward flow is amplified in Figure 1.3-2, and includes activities associated primarily with blocks 1, 2, 3, 4, 6, and 7 in the figure. These activities, primarily related to the manufacturer/producer of goods, are further identified in Figure 1.3-3.

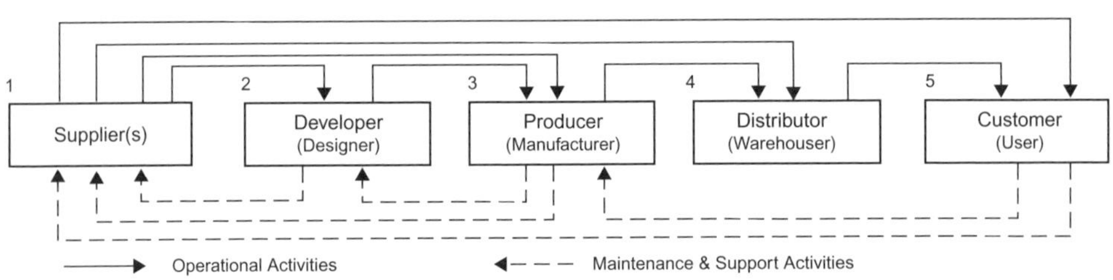

Figure 1.3-1. Flow of Activities – The Supply Chain

[4] This definition was developed by the Council of Logistics Management (CLM), 2805 Butterfield Road, Oak Brook, IL. Refer to the CLM web site, www.clm1.org, "Glossary of Terms."

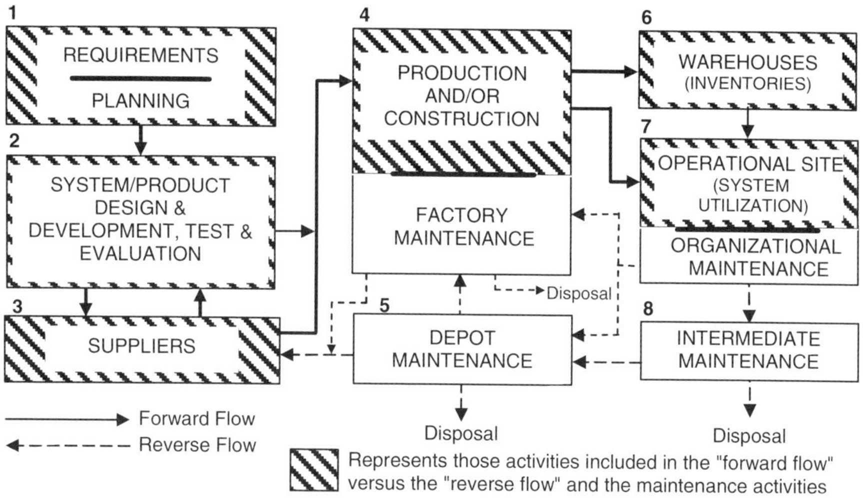

Figure 1.3-2. System Operational/Maintenance Flow

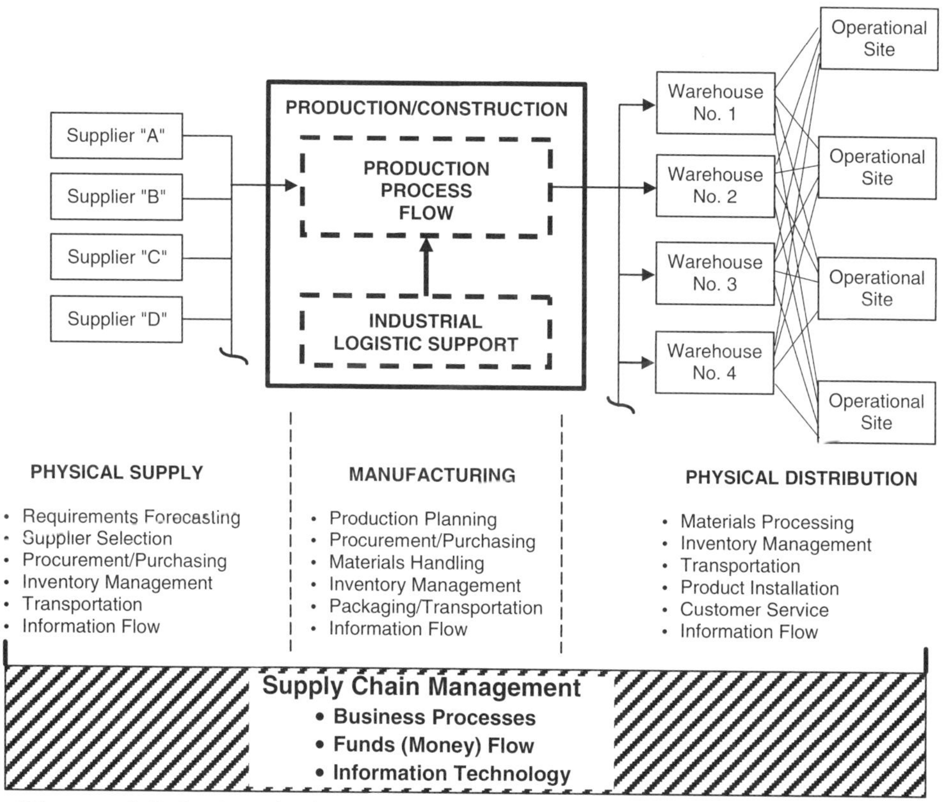

Figure 1.3-3. Logistics Activities in Production/Construction

Logistics in the commercial sector has traditionally been oriented toward managing the physical flow of materials and products among organizations. Activities such as transportation and warehousing have been available to ensure that the movement of products has been continuous and reliable. Purchasing departments have been responsible for the procurement and acquisition of materials, and marketing groups have been responsible for providing information to the customers both before and after each transaction. In essence, the spectrum of logistics has (in many instances) included a number of different organizational elements, working toward a given objective and yet operating somewhat independently.

With the advent of electronic commerce (EC) methods and advanced information technology (IT), the role of logistics has increased in scope and responsibility. The development of bar coding methods, radio frequency identification (RFID) tags, global positioning systems (GPS), and electronic data interchange (EDI) has enabled the rapid and efficient transfer of information on product/material flows. The role of logistics has taken on a more comprehensive and integrated approach; it has grown to include such functions as information technology, marketing, sales, and finance; that is, the economic, finance, and business issues in support of the physical flow of materials and products. At the same time, the current trends toward more outsourcing, increased globalization, and greater international competition have created a need for the establishment of partnerships and coalitions, and for the further expansion of the scope of logistics. These various developments have more recently evolved into the concepts associated with the *supply chain (SC)* and *supply chain management (SCM).*

While there may be slight variations, a good definition of SCM (for our purposes) is as follows.

> The systemic, strategic coordination of the traditional business functions and the tactics across these business functions within a particular company and across businesses within the supply chain, for the purposes of improving long-term performance of the individual companies and supply chains as a whole.[5]

The SC basically includes all those activities associated with inbound logistics (the flow of materials and services from the supplier to the producer/manufacturer), the *material flows* within the factory, and the *outbound logistics* (flow of materials, products, and services from the factory to the customer/consumer/user). These activities are illustrated in Figures 1.3-1, 1.3-2, and 1.3-3, and are critical in the overall development and support of a system throughout its life cycle. The SC may include many different categories of suppliers located throughout the world, one or more producers or manufacturers of products, and consumers operating globally. The formalization of a supply chain is realized through the establishment of *partnerships* or *coalitions*, who work together to service the customer in the best way possible.

[5] Council of Logistics Management (CLM), Oak Brook, IL. Refer to www.clm1.org.

SCM pertains to the *management* of the supply chain, or a group of supply chains, with the objective of providing the required customer services, both effectively and efficiently. It requires a highly integrated approach, employing the appropriate resources (e.g., transportation, warehousing, inventory control, information) and implementing the necessary business processes to ensure complete satisfaction. The objective is to accomplish the activities shown in Figure 1.3-3, with a total business perspective in mind. The results from this overall activity can, in turn, have a significant impact on the subsequent support of a system over its life cycle.

In the *defense* sector, logistics has evolved primarily through the concept of *integrated logistics support (ILS),* which was formally developed in the mid 1960s. At that time, ILS was defined as:

> A composite of all support considerations necessary to assure the effective and economical support of a system or equipment at all levels of maintenance for its programmed life cycle. It is an integral part of all other aspects of system acquisition and operation.[6]

ILS, as initially envisioned, included a "life-cycle approach to the planning, development, acquisition, and operation of systems and equipment to maximize readiness and optimize costs." The principle elements of ILS included the "system design/support interface, reliability and maintainability, maintenance planning, support and test equipment, supply support, transportation and handling, technical data, facilities, personnel and training, logistics support resource funds, logistic support management information, and contractor support services." The emphasis here is on logistics as it pertains to **systems** (versus the supply and distribution of smaller components and consumables emphasized in the commercial sector) and on the effective and efficient **sustaining support** of these systems throughout their planned life cycle. It includes not only the aspects of system maintenance and support of systems during their period of utilization but the **design** of systems for *reliability, maintainability,* and *supportability.*

The principles and concepts of ILS were developed further throughout the 1970s, 1980s, and 1990s, and the definition of ILS was expanded to constitute a:

> Disciplined, unified, and iterative approach to the management and technical activities necessary to (1) integrate support considerations into system and equipment design; (2) develop support requirements that are related consistently to readiness objectives, to design, and to each other; (3) acquire the required support; and (4) provide the required support during the operational phase at minimum cost.[7]

[6] 4100.35G, *Integrated Logistics Support Planning Guide for DoD Systems and Equipment,* Department of Defense, Washington, DC, 1967.

[7] DSMC, *Integrated Support Guide,* Defense Systems Management College, Fort Belvoir, VA, 1994 (DSMC is currently recognized as a major element within the Defense Acquisition University).

The emphasis here is on developing and implementing an effective and efficient system maintenance and support infrastructure and all the activities related thereto. As illustrated in Figure 1.3-4, the required support for an operational system may be provided at different levels depending on the overall system mission requirements, the complexity of the system, its reliability and effectiveness, available resources, cost, etc. Additionally, the various elements of logistics, illustrated in Figure 1.3-5, may be applicable at each of the levels shown in Figure 1.3-4.

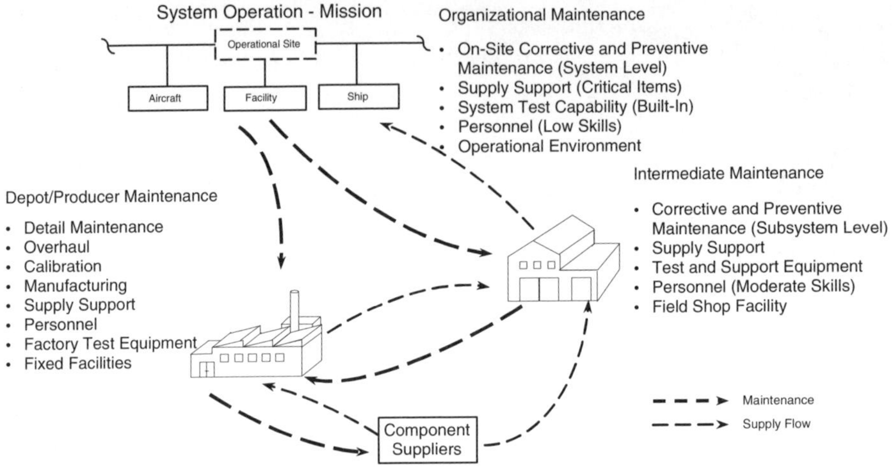

Figure 1.3-4. System Maintenance and Support Infrastructure

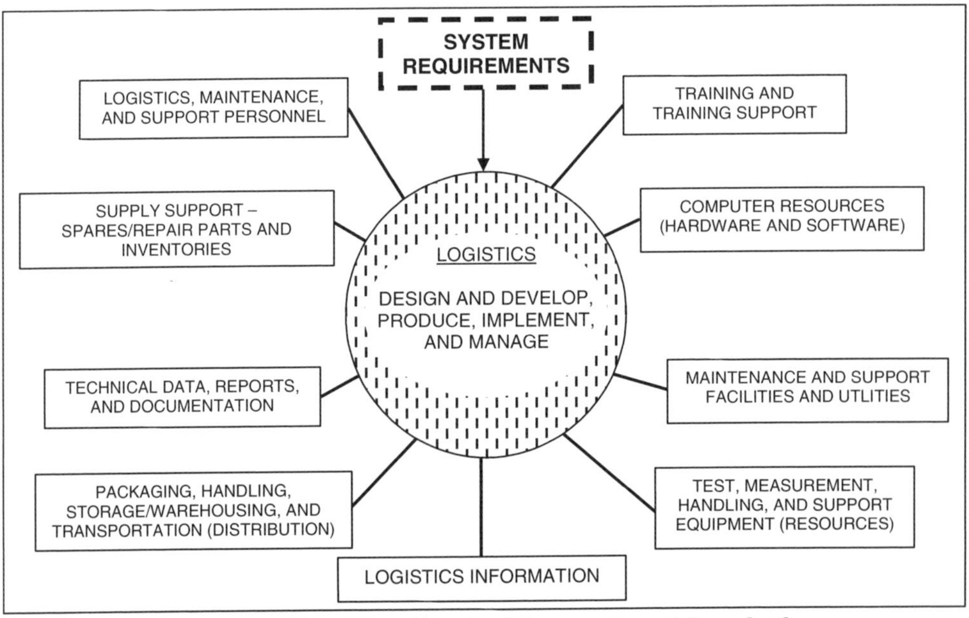

Figure 1.3-5. The Basic Elements of Logistics

Inherent within this definition is the current requirement dealing with the design for supportability. This requirement pertains to the degree to which a system can be effectively supported, both in terms of the built-in design characteristics of the prime mission-related elements of the system and the design characteristics of the overall support infrastructure and the logistics elements shown in Figures 1.3-4 and 1.3-5. The emphasis here is on *design*, the first two items in the preceding definition.

More recently, there has been increased emphasis on logistics (and the "design for supportability") early in the system development process through introduction of the concept of *acquisition logistics*, which can be defined as:

> A multi-functional technical management discipline associated with the design, development, test, production, fielding, sustainment, and improvement modifications of cost-effective systems that achieve the user's peacetime and wartime readiness requirements. The principal objectives of acquisition logistics are to ensure that support considerations are an integral part of the system's design requirements, that the system can be cost-effectively supported throughout its life cycle, and that the infrastructure elements necessary to the initial fielding and operational support of the system are identified, developed, and acquired.[8]

To further stress the need for addressing logistics early in the design process (i.e., the *design for supportability*), the Department of Defense (DoD) recently introduced the concept of *performance-based logistics (PBL).*[9] The objective is to establish some specific metrics associated with the maintenance and support infrastructure, and to include these as quantitative *design-to requirements in the appropriate specifications. Such requirements may be covered through the initial specification of performance requirements at the system level (e.g., operational availability, system effectiveness, life-cycle cost, logistics effectiveness), and then allocated downward to the various elements of the system (e.g., test equipment reliability, transportation times, supply support effectiveness, facility utilization, personnel effectiveness, and cost). The important issue here is to view these logistics-related measures as system* ***performance factors.***

In covering the scope of logistics, one needs to consider the entire spectrum of activity in this area, including both the commercial and defense approaches. Referring to Figure 1.3-2, there is the forward flow commencing with the definition of requirements

[8] MIL-HDBK-502, *Department of Defense Handbook on Acquisition Logistics*, Department of Defense, Washington, DC, 1997, page 4-1. Also, refer to DSMC, *Acquisition Logistics Support Guide,* 3rd Ed., DSMC, Fort Belvoir, VA, 1997.

[9] *Designing and Assessing Supportability in DOD Weapon Systems: A Guide to Increased Reliability and Reduced Logistics Footprint*, Prepared by the Office of Secretary of Defense, 2003. Also, refer to DoD 5000.2-R, *Mandatory Procedures for Major Defense Acquisition Programs (MDAPS) and Major Automated Information System (MAIS) Acquisition Programs*, Department of Defense, Washington, DC, 2002.

and initial planning (block 1), to system design and development (block 2), manufacturing and production (blocks 3 and 4), and the distribution of products to warehouses and operational sites (blocks 6 and 7). At the same time, there is a *reverse flow* beginning at the user's operational site (block 7) and back to various levels of maintenance (blocks 5, 7, and 8), the original producer (blocks 3 and 4), and/ or a third-party maintenance provider (block 3). Both "flows" are critical in the development and sustaining support of systems.

Finally, it should be emphasized that this entire spectrum of logistics activity must be considered as an integrated entity and be addressed early in the design process. While the requirements pertaining to the *design for supportability* have primarily been related to the infrastructure shown in Figure 1.3-4, it is the entire two-way flow process shown in Figures 1.3-1 and 1.3-2 that must be addressed in the design process and covered through the specification of PBL factors.

Topic 1.4: Supportability Requirements in the System Life Cycle

Logistics (and the design for supportability), as viewed from a total system's perspective, includes activities throughout each phase of the system life cycle. While there may be some variations, the "model" illustrated in Figure 1.4-1 provides the general framework for discussion throughout this Toolkit.[10]

Initially, the identification of need, the conduct of a feasibility analysis, development of system operational requirements and the maintenance support concept, and early planning lead to the definition of logistics and supportability requirements. PBL factors are identified and specified through requirements analysis, which is an inherent step within the system engineering process.[11] This leads to the functional analysis and allocation of requirements. The system is defined in "functional" terms and system-level quantitative design-to requirements are allocated to the applicable elements of the system; i.e., down to the equipment, software, people, facilities, data, etc. Quantitative supportability requirements, established at the system-level, are allocated or apportioned, down to the applicable elements of the system (to include the mission-related elements and the elements of the logistics and maintenance support infrastructure). The applicable "design-to" supportability requirements are included in the appropriate specification(s) as an **input to design. Referring to Figure 1.4-1, these activities are inherent within the Concept Refinement and Technology Development Phases.**

Subsequently, there is an on-going iterative process of synthesis, analysis, and design optimization. *Synthesis* refers to the combining and structuring of system components in such a way as to represent a feasible system configuration. The basic requirements for the system have been established, some preliminary trade-off studies have been completed, and a baseline configuration needs to be developed. Synthesis is *design.* Initially, synthesis is employed to develop preliminary concepts and to establish basic relationships among various elements and components of the system. Later, when sufficient functional definition and decomposition have occurred, synthesis is used to further define the *hows* in response to the *what* requirements. Synthesis involves the selection of a configuration that could be representative of the form that the system will ultimately take, although a final configuration is certainly not to be assumed at this point.

---

[10] *Designing and Assessing Supportability in DOD Weapon Systems: A Guide to Increased Reliability and Reduced Logistics Footprint,* OSD, October 24, 2003, page 5.

[11] MIL-HDBK-502, "Department of Defense Handbook on Acquisition Logistics," Department of Defense, Washington, DC, May 1977, pages 4-13.

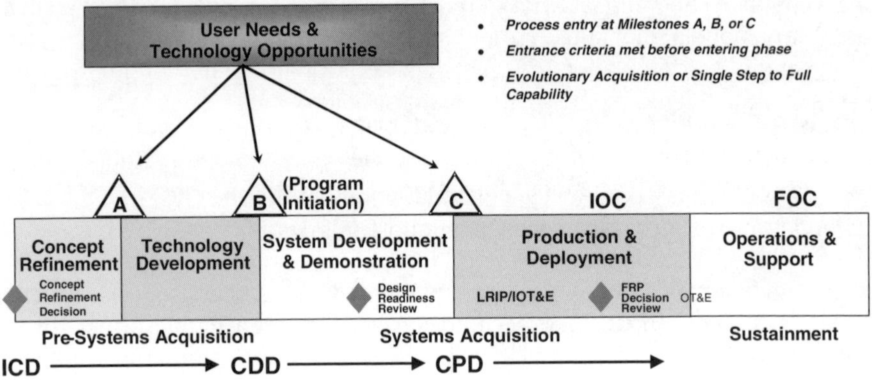

ICD: Initial Capabilities Document
CDD: Capability Development Document
CPD: Capabilities Production Document
IOC: Initial Operating Capability
FOC: Full Operating Capability
LRIP: Low Rate Initial Production
OT&E: Operational Test and Evaluation
FRP: Full Rate Production
IOT&E: Initial Operational Test and Evaluation

Figure 1.4-1. DoD 5000 Defense Acquisition Management Framework

(Source: *Designing and Assessing Supportability in DOD Weapon Systems: A Guide to Increased Reliability and Reduced Logistics Footprint*,[9] Figure 1.1, page 5)

Within this iterative process of synthesis and analysis, there are numerous alternative design approaches that must be evaluated with regard to supportability requirements. This is basically accomplished through application of the supportability analysis (SA), utilizing various combinations of analytical methods/tools (e.g., reliability and maintainability predictions, maintenance task analyses, level-of-repair analyses, life-cycle cost analyses, failure mode, effects, and criticality analyses, etc.). The supportability analysis (SA) constitutes an on-going analytical process, which supports the day-to-day design review and evaluation effort in system development, and which ultimately leads to the design of the logistics and maintenance support infrastructure and its elements. This continuous process of defining (and refining) supportability requirements is accomplished throughout the Technology Development and the System Development and Demonstration Phases identified in Figure 1.4-1.

With a proposed system configuration established, the next step includes that of *validation.* This question is – *will the system configuration as designed (to include its support capability) completely fulfill the requirements initially specified?* In response, a comprehensive system test and evaluation effort must be established for the purposes of assessment and follow-on corrective-action as required. Such an effort may commence through the application of selected analytical methods (e.g., simulation) to measure and evaluate the various elements of the system and its characteristics, and subsequently continue with test and evaluation through the use of laboratory models, mockups, prototypes, production/operational system components, and/or various combinations of these as appropriate. The various elements of the logistics and maintenance support infrastructure must be evaluated, in terms of compliance with the initially specified PBL requirements, and areas of deficiency must be identified and result in some form of corrective action. Referring to Figure 1.4-1, the test and evaluation effort is continuous throughout the System Development and Demonstration Phase, the Production and Deployment Phase, and during the initial stages of the Operations and Support Phase.

Given that the proposed system configuration is compliant with all of the initially specified requirements (to include complete customer satisfaction), there is an on-going effort of *continuous product and process improvement (CPPI)* that should be implemented. The questions are -- *can the system be improved in terms of effectiveness and cost? Can the reliability of the system be increased? Can the system life-cycle cost be decreased? Can the logistics footprint be reduced?* Through the implementation of an on-going data collection, analysis, and feedback capability, one can assess and identify areas where selected design changes may be considered for the purposes of improvement. Such an evaluation effort should cover not only the equipment, software, and specific elements of the system, but also the processes required for system operations and support. This level of activity is applicable throughout the System Operations and Support Phase. Additionally, the processes required and applicable in the follow-on System Retirement and Material Recycling/ Disposal Phase (not shown in the figure) should also be evaluated for improvement.

The objective for including this topic, in this first section of the Toolkit, is to emphasize that logistics and supportability activities are life-cycle-oriented, and are an inherent requirement in each of the program phases in Figure 1.4-1. This objective will become even more evident as one proceeds through the sections that follow.

Topic 1.5: Performance-Based Logistics (PBL) and Supportability

The advent of Performance-Based Logistics (PBL) evolved with the objective of (1) emphasizing the importance of considering logistics and supportability factors in the same context with other system **performance** factors, and (2) specifying these factors as part of the early requirements analysis process as specific "design-to" criteria being defined during the conceptual design phase. More specifically, PBL is:

> DOD's preferred approach for implementing product support. PBL is a strategy for weapon system life cycle support that brings higher levels of system readiness through efficient management and direct accountability. It describes performance goals for a weapon system's readiness, and encourages the creation of incentives for attaining the goals through clear lines of authority and responsibility.[12]

Referring to Figure 1.5-1, one can see that there are many different categories and levels of performance measures that can be applied in the area of logistics and supportability. The emphasis through application of the PBL objective is to identify those technical performance measures (TPMs) that could significantly impact the overall supportability of the system and, thus, the system's ability to directly accomplish its mission; e.g., cost effectiveness (CE), system effectiveness (SE), operational availability (A_o), mean time between maintenance (MTBM), maintenance downtime (MDT), life-cycle cost (LCC), and so on. Each of these system-level measures will, of course, be heavily dependent on those supporting measures at the sub-system level and below. A few specific examples of PBL factors, which could be specified as "design-to" requirements, are noted as follows:

1. The operational availability (A_o) for the system shall be 98%, or greater.
2. The system mean time between maintenance (MTBM) shall be 1,000 hours, or greater.
3. The mean downtime (MDT) for the system shall be 4 hours, or less.
4. The manufacturing process shall be capable of producing *x* products, in *y* time, and at a *z* unit cost, with a defect rate not to exceed 1%.
5. The response time for the logistics and maintenance support infrastructure shall not exceed 4 hours.

[12] "Product Support: A Program Manager's Guide to Buying Performance," Department of Defense, Washington, DC, November 2001.

6. The process time for removing an obsolete item from the operational inventory shall not exceed 12 hours, the cost per item processed shall be *x* dollars or less, and the resulting environmental degradation shall be zero.
7. The system shall be designed such that at least 95% of all of its components can be recyclable for other applications.
8. The unit life-cycle cost for the system shall not exceed *x* dollars, based on a deployment of *y* quantities and a 10-year planned life cycle.

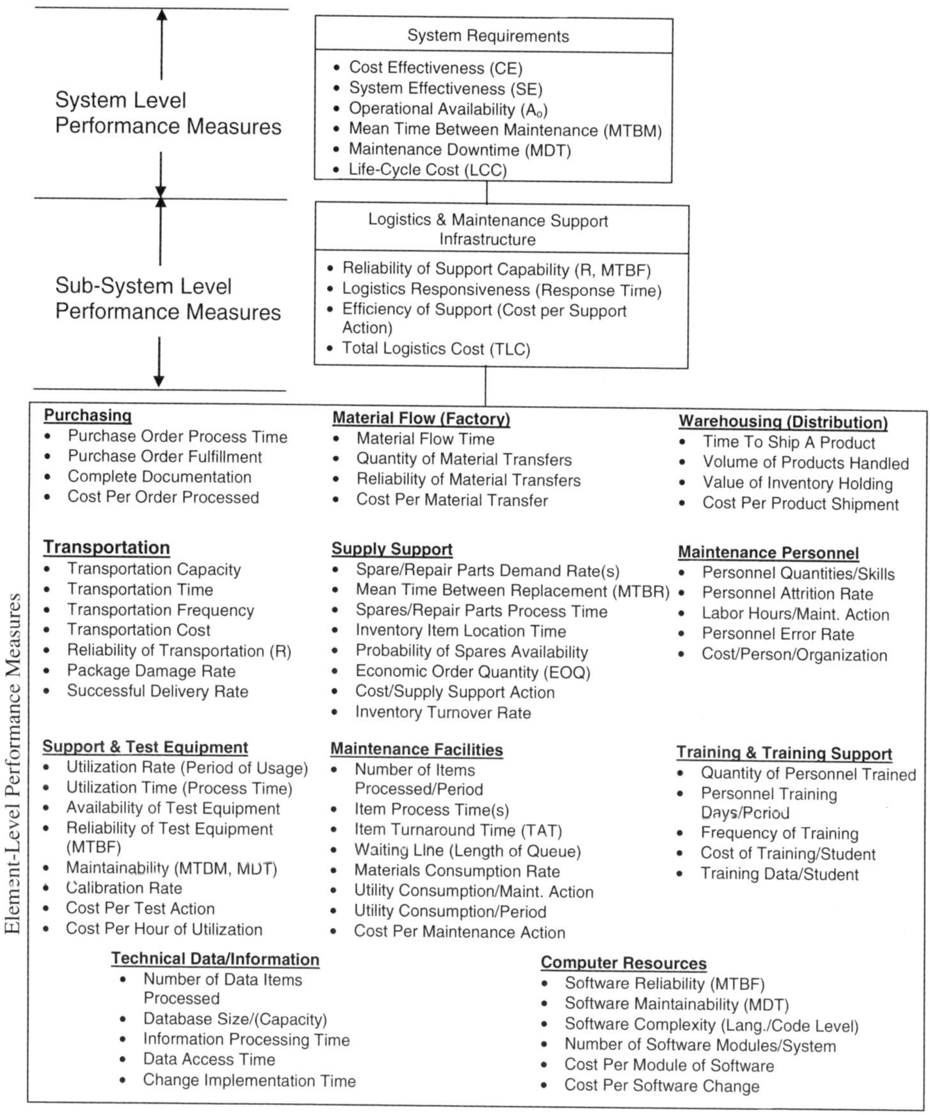

Figure 1.5-1. Selected Technical Performance Measures for Logistics and Maintenance Support Infrastructure

Fulfilling the PBL objectives involves the following basic steps: (1) Define the appropriate system-level requirements, as part of the early conceptual definition process, that reflect supportability and include these as "performance" factors in the basic system specification (Type "A," as described in MIL-STD-490A, or equivalent); (2) Ensure that these requirements are properly addressed throughout the system design and development process through the top-down systems engineering process; and (3) Ensure that the proper program incentives are identified and enforced through the appropriate planning process and associated contracting structure.

From a technical perspective, proper implementation of the systems engineering process in the design and development of new systems, and in the re-engineering and modification of existing systems, directly responds to these PBL objectives through the requirements analysis, functional analysis, and the subsequent allocation of top-level requirements down to the subsystem and the various elements of logistics.

Topic 1.6: Related Terms and Definitions

With the objective of further clarifying the scope of logistics and its many interfaces, it seems appropriate to direct some attention to a few additional terms and definitions. While not to be considered as being all-inclusive, an understanding of the areas noted in this section is of particular importance.

Topic 1.6.1: Systems Engineering

Systems engineering may be defined differently depending on one's background and personal experience. The International Council on Systems Engineering (INCOSE) defines it as:

> *An engineering discipline whose responsibility is to create and execute an interdisciplinary process to ensure that the customer and stakeholders' needs are satisfied in a high-quality, trustworthy, and cost and schedule efficient manner throughout a system's entire life cycle. The process is usually comprised of the following seven tasks:* **S***tate the problem;* **I***nvestigate alternatives;* **M***odel the system;* **I***ntegrate;* **L***aunch the system;* **A***ssess the performance; and* **R***e-evaluate (***SIMILAR***). The system engineering process is not sequential. The functions are performed in a parallel and iterative manner.*[13]

The Department of Defense defines systems engineering as:

> *An approach to translate approved operational needs and requirements into operationally suitable blocks of systems. The approach shall consist of a top-down, iterative process of requirements analysis, functional analysis and allocation, design synthesis and verification, and system analysis and control. Systems engineering shall permeate design, manufacturing, test and evaluation, and support of the product. Systems engineering principles shall influence the balance between performance, risk, cost, and schedule.*[14]

A slightly different definition (and one preferred by the authors) is:

> *The application of scientific and engineering efforts to: (1) transform an operational need into a description of system performance parameters and a system configuration through the use of an iterative process of definition, synthesis, analysis, design, test and evaluation, and validation; (2) integrate related technical parameters and ensure the compatibility of all physical, functional, and program interfaces in a manner that optimizes the total definition and design; and (3) integrate reliability, maintainability, usability (human factors), safety and security, producibility, supportability*

---

13 "A Guide to the System Engineering Body of Knowledge (SEBok) - Introduction," *INSIGHT*, Vol. 5, No. 1, International Council on Systems Engineering (INCOSE), 2150 N. 107th Street, Suite 205, Seattle, WA 98133-9009.

14 DOD Regulation 5000.2-R, "Mandatory Procedures for Major Defense Acquisition Programs (MDAPS) and Major Automated Information System (MAIS) Programs," Department of Defense, Washington, DC, Chapter 5, Paragraph C5.2, April 5, 2002.

(serviceability), disposability, and other such factors into a total engineering effort to meet cost, schedule, and technical performance objectives.[15]

Basically, systems engineering is *good* engineering with certain designated areas of emphasis, a few of which are noted as follows:

1. A top-down/bottom-up approach is required (versus bottom-up only), viewing the system as a *whole.* Although engineering activities in the past have very adequately covered the design of various system components, the necessary *overview* and an understanding of how these components effectively fit together have not always been present.
2. A *life-cycle* orientation is required, addressing all phases to include system design and development, production and/or construction, distribution, system operation (utilization), sustaining maintenance and support, and retirement and material recycling/disposal. Emphasis in the past has been placed primarily on system design activities, with little (if any) consideration during the design process toward their impact on production, operations, support, and disposal.
3. A better and more complete effort is necessary relative to the initial *identification of system requirements*, relating these requirements to specific design goals, the development of appropriate design criteria, and the follow-on analysis effort to ensure the effectiveness of early decision making in the design process. In the past, the early front-end analysis effort, as applied to many new systems, has been minimal. Consequently, greater individual design efforts have been required downstream in the life cycle, many of which were not well integrated with other design activities and required modification(s) later on.
4. An *interdisciplinary* effort (or team approach) is required throughout the system design and development process to ensure that **all** design objectives are met in a satisfactory manner. This necessitates a complete understanding of many different design disciplines and their interrelationships, particularly for large projects and when there are many different organizations involved.

Referring to Figure 1.6.1-1, a top-down/bottom-up illustration of the systems engineering process is presented. The emphasis throughout this Toolkit pertains to the top-down shaded area, with the necessary "feedback" provisions noted throughout! When referring to the life cycle, one should view not only the prime mission-related elements of the system (e.g., a radar set, a communications network, fire control equipment) but the applicable production process, the support infrastructure, and the retirement and material recycling/disposal process as well.

[15] Blanchard, B.S., *System Engineering Management*, 3rd Ed., John Wiley & Sons, Hoboken, NJ, pages 15-22, 2004. This is a modified version of the definition of systems engineering that was included in the original version of MIL-STD-499, "Systems Engineering," Department of Defense, Washington, DC, July 1969.

These activities are all interrelated, the interfaces are numerous, and they must be addressed concurrently.

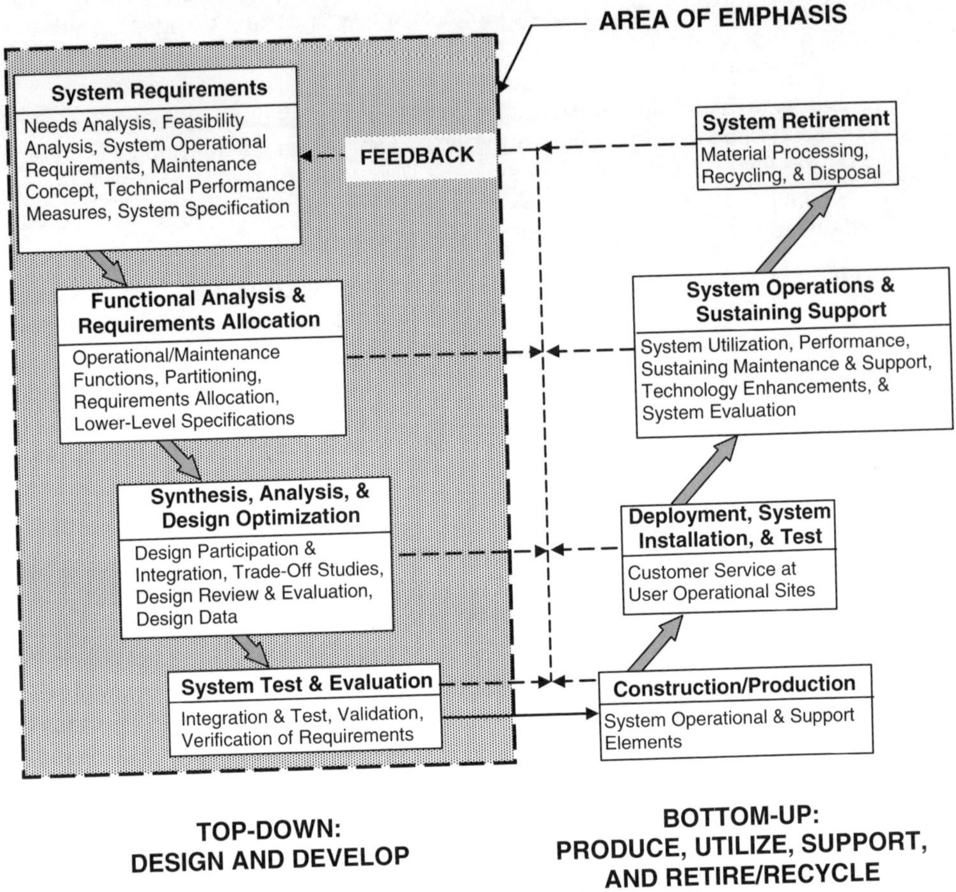

Figure 1.6.1-1. Top-Down/Bottom-Up System Development Process

Referring to Figure 1.6.1-2, the requirements for logistics and system support are initially determined during the early stages of the top life cycle, and the design of the prime mission-related elements of the system must reflect the incorporation of the appropriate logistics and supportability characteristics. As one determines production/construction requirements (second life cycle), the ultimate design configuration must include the requirements pertaining to SCM and the activities presented in Figure 1.3-3. In the development of the maintenance and support infrastructure (third life cycle), the results need to reflect a network similar to that presented in Figure 1.3-4. Additionally, there should be an equivalent flow covering the activities associated with the processing of system components for recycling and/or disposal (to include the activities dealing with "reverse" logistics in the fourth life cycle). This constitutes an integrated "back-and-forth" process of evolutionary development, with the necessary feedback provisions for continuous product and process improvement (CPPI).[16]

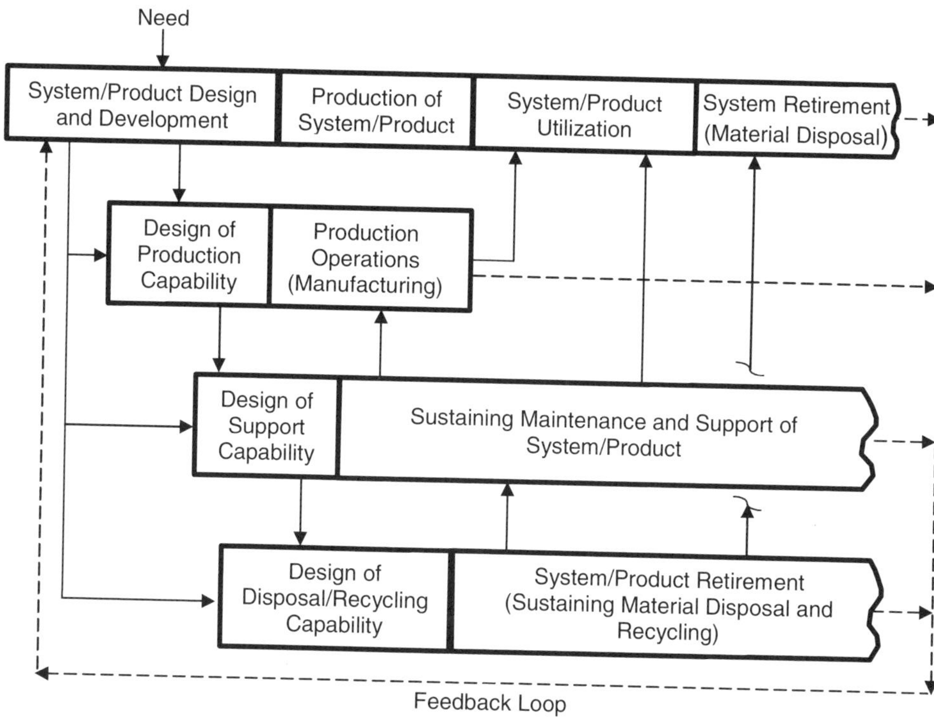

Figure 1.6.1-2. The Life Cycles and Their Relationships

The systems engineering process constitutes the "vehicle" through which the *design for supportability* requirements can be properly addressed and implemented in the overall system development process. Referring to Figure 1.6.1-1, supportability requirements are initially determined from the beginning, the appropriate PBL factors are included in the identification and prioritization of technical performance measures (TPMs) for the system, system-level requirements are allocated to the various elements of the system, the iterative process of synthesis and analysis takes place, design trade-off studies are conducted, and the ultimate design configuration evolves. This process forms the basis for discussion in most of the sections throughout this Toolkit.

[16] Blanchard, B.S., *Logistics Engineering and Management,* 6th Ed., Pearson Prentice Hall, Upper Saddle River, NJ, pages 16-17, 2004.

Topic 1.6.2: System Analysis

Inherent within the system engineering process (i.e., involving the steps associated with synthesis, analysis, evaluation, and design optimization) is an ongoing analytical effort. In a somewhat purist sense, *analysis* refers to a separation of the whole into its component parts, an examination of these parts and their interrelationships, and a follow-on decision relative to a future course of action.

More specifically, throughout system design and development there are many different alternatives (or trade-offs) requiring some form of evaluation. For instance, there may be alternative operational scenarios to consider, alternative logistics and maintenance support concepts, alternative equipment packaging schemes, alternatives associated with the selection of components (e.g., COTS items), alternative diagnostic routines, and alternative manual versus automation applications. The process of investigating these various alternations in response to an identified problem, and the evaluation of each in terms of some criteria, constitutes an ongoing analytical effort. This overall system analysis process, presented in a rather generic sense, is illustrated in Figure 1.6.2-1.

To accomplish this activity in an effective manner, the design engineer (or analyst) relies on the use of a variety of available analytic techniques/models to include operations research methods such as simulation, linear and dynamic programming, integer programming, optimization methods, queuing theory, control theory, accounting and networking methods, and so on, to help solve problems. Further, mathematical models are used to help facilitate the quantitative analysis effort.

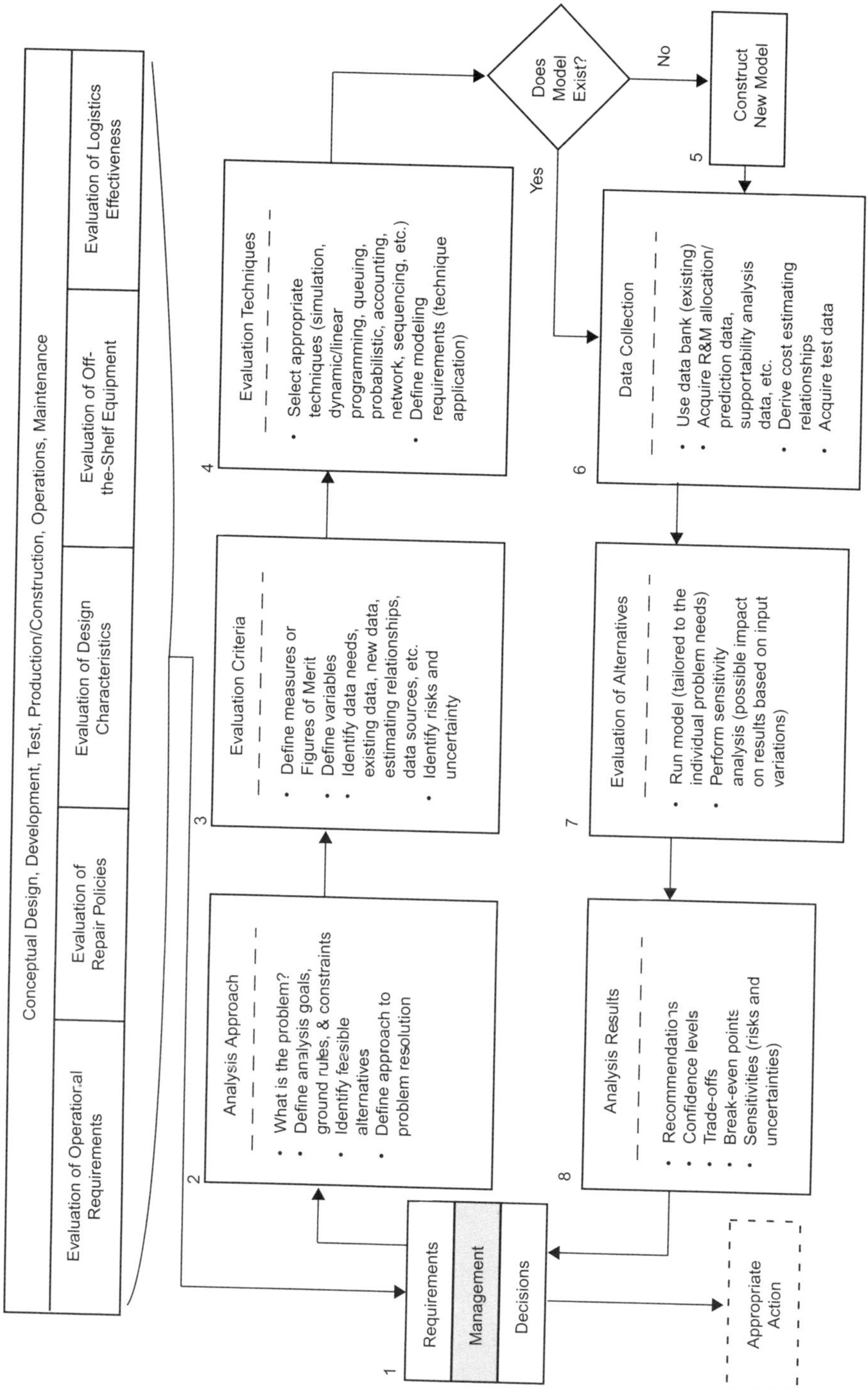

Figure 1.6.2-1. The System Analysis Approach

Topic 1.6.3: Supportability Analysis (SA)

The *supportability analysis (SA)* is an iterative analytical process, accomplished as part of the overall system analysis activity, by which supportability design criteria and the logistic support for a new (or re-engineered) system are first identified and then evaluated. The SA constitutes the application of selected quantitative methods, combined to help solve a wide variety of problems, with the objective of designing and developing an effective and efficient logistic and maintenance support infrastructure. More specifically, the SA addresses itself to the following objectives:[17]

1. Aid in the initial establishment of supportability requirements during conceptual design. In the evaluation of alternative technologies, accomplished as part of an early feasibility analysis, supportability-related factors must be addressed in the decision-making process. Alternatives in selecting a "technical" approach must be evaluated from a total life-cycle perspective.

2. Aid in the early establishment of supportability design-to criteria through the definition of system operational requirements and the maintenance support concept, the identification and prioritization of technical performance measures (TPMs) and performance-based logistics (PBL) factors, the performance of a functional analysis, and in the allocation of requirements. System-level design-to factors are established that lead to the logistics and maintenance support infrastructure, and to the subsequent allocation of requirements to the appropriate lower-level elements of the system. It is this stage that leads to establishment of a hierarchical set of factors as shown in Figure 1.5-1.

3. Aid in the synthesis, analysis, and design optimization effort through the conducting of trade-off studies and the evaluation of various design alternatives. Specific applications may entail the evaluation of:

   a. Alternative transportation and distribution methods (e.g., packing and shipping schemes, transportation modal and inter-modal options, warehousing approaches).

   b. Alternative repair policies (e.g., repair versus discard-at-failure and, if repair, the level of maintenance at which the repair is to be accomplished).

   c. Specific reliability and maintainability characteristics in system/ equipment design (e.g., alternative packaging schemes, test approaches

---

[17] For defense systems, refer to MIL-HDBK-502, "DOD Handbook-Acquisition Logistics," Department of Defense, Washington, DC, May 1997. The principles and concepts associated with supportability analysis (SA), presented throughout this Toolkit, are not new and have been previously covered under such titles as *logistic support analysis (LSA), maintenance engineering analysis (MEA), maintenance level analysis (MLA), maintenance task analysis (MTA), maintenance engineering analysis record (MEAR), maintenance analysis data system (MADS),* and so on. Although the specific nomenclature may change from time to time, the objectives and processes utilized in each instance are basically the same.

and the extent to which diagnostic provisions are incorporated, built-in versus external test methods, the incorporation of manual versus automated maintenance provisions, material handling provisions, levels of accessibility, degrees of standardization, and so on).

d. Two or more COTS items being considered for a single application. Assuming that new design is not appropriate (due to cost, production lead time, supplier availability, etc.), which item, when installed as part of the system, will reflect the smallest overall logistics burden in the long term? The objective is to influence system design for supportability through the appropriate selection of system components and potential suppliers. The results will likely impact future procurement decisions, distribution and support policies, and the type of contractual structure imposed.

4. Aid in the evaluation of a given design configuration (whether interim or final) relative to the determination of specific logistics and maintenance support resource requirements; i.e., a new design configuration described through a data package or an existing system configuration already being utilized. Once the system requirements are known and system data are available, it is possible to determine the requirements for spares and repair parts, test and support equipment, personnel quantities and skill levels, training requirements, facilities, transportation and handling requirements, computer resources, information, and so on. It is this step in the supportability analysis (SA) that leads to the identification of the various logistics elements shown in Figure 1.3-5 and this, in itself, constitutes an iterative process ultimately leading to the definition of the logistics and maintenance support infrastructure.

5. Aid in the measurement and evaluation, or *assessment*, of an operating system to determine its effectiveness and the degree to which the system can be supported in the user's environment. Given a fully operational capability, can the system be effectively and efficiently supported throughout its planned life cycle? Have the applicable PBL requirements for the system been fulfilled (refer to Figure 1.5-1)? Field data are collected, evaluated, and the results are analyzed and compared against the initially specified requirements for the system. The objective is not only to verify that the system is successfully accomplishing its mission, but also to identify the high-cost/high-risk areas and to recommend and incorporate modifications as part of a *continuous product/process improvement (CPPI)* effort.

6. Aid in the initial determination of the requirements for, and in the subsequent assessment of, logistics resources necessary during and throughout the system retirement and material recycling/disposal phase of the life cycle. This basically addresses the activities reflected by the fourth life cycle shown in Figure 1.6.1-2.

Referring to Figure 1.6.3-1, the intent here is to provide an illustration as to how the SA can be related to the overall system development effort. The blocks on the left (e.g., blocks 1, 3, 5, 7, 9, etc.) represent the basic system design and development process conveyed in Figure 1.6.1-1. The blocks on the right (e.g., blocks 2, 4, 6, 8, 10, ---, 15) reflect the major areas of emphasis with regard to the logistics and maintenance support infrastructure. In the early phases of system development the emphasis is on *design influence,* whereas the objective later on is on *resource identification and validation.* Further, as the system is utilized, the emphasis is on *supportability assessment*, along with the necessary feedback provisions for the purposes of corrective action and/or product/process improvement as required.

Inherent within the overall spectrum of the supportability analysis (SA) effort is the application of a wide variety of analytical methods and models, applied to varying degrees depending on the depth of system design definition at the time, and directed toward facilitating accomplishment of specific SA tasks. For example, reliability and maintainability prediction data are required as an "input" to the SA, leading to the identification of projected maintenance frequencies and downtimes; the Failure Mode, Effects, and Criticality Analysis (FMECA) may be required to determine critical failure modes and cause-and-effect relationships; the maintenance task analysis (MTA) is required in determining personnel quantities and skill-level requirements; a level-of-repair analysis (LORA) is required to determine the depth of maintenance support that will be required throughout the system utilization phase; a life-cycle cost analysis (LCCA) is required in the evaluation of alternatives from a total life-cycle perspective; and so on. Referring to Figure 1.6.3-2, a few of these analytical methods and tools are presented for illustrative purposes. It is through the proper "combined" application of these tools, utilized in a timely manner, that enables the accomplishment of the SA in an effective and efficient manner.

Referring to the six items identified in the beginning of 1.6.3, the basic objective(s) in the performance of the SA (or equivalent) has essentially resulted in placing a great deal of emphasis on the completion of the activity described by Item 4. More specifically, there has (through the years) been a great deal of effort expended in the preparation and processing of various data in support of justifying the logistics resources required for a given system. The accomplishment of such has basically resulted in an "after-the-fact" identification and evaluation of a given design configuration. On the other hand, there has been very little emphasis on the activities described under Items 1, 2, and 3 which, when completed, lead to the ultimate design configuration of the logistics and maintenance support infrastructure. If one is to successfully attain the goals pertaining to the *design for supportability*, much more emphasis needs to be exerted in the "front-end;" i.e., Items 1, 2, and 3.[18]

[18] The supportability analysis (SA) is discussed in detail in Section 5.

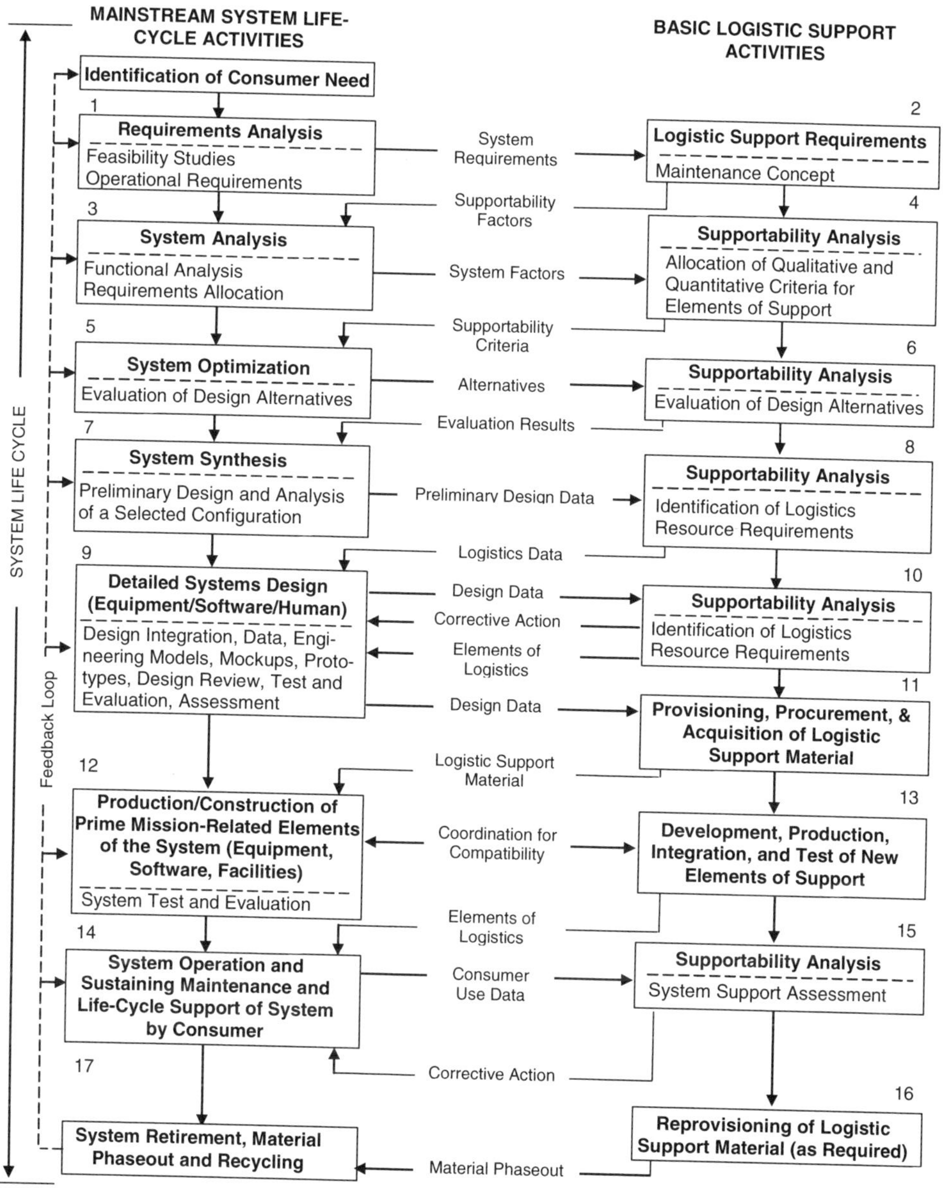

Figure 1.6.3-1. The Interface Relationships Between Basic Design and Logistics Functions

(Source: Blanchard, B.S., *Logistics Engineering and Management*, 6th Edition, Pearson Prentice Hall, Upper Saddle River, NJ, Figure 1.9, 2004)

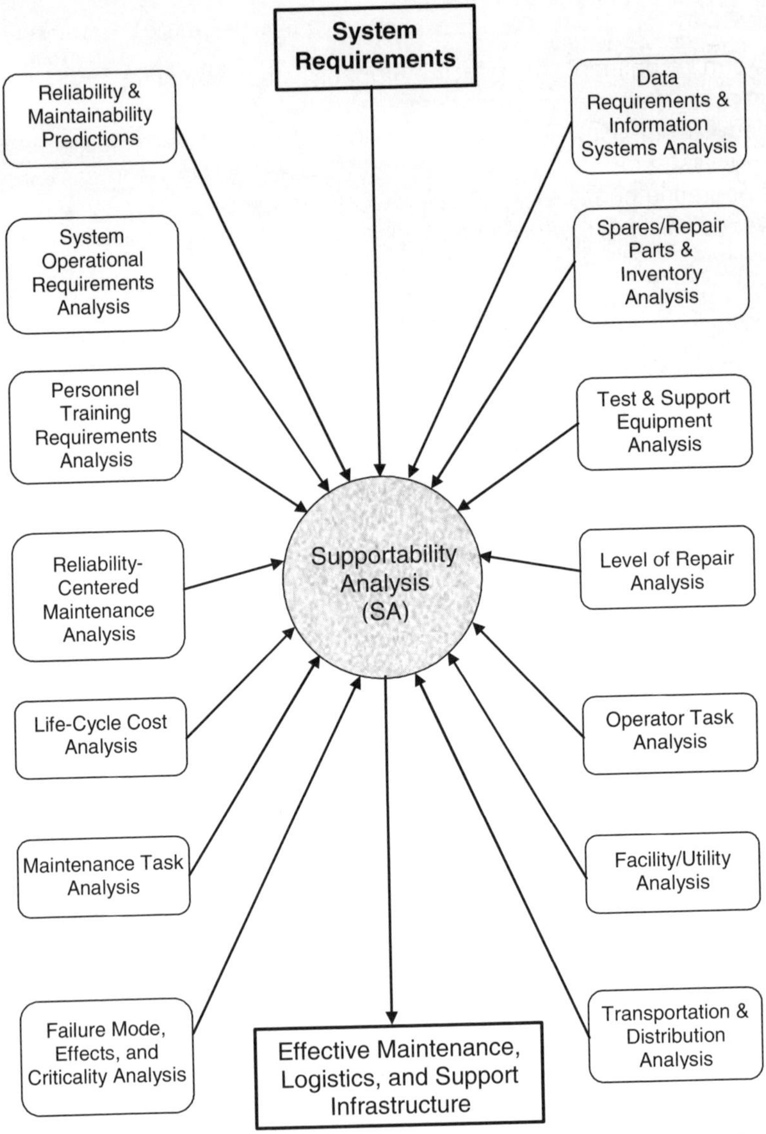

Figure 1.6.3-2. Supportability and Supplemental Analysis Efforts

Topic 1.6.4: Reliability (R)

Reliability can be defined simply as the probability that a system, product, or item will perform in a satisfactory manner for a given period of time, or in the accomplishment of a defined mission scenario, when used under specified operating conditions. This definition stresses the elements of probability, satisfactory performance, time, and specified operating conditions. These four elements are extremely important, since each plays a significant role in determining system/product reliability and its impact on supportability.

1. *Probability*, the first element in the reliability definition, is usually stated as a fraction or a percentage signifying the number of times that an event occurs (successes), divided the total number of trials. For example, a statement that the probability of survival (*Ps*) of an item for 80 hours is 0.75 (or 75%) indicates that we can expect that the item will function properly for at least 80 hours, 75 times out of 100 trials. When there are a number of supposedly identical items operating under similar conditions, it can be expected that failures will occur at different points in time; thus, failures are described in probabilistic terms. The fundamental definition of reliability is heavily dependent on the concepts derived from probability theory.
2. *Satisfactory performance*, the second element of the reliability definition, indicates that system mission requirements have been defined and specific criteria have been established that describe what is to be considered as satisfactory system "operation." A combination of qualitative and quantitative factors defining the functions that the system or product is to accomplish, usually presented in the context of a system specification, are required. When addressing the issue of "performance," this includes the identification of any PBL factors that relate directly to the successful accomplishment of the mission (i.e., function) that the system is to perform.
3. The third element, *time*, is most important since it represents a measure against which the degree of performance can be related. For example, one must know the time parameter in order to assess the probability of completing a mission or a given function as scheduled. Of particular interest is being able to predict the probability that an item will survive (without failure) for a designated period of time (sometimes designated as *R* or *P*). Also, reliability is frequently defined in terms of *mean time between failure (MTBF)*, *mean time to failure (MTTF)*, or *failure rate (lambda)*; thus, the aspect of time is critical in reliability measurement.
4. The *specified operating conditions* under which a system or product is expected to function constitute the fourth significant element in the reliability definition. These conditions include environmental factors such as geographical location(s) where the system is expected to operate, the operational profile, the transportation profile, temperature cycles, humidity, vibration, shock, noise levels, external factors such as toxicity, and so on.

> Such factors must address not only the conditions for the period when the system or product is operating, but also the conditions for the periods when the system (or a segment of such) is in a storage mode or being transported from one location to another. Experience has indicated that the transportation, handling, and storage modes are sometimes more critical from a reliability standpoint than the conditions experienced during actual system operational use.

These four elements are critical in determining the reliability of a system or product. System reliability (or unreliability) is a key factor in determining the frequency of maintenance, and the maintenance frequency obviously has a significant impact on logistics and maintenance support requirements. Thus, reliability is a key factor in fulfilling the requirements in the *design for supportability.*[19]

[19] Reliability is discussed extensively throughout this Toolkit, and knowledge of reliability principles and concepts is essential in meeting objectives pertaining to system supportability. The measures of reliability are discussed further in Section 2.1, reliability design requirements are covered in Section 4.2.1, and reliability test and evaluation requirements are presented in Section 6.2.1. Additionally, the subject is covered extensively in "Reliability Toolkit: Commercial Practices Edition," Reliability Analysis Center (RAC), 201 Mill Street, Rome, NY 13440-6916 (latest edition).

Topic 1.6.5: Maintainability (M)

Maintainability, like reliability, is an inherent characteristic of system/product design. It pertains to the ease, accuracy, safety, and economy in the performance of maintenance actions. A system should be designed such that it can be maintained without large investments of time, money, and other resources (e.g., personnel, materials, facilities, test equipment) and without adversely affecting the mission of that system. Maintainability is the *ability* of an item to be maintained, whereas maintenance constitutes a series of actions to be taken to restore an item or retain it in an effective operational state. Maintainability is a *design parameter.* Maintenance is the result of design.

Maintainability can also be defined as a characteristics in design that can be expressed in terms of maintenance frequency factors, maintenance times (elapsed times and labor hours), and maintenance cost. These terms may be presented as different figures-of-merit; therefore, maintainability may be defined on the basis of a combination of factors such as:

1. A characteristic of design and installation expressed as the probability that an item will be retained in or restored to a specified condition within a given period of time, when maintenance is performed in accordance with prescribed procedures and resources.
2. A characteristic of design and installation expressed as the probability that maintenance will not be required more than *x* times in a given period, when the system is operated in accordance with prescribed procedures. This may be analogous to reliability when the latter deals with the overall frequency of maintenance.
3. A characteristic of design and installation expressed as the probability that the maintenance cost for a system will not exceed *y* dollars per designated period of time, when the system is operated and maintained in accordance with prescribed procedures.

Maintainability requires the consideration of many different factors involving all aspects of the system, and the measures of maintainability often include a combination of factors such as those identified in Table 1.6.5-1 (as applicable).[20]

[20] Maintainability is discussed extensively throughout this Toolkit, and knowledge of maintainability principles and concepts is essential in meeting the objectives pertaining to system supportability. Some of the measures of maintainability are discussed further in Section 2.2, maintainability design requirements are covered in Section 4.2.2, and maintainability test and demonstration requirements are presented in Section 6.2.2. Additionally, the subject is covered extensively in "Maintainability Toolkit: A Practical Guide for Designing and Developing Maintainable Products and Systems," Reliability Analysis Center (RAC), 201 Mill Street, Rome, NY 13440-6916 (latest edition).

Table 1.6.5-1. Typical Measures of Maintainability

1. MTBM: Mean Time Between Maintenance, including both preventive (scheduled or condition-based) and corrective (unscheduled) actions. It reflects reliability (MTBF and MTBR). May also be considered a measure of reliability.
2. MTBR: Mean Time Between Replacement of an item as part of a maintenance action (usually generates a spare-part requirement).
3. $\bar{M}$: Mean active maintenance time (mean time required for active maintenance excluding any delays). It is a function of $\bar{M}_{ct}$ and $\bar{M}_{pt}$.
4. $\bar{M}_{ct}$: Mean corrective maintenance time.
5. $\bar{M}_{pt}$: Mean preventive maintenance time.
6. $\tilde{M}_{ct}$: Median active corrective maintenance time.
7. $\tilde{M}_{pt}$: Median active preventive maintenance time.
8. $MTTR_g$: The geometric mean time to repair.
9. M_{max}: Maximum active corrective maintenance time (usually specified at the 90% or 95% confidence level).
10. MDT: Maintenance Downtime. The total time during which a system is not in condition to perform its intended function. MDT includes active maintenance time ($\bar{M}$), logistic delay time (LDT), and administrative delay time (ADT).
11. MLH/OH: Maintenance Labor Hours per Operating Hour. Maintenance labor hours per system operating hour.
12. Cost/OH: Cost per Operating Hour. Maintenance cost per system operating hour.
13. Cost/MA: Cost per Maintenance Action. Maintenance cost per maintenance action.
14. TAT: Turnaround Time. That element of maintenance time needed to service, repair, and check out an item for recommitment. TAT is the time it takes for an item to go through a complete cycle from operational installation through a maintenance shop and into the spare inventory ready for use.
15. Self-test Thoroughness: The scope, depth, and accuracy of testing.
16. Fault-Isolation Accuracy: The accuracy of system diagnostic routines in isolating the cause of failure (measured in percent).

Topic 1.6.6: Human Factors (HF)

Human factors (or *ergonomics*) pertain to the human element of the system and the interfaces between the human being, the machine (equipment), facilities, and associated software. Quite often in a development of a system, the emphasis is on the design of hardware and software and the *human* element tends to be ignored. The objective here is to assure complete compatibility between the system's physical and functional design features and the human element in the operation, maintenance, and support of the system. In the development of a system (and its elements), specific considerations in design must include the following factors:

1. *Anthropometric factors.* Anthropometry deals with the measurement of the dimensions and the physical characteristics of the human body (e.g., standing height, sitting height, arm reach, breadth, buttock-knee length, hand size, and weight). When establishing the basic design requirements involving the human being (for work space application, work surface design, control panel layout, accessibility for maintenance), one obviously must take into consideration the physical dimensions of the human body. Both "structural" dimensions (when the body is fixed and in a static state) and "functional" dimensions (when the body is engaged in some physical activity and in a dynamic state) must be measured and used in designing for the performance of operational functions, logistics functions, and maintenance support functions. Further, the design engineer must consider both *male* and *female* dimensions, with the appropriate ranges of variability (usually from the 5th to the 95th percentiles). For example, the height of a male may range from 63.6 inches (5th percentile), to 68.3 inches (50th percentile), or 72.8 inches (95th percentile). The height of the female may vary from 59.0 inches (5th percentile), to 62.9 inches (50th percentile), or 67.1 inches (95th percentile). Although the average values may be used, the design workspaces, surfaces, and so on, must consider variations for both male and female operators and maintainers.
2. *Human sensory factors.* This category relates to the human sensory capacities, particularly sight or vision, hearing, feel or touch, smell, and so on. In the design of work stations, surfaces, operator consoles and panels, and maintenance control facilities, the design engineer must be cognizant of the human's capability relative to sight as it pertains to vertical and horizontal fields of view, angular fields of view, the detection of certain objects from different angles, the detection of certain colors and varying degrees of brightness from different angles, and so on. The placement of panel displays and controls as a function of use and the employment of different color combinations to facilitate the accomplishment of manual tasks require knowledge of the human being's capability for seeing. In addition, the designer needs to understand the human's capacity for hearing in terms of both frequency and intensity (or amplitude). The design of work areas for oral communications and the use of auditory displays require knowledge relative

to the effects of noise on the performance of work. For instance, as the noise level increases, a human begins to experience discomfort and both productivity and efficiency decrease. If the noise level approaches 120 to 130dB, then a physical sensation in some form, or pain, will likely occur. In essence, the system designer needs to integrate the capabilities of the human into the final product.

3. *Physiological factors.* Although the study of physiology is obviously well beyond the scope of this Toolkit, it is appropriate to recognize the effects of environmental stresses on the human body during the performance of manual tasks. *Stress* refers to any type of external activity, or environment, that acts on an individual in such a manner as to cause a degrading impact. Some typical causes of stress are: (1) high and low temperatures, or temperature extremes; (2) high humidity; (3) high levels of vibration; (4) high levels of noise; and (5) large amounts of radiation or toxic substances in the air. To vary degrees, these environmental effects will negatively impact on human performance; that is, physical fatigue will occur, motor response will be slower, mental processes will slow down, and the likelihood of error will increase. These externally related stress factors will normally result in individual human "strain." *Strain* may, in turn, have an impact on any one or more of the human's biological functions (e.g., the circulatory system, digestive system, nervous system, and respiratory system). Measures of strain may include parameters such as blood pressure, body temperature, pulse rate, and oxygen consumption. These factors of strain, caused by external stresses, will definitely have an impact on the performance of human operator and maintenance functions if the design fails to address the physiological effects on the human.

4. *Psychological factors.* This category relates to factors that pertain to the human mind; that is, the emotions, traits, attitudinal responses, and behavioral patterns as they relate to job performance. All other conditions may be perfect relative to completing a task in an effective manner. However, if the individual operator (or maintenance technician, or logistics practitioner, or whomever) lacks the proper motivation, initiative, dependability, self-confidence, communication skills, and so on, the likelihood of performing a required task in an effective manner is low. Generally, a person's attitude, initiative, motivation, etc., are based on the needs and expectations of the individual. This, in turn, is a function of system design and the organizational environment within which the individual performs. If the tasks to be completed are perceived as being too complex, the individual may become frustrated, a poor attitude may develop, and errors will occur. On the other hand, if the tasks are too simple and routine, there is little challenge, boredom prevails, and errors will occur as a result of attitude. Further, as an external factor, the management style of the supervisor may cause an attitudinal problem. In any event, it is appropriate to consider the possible psychological effects on the human in the design and development of a system.

In addition to considering the aforementioned general characteristics associated with the human, it is necessary to have some understanding of the human's ability to deal with and process information. Whether a function should be automated or accomplished by the human and, if accomplished by the human, to what extent, is dependent on the human's ability to detect, react, process information, and provide a response in a timely manner.

While the subject of human factors has not (in the past) received a great deal of attention, other than for the design of operator stations/consoles (for airplanes, ships, missiles, communication centers), there needs to be a greater degree of emphasis in this area as it applies to all elements of the system. Many of the failures that have occurred in systems in the past have been *operator-induced* and/or *maintenance-induced*; i.e., system failures caused by the human during the performance of operational, maintenance, and related support tasks. Often, these failures tend to be ignored; however, the human constitutes a major element of the system and these are, indeed, *system failures.*[21]

[21] Refer to Section 2 for some specific measures associated with human factors, and Section 4.2.3 for human factors activities in system design.

Topic 1.6.7: Safety and Security

Safety is a system design characteristic. The selection of certain materials in the design and construction of a system could produce harmful toxic effects on people; the placement and mounting of components could cause injuries to the operator and/or the maintainer; the use of certain fuels, hydraulic fluids, and/or cleansing liquids could result in an explosive environment; the location of certain electronic components close together may cause the generation of an electrical hazard; the performance of a series of strenuous tasks during the operation and maintenance of the system could cause personal injury; and so on.

Safety is important both from the standpoint of the human operator and/or maintainer, and from the standpoint of other elements of the system. Through faulty design, one can create problems that could result in human injury. Also, problems can be created that will result in damage to other elements of the system. Inherent within supportability objectives is the *design for system safety.*

In a similar context, the issue of *security* (and the *design for security*) needs to be addressed, particularly in view of the continuing threats of terrorism. This pertains to the design of a system such as to preclude the planned introduction of faults/failures that will cause the system (or any portion thereof) to be completely destroyed, resulting in the damage of materials, facilities, and/or the loss of life. The objective, of course, is to prevent an individual (or group of individuals) from intentionally sabotaging a system for one reason of another. In response, the design should consider the incorporation of an external security alarm capability that will detect an attempt by unauthorized personnel to operate/maintain the system, a condition-based monitoring capability that will enable a continuing check on the status of the system, a built-in mechanism that will detect and initiate an alarm and will prevent a chain reaction of failures that will lead to system damage or destruction, and/or any combination of these. Without a safe and secure logistics and maintenance support infrastructure, there is no guarantee that the system will successfully accomplish its intended mission.

Topic 1.6.8: Producibility

Producibility is a measure of the relative ease and economy of producing a system or a product. The characteristics of design must be such that an item can be produced easily and economically, using conventional and flexible manufacturing methods and processes without sacrificing function, performance, effectiveness, or quality. *Simplicity* and *flexibility* are underlying objectives, and it is a goal to minimize the use of critical and non-standard materials and processes, the use of proprietary items, the use of special production (manufacturing) tooling and facilities, the application of unrealistic tolerances in fabrication and assembly, the use of special test systems, and the need for highly skilled personnel in manufacturing. Additionally, production and procurement lead times and material flow times should be minimized to the extent possible.

Producibility objectives (and the *design for producibility*) should apply to the elements of the logistics and maintenance support infrastructure as well as to the prime mission-related components of the system. Further, the application of "producibility" characteristics in the design of a manufacturing capability tends to have a highly beneficial impact on a design of the system overall relative to such areas as accessibility, functional packaging and modularization, labeling, the incorporation of good diagnostics, and so on, in addition to a potential realization of reduced life-cycle cost.[22]

[22] "Designing and Assessing Supportability on DOD Weapon Systems: A Guide to Increased Reliability and Reduced Logistics Footprint," Prepared by the Office of the Secretary of Defense, October 24, 2003, pages 12,14.

Topic 1.6.9: Disposability

Disposability pertains to the degree to which an item can be removed from the operational inventory, recycled for some other use, or disposed of economically and without causing any degradation to the environment; that is, resulting in the generation of solid waste, toxic substances (air pollution), water pollution, noise pollution, and so on (refer to the fourth life cycle shown in Figure 1.6.1-2). Should this area not be addressed in system design from the beginning, the logistics requirements may turn out to be rather extensive and costly when complying with the environmental requirements currently being imposed, both nationally and internationally. For example, a large incineration facility may be required for material decomposition and this facility, in turn, may include large amounts of capital equipment that requires maintenance and can be very costly to support.[23]

[23] *Logistics Engineering and Management*, 6th Ed., Pearson Prentice Hall, Upper Saddle River, NJ, 2004, pages 39, 295-296, and 368.

Topic 1.6.10: Quality

The term, *Quality*, is often applied in different contexts, and the definition of such may vary somewhat depending on the situation. Within the context of this Toolkit, "quality" is viewed from a rather broad perspective. More specifically, and referring to D. Garvin's definition discussed in "Quality Toolkit," quality can be defined in terms of five basic categories:[24]

1. Transcendent: a subjective feeling of "goodness." This relates to the customer's subjective feeling(s) about the product.
2. Product-based: measured attributes of the product.
3. Manufacturing-based: conformance with specifications.
4. Value-based: "goodness" for the price.
5. User-based: defined by the ultimate customer.

While the resultant product output may be in compliance with all of the required standards and specifications, there may be a number of subjective issues that also need to be addressed; e.g., product features and characteristics, the feeling of "goodness" about the product, general overall customer satisfaction. One might start by asking the following questions: Does the customer feel "good" about the product? Does the customer feel that he/she is getting a good deal from an economic perspective? Is the customer completely satisfied with the product in all aspects?

With a broad objective in mind, the concept of *Total Quality Management (TQM)* was introduced in the 1970s-1980s, primarily in the defense sector. TQM can be described as a total integrated management approach that addresses system/product quality during all phases of the system life cycle and at each level in the overall system hierarchy. It provides a "before-the-fact" orientation to quality, and it focuses on system design and development activities as well as on production, manufacturing, assembly, construction, logistics and maintenance support, and related functions. Some specific characteristics of TQM are noted:

1. Total *customer satisfaction* is the primary objective, as compared with the practice of accomplishing as little as possible in conforming to the minimum requirements.
2. Emphasis is placed on the iterative practice of *continuous product/process improvement (CPPI)* as applied to engineering, production, and logistics and support processes. The objective is to seek improvement on a day-to-day basis, as compared with the often-imposed last-minute single thrust initiated to force compliance with a standard (the Japanese version of this approach, known as *Kaizen*, is growing in popularity internationally).

[24] "Quality Toolkit: A Practical Guide for Developing and Implementing a Quality Program for Products and Services," Reliability Analysis Center (RAC), 201 Mill Street, Rome, NY 13440-6916 (latest edition), pages 4-11.

3. In support of Item 2, an individual understanding of processes, the effects of variation, the application of process control methods, and so on, is required. If the individual employees in a given firm are to be contributors to "continuous improvement," they must know the various processes and their inherent characteristics.
4. Emphasis is placed on a total organizational approach, involving every group in the organization, and not just the Quality Control or Quality Assurance Group. Individual employees are motivated from within and are recognized as being key contributors in meeting TQM objectives.

Included within the function of system design and development is the consideration of: (1) the inherent characteristics of the various components selected for the ultimate system configuration, and (2) the characteristics of the processes that will be utilized to produce the prime system components and the associated elements of support (refer to the four life cycles shown in Figure 1.6.1-2). Quality in design pertains to many of the issues discussed earlier in this section, with emphasis directed toward simplicity, flexibility, standardization, and so on. Of a more specific nature are the concerns of *variability*, in which a reduction in the variation of the dimensions for specific component designs, or tolerances in process design, might result in an overall improvement. Taguchi's approach to *robust design* is to provide a design that is insensitive to the variations normally encountered in production (manufacturing) and/or when the system is in operational use. The more robust the design, the fewer the support requirements, the lower the life-cycle cost, and the higher degree of system effectiveness. Overall design improvement is anticipated through a combination of careful component evaluation and selection, the use of appropriate *statistical process control (SPC)* methods, and the application of experimental (i.e., *design of experiments*) testing approaches, applied on a continuing basis.[25]

More recently, there has been a great deal of emphasis on the application of *six sigma* methods. Basically, six sigma is a management philosophy and approach, supplemented with a set of tools, designed to eliminate product defects, increase customer satisfaction, and increase the profits of various production (manufacturing) enterprises. The expectations include lower product cost, lower cycle time, lower occurrence of product/process variation, lower waste, lower logistics footprint, lower system life-cycle cost, higher quality, higher product reliability, and higher customer satisfaction. Essentially, six sigma can be considered as a metric for quality.[26]

[25] Dr. Genichi Taguchi has developed mathematical and related techniques applicable to the evaluation of design variables. Refer to Rose, P.J., *Taguchi Techniques for Quality Engineering*, McGraw-Hill, New York, NY, 1988.

[26] Two recommended references for six sigma are: (1) Pande, P., and L. Holpp, *What is Six Sigma?*, McGraw-Hill, New York, NY, 2002; and (2) Tennant, G., *Design for Six Sigma: Launching New Products and Services Without Failure*, Gower Publishing Ltd., 131 Main Street, Burlington, VT 05401, 2002.

Topic 1.6.11: System Effectiveness (SE)

In addressing the overall effectiveness for a system, one needs to consider both the *technical* and the *economic* sides of the spectrum and the appropriate balance between the two, as illustrated in Figure 1.6.11-1.

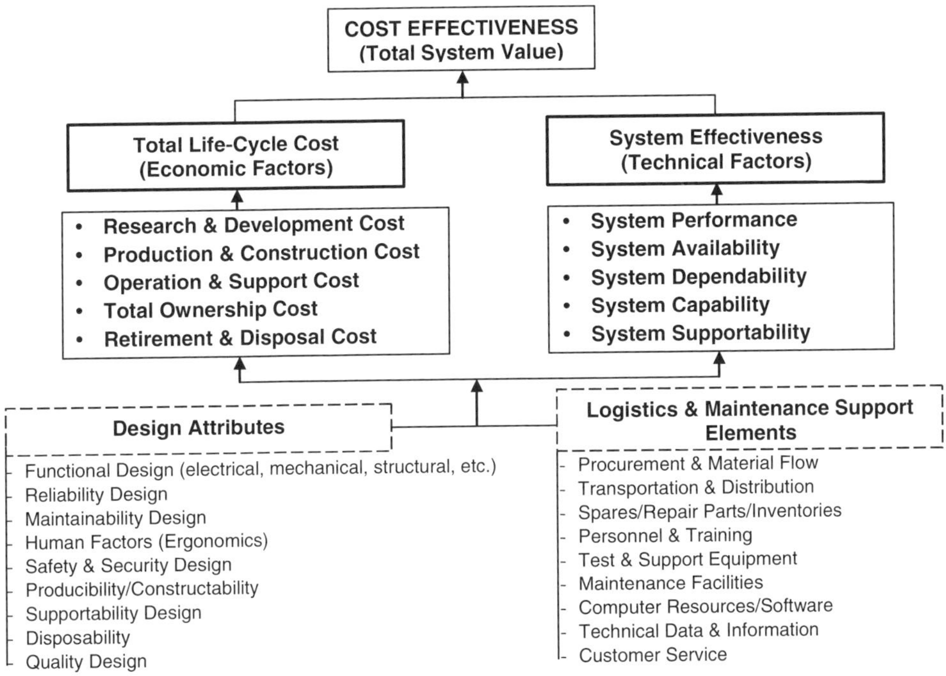

Figure 1.6.11-1. Basic Ingredients of Cost Effectiveness

In the figure, which represents a generic approach, the "technical" factors (when combined and related to a specific mission scenario) have been identified, included within, and designated as a measure of *system effectiveness (SE)*. Additionally, one needs to consider applicable economic factors in order to address the issue of *efficiency.* Thus, there is an overall balance between the technical side of the spectrum and the economic side of the spectrum, and the applicable factors on each side can have a strong interactive impact on each other, the results of which can be a measure of *cost effectiveness.*

Historically, different measures have been used to determine the SE for a given system. For example, in the early 1970s a joint government-industry committee activity was established to study and determine the appropriate "measures of effectiveness (MOEs)" for defense systems. Three different approaches resulted, one for each the Army, Navy, and the Air Force. The Air Force defined SE as being a product of *availability (A)*, *dependability (D)*, and *capability (C)*, or SE = (A)(D)(C); the

Army used the terms *operational readiness, mission reliability,* and *design adequacy*; and the Navy defined it as a function of *performance, availability*, and *utilization.* In essence, the objectives were the same in each case, although the nomenclature and factors used were different.[27]

For the purposes of further discussion herein, it is assumed that SE can be defined in terms of the following:

1. *System performance (P)* – the capacity of a power plant, range or weight of an airplane, acceleration and speed of a ground vehicle, destructive capability of a weapon, quantity of letters processed through a postal system, number of items manufactured per element of time in a production plant, amount of cargo delivered by a transportation system, accuracy of a radar capability, and so on.
2. *Availability (A)* – the degree to which a system is in the operable and committable state at the start of a mission when the mission is called for at an unknown random point in time. This factor is also known as *operational readiness.* Availability is a function of operating time (reliability) and downtime (maintainability and supportability).
3. *Dependability (D)* – the measure of the system operating condition at one or more points during the mission, given the system condition at the start of the mission (i.e., availability). Dependability is a function of operating time (reliability) and downtime (maintainability and supportability).

Another good example pertains to the measure, *Overall Equipment Effectiveness (OEE)*, which is often applied as a MOE for a manufacturing plant. In this case, *OEE* is a function of *availability (A), performance rate (P),* and *quality rate (Q)*, or OEE = (A)(P)(Q).[28]

A combination of these (and perhaps other) measures represent the technical characteristics of a system, as opposed to the cost and economic aspects. By inspection, one can see that the logistics and maintenance support infrastructure can have a significant impact on the overall effectiveness of a given system. The Performance-Based Logistics (PBL) metrics, described in Section 1.5 (also refer to Figure 1.5-1), are inherent within the measures of performance, availability, and

[27] A Weapon System Effectiveness Industry Advisory Committee (WSEIAC), to include a number of different subcommittees, was established in the early 1970s for the purposes of developing a set of effectiveness measures for various categories of defense systems. The results ended up with a slightly different approach for each of the services.

[28] In the early 1970s, the Japanese developed the concept of *Total Productive Maintenance (TPM)*, with the objective of improving productivity in the factory, improving production line reliability, reducing maintenance and associated support costs in the factory, and ultimately reducing the cost of producing a product. The metric used in measuring the effectiveness of the manufacturing plant is *OEE.*

dependability. System operation is highly dependent on various logistics-related activities shown in Figures 1.3-2 and 1.3-3. System availability and dependability are highly dependent on the maintenance support infrastructure shown in Figure 1.3-4 and its capability should a failure occur. Basically, the logistics and maintenance support infrastructure is measured through the various parameters of system effectiveness.

Topic 1.6.12: Life-Cycle Cost (LCC)

In Figure 1.6.11-1, the other side of the balance pertains to the economic issues, which must be addressed along with the technical factors in making decisions throughout the system development process. As system design and development evolve, design decisions of a technical nature can have a great impact on cost, and vice versa. If one is to properly assess the risks associated with the day-to-day decision-making process, both sides of the balance need to be addressed as the interaction effects could be numerous.

Of particular significance on the economic side is the total *life-cycle cost (LCC)* of a given system, which may be broken down into different categories shown in the example in Figure 1.6.12-1 and described as follows:

1. *Research, design, and development cost*: the cost of early planning; feasibility studies; system requirements analyses; detail design and development; development of prototype models; system test and evaluation; design data/documentation; and all related customer, contractor, and supplier activity pertaining to the development of a new system and/or the re-engineering of an existing system. The costs associated with the *design for supportability*, the supportability analysis (SA), design and test of the logistics and maintenance support infrastructure, etc., are included.
2. *Production and/or construction cost*: the cost of production planning; design and construction of the production (manufacturing) facility; fabrication, assembly, and test of operational systems (early production and full-scale production models); the incorporation of system modifications; sustaining operation and maintenance of the production capability; and supporting supplier activity. The costs associated with the logistics and supply chain activities identified in Figure 1.3-3 are included; i.e., the costs associated with physical supply, materials flow in the factory, and physical distribution.
3. *Operation and support cost*: the cost of system operation (utilization); the incorporation of system modifications; sustaining system maintenance and support; and related supplier activity. The costs associated with logistics and maintenance support activities identified in Figure 1.3-4 are included; i.e., the costs associated with maintenance personnel, spares/repair parts and related inventories, test and support equipment, transportation and handling, maintenance facilities, computer resources, technical data, and information.
4. *System retirement and material recycling/disposal costs*: the cost of phasing the system and its components out of the operational inventory due to obsolescence or wearout; recycling and the processing of items for other uses; the disposal of items that can not be recycled; and related supply activities. The cost of logistics activities required for the material recycling and/or disposal are included (refer to the fourth life cycle illustrated in Figure 1.6.1-2).

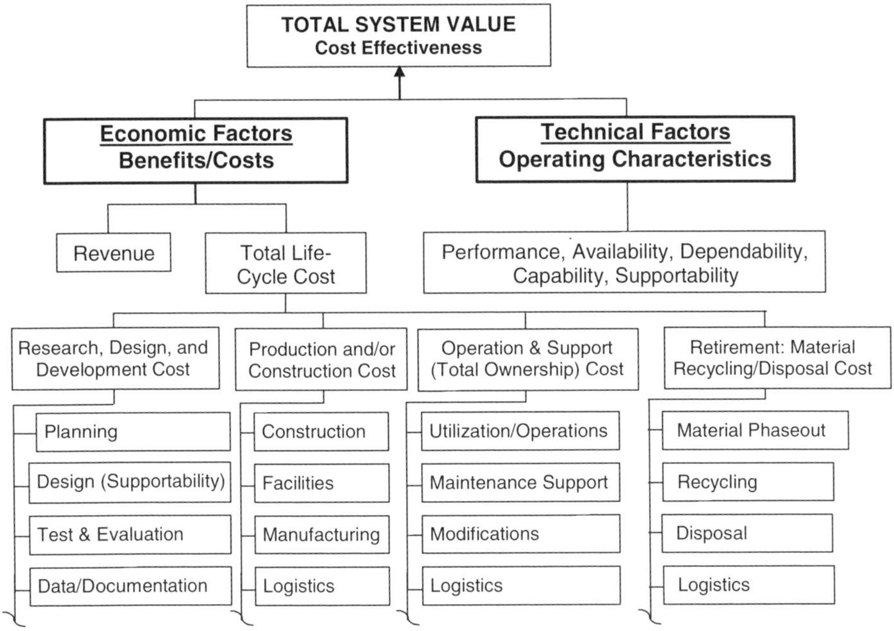

Figure 1.6.12-1. Life-Cycle Cost and Major Cost Categories (Partial Example)

Life-cycle costs may be categorized in many different ways, depending on the type of system, the nature of its mission, and on the visibility and sensitivities desired in a cost-effectiveness measurement. The key objective is to ensure that **ALL** costs are included, whether they are considered as being relevant or not. It is not uncommon to find that, in accomplishing a life-cycle cost analysis (LCCA), certain costs have been left out for one reason or another (e.g., because of not being considered as being important, for political reasons, etc.). This, of course, precludes the possibility of gaining full visibility of the total cost, identifying the high-cost contributors, and being able to identify the true risks associated with the day-to-day decision-making process.[29]

[29] The term "Total Ownership Cost (TOC)" is used throughout the literature as it applies to major defense systems. While it is presumed that TOC basically equates to System Operation and Support Cost (Item 3 – also refer to Figure 1.6.12-1), there have been some variations in the past as to what is and is not included within the TOC category. Additionally, there are some activities and the costs of such (e.g., system enhancements with the costs identified under design and development) that will have a great impact on TOC; however, the total impact of design and support costs are not addressed as an integrated entity. In accomplishing a LCCA, it is critical that a complete cost breakdown structure (CBS), describing what is included/excluded, be identified, and that full-cost visibility be possible across the entire spectrum shown in Figure 1.6.12-1.

Topic 1.6.13: Cost Effectiveness (CE)

As stated earlier, the ultimate objective is to design and develop a system that is cost-effective; i.e., one that will meet all of the technical requirements and at the right cost. While there may be some variations in measuring cost-effectiveness, the relationships presented in Figure 1.6.11-1 are assumed herein for discussion purposes. Referring to the figure, *cost effectiveness (CE)* relates to the measure of a system in terms of mission fulfillment (system effectiveness) and total life-cycle cost.

With regard to supportability (and the logistics and maintenance support infrastructure), there are requirements on both sides of the spectrum. On one end, the objective is to design, develop, and provide an effective support capability that will be ready and available when needed. On the other end, such a capability needs to be economically feasible from a total life-cycle perspective. Thus, both sides of the balance need to be addressed throughout the system development process.

Topic 1.7: Management of Logistics and Supportability

The successful implementation of logistics and supportability objectives depends not only on the *technology* issues and the process with all of its associated tools/techniques, but on the *management* issues as well. There are two sides of the spectrum, each dependent on the other. The best tools/models may be available to implement the processes illustrated in Figures 1.6.1-1 and 1.6.3-1; however, there is no guarantee for success unless the proper organizational environment has been created and an effective management structure is in place. The requirements for *logistics* and the *design for supportability* must evolve from the top, and top-management must first believe in and then provide the necessary support to enable the proper application of the tools, techniques, models, and processes to various programs. Specific objectives must be defined, policies and procedures must be developed and properly implemented, and an effective reward structure must be supported. The challenge is *proper implementation.*

While Section 9 includes an extensive coverage of logistics and supportability management (i.e., program requirements, planning and organization, management and control activities, etc.), it is appropriate at this point to establish a baseline for early planning purposes. Although there may be some variations in application and specific nomenclature, the "DOD 5000 Defense Acquisition Management Framework," shown in Figure 1.4-1, is presented herein as an initial baseline. This baseline has been expanded to reflect the basic system engineering activities shown in Figure 1.6.1-1 and the logistics activities presented in Figure 1.6.3-1. The integration of these activities into a total "life-cycle" perspective is presented in Figure 1.7-1.[30]

[30] The DOD 5000 Model is included and described in *Designing and Assessing Supportability in DOD Weapon Systems: A Guide to Increased Reliability and Reduced Logistics Footprint*, Office of the Secretary of Defense, October 24, 2003, page 5.

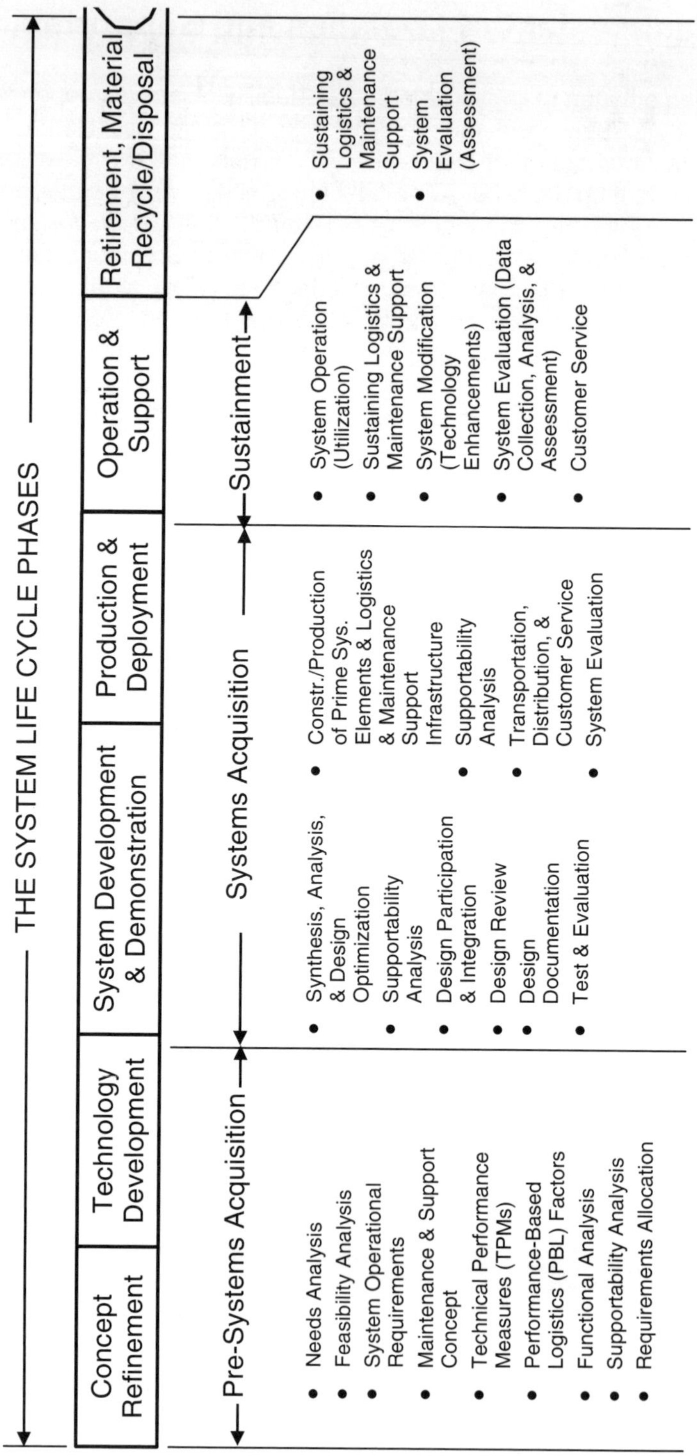

Figure 1.7-1. Logistics and Supportability Throughout the System Life Cycle

Topic 1.7.1: Concept Refinement and Technology Development Phases

Referring to Figure 1.7-1, the basic **requirements** for supportability are initially established for the overall system during the Concept Refinement Phase. The needs analysis is accomplished and the requirement for a logistics and maintenance support infrastructure is described. During the subsequent feasibility analysis, alternative technologies (i.e., a "technical approach" as opposed to a specific design solution) are evaluated and a proposed approach to solving the problem at hand is recommended. At this point, each of the possible alternatives being evaluated must be viewed in the context of the entire system life cycle. Thus, the consideration of logistics and supportability requirements is inherent within the accomplishment of a feasibility analysis.

System operational requirements are then defined. This constitutes a description of the system's proposed mission(s) and related scenarios, specific technical performance requirements, geographical distribution, effectiveness requirements, proposed system life cycle, and the anticipated operational environment. The applicable performance-based logistics (PBL) factors, as they are applied to the prime mission-related elements of the system and the *design for supportability* requirements, are defined at this point.

From the definition of operational requirements, the system maintenance and support infrastructure is developed, with the results being presented in a structure similar to that illustrated in Figure 1.3-4. This includes a definition of the various levels of support, the expected organizational responsibilities for maintenance, the anticipated duration of support, applicable effectiveness requirements (PBL factors as they apply to the support infrastructure and its elements), and the anticipated maintenance environment. At this point, the objective is to develop an initial "concept" for the establishment of supportability "criteria" as an **input** to the overall infrastructure design. This concept (i.e., illustrated structure) will be updated and refined, though application of the supportability analysis (SA), as the design process evolves.

Given the description of system operational requirements and the defined maintenance and support concept, the appropriate technical performance measures (TPMs) and performance-based logistics (PBL) factors for the "system" are identified and prioritized in terms of anticipated degrees of importance. While there may be many different requirements (i.e., measures), the question is -- which TPM/PBL factors are more important than others, when considering mission and related criticalities, and what are the areas in design that should receive more emphasis than others? This process aids in the establishment of specific design priorities.

The process thus far (identified by first block in Figure 1.6.1-1) has resulted primarily in the definition of requirements at the "system" level, and described through the

System Specification (Type "A") and supporting documentation. The next step is to define the system in "functional" terms (through the functional analysis), accomplish the necessary partitioning and functional packaging of the system, identify the basic elements of the system (i.e., major sub-systems and below), determine the basic resource requirements for the system (e.g., hardware, software, people, facilities, data), and allocate the appropriate requirements (i.e., TPM/PBL factors) from the top level down to the applicable elements/components of the system. The question is -- given the design-to requirements for the "system," what design-to requirements should be specified for each sub-system and below?

While this overall process is somewhat iterative in nature and commences during the Concept Refinement Phase, the major thrust of activity as one evolves from the system-level and down is accomplished as part of the Technology Development Phase. Supportability requirements evolve from the definition of system operational requirements and the maintenance support concept, the identification of maintenance functions (as part of the functional analysis), the allocation of PBL factors to different elements of support (refer to Figure 1.3-5), and the subsequent accomplishment of supportability analysis (SA).

Topic 1.7.2: System Development and Demonstration Phase

Stemming from the requirements analysis activity in the earlier program phases, there is the on-going process of synthesis, analysis, and design optimization that leads to the ultimate system configuration that is to be produced and delivered to the customer for operational use. This includes the accomplishment of numerous trade-off studies, the day-to-day design participation by representatives from various technical disciplines, the conductance of formal design reviews, system test and evaluation, and feedback, and the incorporation of changes as necessary for corrective action and/or product improvement. Essentially, this activity is reflected by the third and fourth blocks (left side) in Figure 1.6.1-1, and blocks 5, 7, and 9 in Figure 1.6.3-1.

With regard to logistics and system support, the primary activity thrust is represented by blocks 6, 8, 10, and 11 in Figure 1.6.3-1. Within the spectrum of the supportability analysis (SA), various design alternatives are evaluated employing such techniques/ tools as reliability and maintainability predictions and analyses, life-cycle cost analysis (LCCA), FMECA/Fault Tree Analysis (FTA), reliability-centered maintenance (RCM) analysis, maintenance task analysis (MTA), level of repair analysis (LORA), and so on. Additionally, the requirements pertaining to the *design for supportability* must be addressed in the day-to-day design participation process (as applicable), as part of the formal design review process, and verified through system test and evaluation.

Topic 1.7.3: Production and Deployment Phase

Logistics requirements and activities during the Production and Deployment Phase may fall into two basic categories:

1. The physical supply, material flow, distribution, and associated supply chain activities, illustrated in Figure 1.3-3. This includes all supplier, producer, warehouses, distributor, and related activities associated with the initial production, distribution, and installation of the system/product at the customer's operational site.
2. The acquisition, through direct procurement and/or production, of the elements of the logistics and maintenance support infrastructure, illustrated in Figures 1.3-4 and 1.3-5. This includes spares and repair parts, test and support equipment, facilities, maintenance software, data, etc., identified through the supportability analysis (SA).

The first category refers to the various facets of commercial logistics needed to support production (manufacturing) operations, and the second category pertains to those items that are necessary and being acquired for the maintenance support of the system throughout its planned life cycle.

Topic 1.7.4: System Operations and Sustaining Support Phase

This phase pertains to system utilization and the fulfillment of its mission objectives throughout the planned operational life cycle. Logistics and support activities include those represented through the system maintenance and support infrastructure illustrated in Figure 1.3-4. Additionally, there is a need for an on-going data collection, analysis, and reporting capability that will provide the necessary feedback and assessment of the true effectiveness and efficiency of the system during its period of operational use. Problem areas are noted and corrective action is taken as appropriate. This provides the first accurate assessment of the system with regard to its *supportability.* The question is -- does the system, in its operational configuration, meet all of the *design for supportability* requirements that were specified during the Concept Refinement and Technology Development Phases?

Topic 1.7.5: System Retirement and Material Recycling/Disposal

A final, and very important, phase of the system life cycle pertains to the activities associated with the retirement of the system from the operational inventory, and the processing of obsolete materials for recycling and/or disposal. While this phase is often ignored, the logistics and related support requirements can be rather extensive; e.g., extensive packaging and transportation requirements, special facility requirements for material decomposition, special tools for disassembly and disposal tasks, and so on. Thus, it is important that emphasis be placed on the *design for disposability*, which, in turn, can significantly impact the *design for supportability*.

For More Information:

1. Blanchard, B.S., *Logistics Engineering and Management*, 6th Ed., Pearson Prentice Hall, Englewood Cliffs, NJ, 2004.
2. Blanchard B.S., *System Engineering Management*, 3rd Ed., John Wiley & Sons, Hoboken, NJ, 2004.
3. Langford, J.W., *Logistics Principles and Applications*, McGraw-Hill, New York, NY, 1995.
4. MIL-HDBK-502, "Department of Defense Handbook on Acquisition Logistics," Department of Defense, Washington, DC, 1997.
5. "Designing and Assessing Supportability in DOD Weapon Systems: A Guide to Increased Reliability and Reduced Logistics Footprint," Prepared by the Office of Secretary of Defense, October 24, 2003.
6. "Reliability Toolkit: Commercial Practices Edition," Reliability Analysis Center, 201 Mill Street, Rome, NY.
7. "Maintainability Toolkit: A Practical Guide for Designing and Developing Maintainable Products and Systems," Reliability Analysis Center, 201 Mill Street, Rome, NY.
8. "Quality Toolkit: A Practical Guide for Developing and Implementing a Quality Program for Products and Services," Reliability Analysis Center, Rome, NY.

Section 2: The Measures of Supportability

Insight

The logistics and engineering factors which indicate the degree to which a system is able to achieve its stipulated operational and support requirements in a cost effective manner are those measures associated with *supportability*. A system must, in compliance with its specified performance parameters, be *available, reliable, maintainable, and affordable over the projected system life cycle*. The purpose of this Section is to define the measures of supportability, emphasizing those directly associated with the various elements of logistics and their ancillary engineering disciplines. There are a variety of applicable mathematical methodologies used as tools for analyzing the attributes of supportability. Availability, Reliability, and Maintainability entail focus on statistical algorithms; Life Cycle Cost analysis involves the pertinent financial analysis formulae. All such quantitative tools of analysis must be integrated to provide a valid picture of the supportability of a system.

Topic 2.1: Reliability Measures

System *Reliability* is described as the probability that a system (or product) will successfully perform in a satisfactory manner for a prescribed period of time in the accomplishment of a mission when used under specified operational conditions. The Reliability function, R(t), may be expressed as:

$$R(t) = 1 - F(t) \tag{2.1}$$

Where F(t) is the probability that the system will fail by time t. F(t) is basically the failure distribution function, or the unreliability function. If the random variable t has a density function f(t), then the expression for reliability is:

$$R(t) = 1 - F(t) = \int_t^{\infty} f(t)dt \tag{2.2}$$

Assuming that the mean time to failure is described by an exponential density function, then:

$$f(t) = \frac{1}{\theta}e^{-t/\theta} \tag{2.3}$$

where θ is the mean life, t is the time period of interest, and e is the natural logarithm base (2.7183). The reliability at time t is:

$$R(t) = \int_t^{\infty} \frac{1}{\theta}e^{-t/\theta}dt = e^{-t/\theta} \tag{2.4}$$

This basic derivative principle provides the basis for calculating the reliability for *electrical/electronic* systems and *mechanical* systems, as described in the following topics.

Topic 2.1.1: Reliability for Electrical/Electronic Systems

Based on $R(t) = e^{-t/\theta}$, electrical/electronic systems employ the single-point *reliability* algorithm, which is expressed as:

$$R(t) = e^{-\lambda t}, \text{ or simply } R = e^{-\lambda t} \qquad (2.5)$$

where:

R	=	Reliability (Probability of success).
λ	=	Failure rate.
t	=	Specified operating cycle.

The Failure Rate (λ) is the reciprocal of the Mean Time Between Failure (MTBF); i.e.,

$$\lambda = \frac{1}{MTBF} \text{ and } MTBF = \frac{1}{\lambda} \qquad (2.6)$$

Within the perspective of *supportability*, the Mean Time Between Failure (MTBF) denotes the average interval of operation between unscheduled events caused by system component failure, operator-induced failure, software failure, etc., after which repair (corrective maintenance) is required to restore the system to operational status. The repair interval (MTBF) is used to project the frequency required for corrective maintenance actions over the life cycle of the system. This factor governs the planning for manpower, training, supply support, facilities, equipment, etc., which are translated to financial resources which must be dedicated for system repair and restoration. The plot guide for determining $R = e^{-\lambda t}$ is provided in the nomograph at Figure 2.1.1-1.[1]

It is noted that for non-repairable and one-shot devices, the term *Mean Time to Failure (MTTF)* is frequently used in lieu of *MTBF* (see Equation 2.6).

[1] Nomograph derived from NAVAIR 01-1A-32, *Reliability Engineering Handbook*, Naval Air Systems Command, U.S. Navy, Washington DC, 1977.

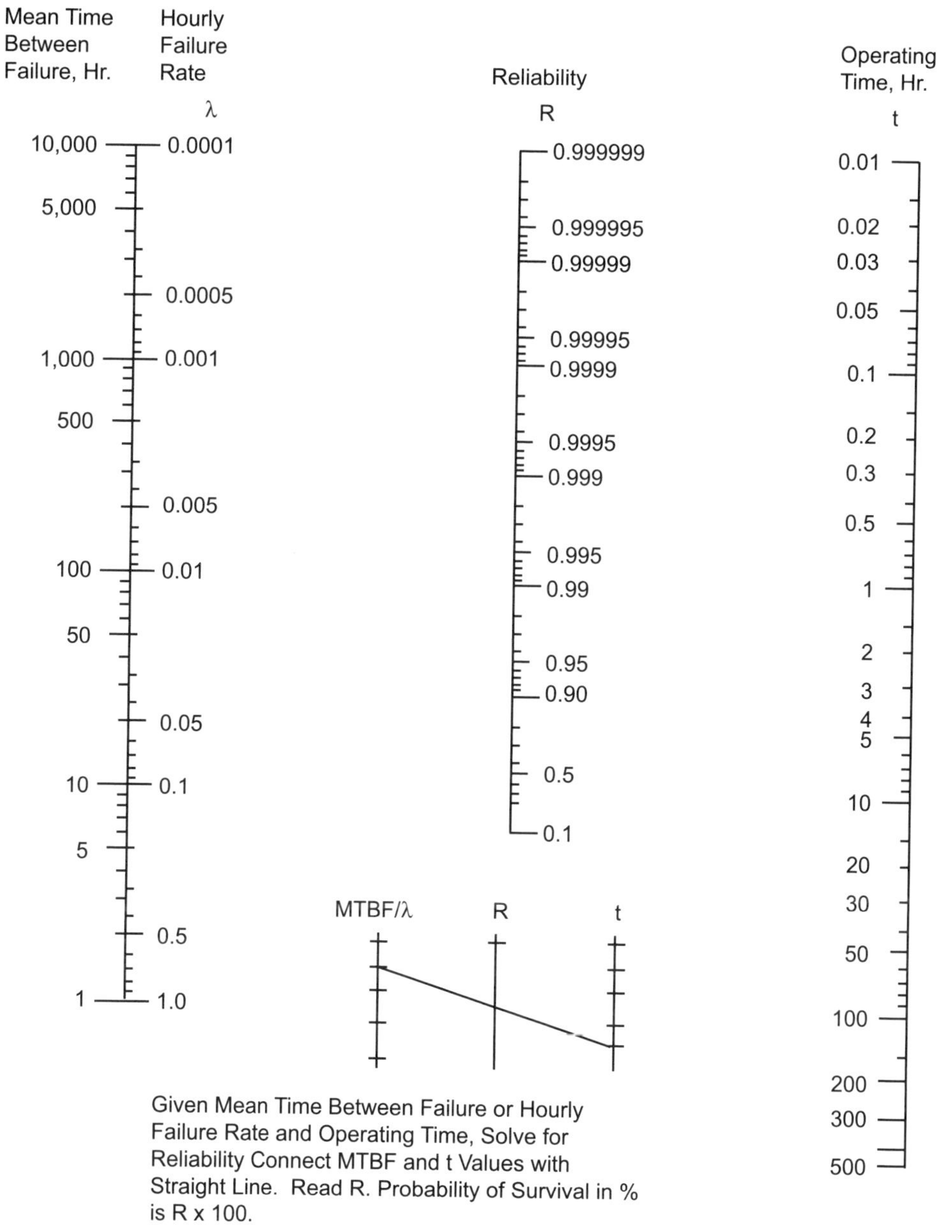

Reliability nomograph for the exponential failure distribution. (NAVAIR 01-1A-32, Reliability Engineering Handbook, Naval Air Systems Command, U.S. Navy, Washington, DC, 1977.)

Figure 2.1.1-1. Reliability Nomograph

Topic 2.1.2: Reliability for Mechanical Systems

Reliability for mechanical systems is primarily based on the *Weibull* distribution. The Weibull distribution is described by the *shape parameter*, β, the *scale parameter*, θ (sometimes called η), and the *location* parameter, γ (See Figure 2.1.2-1). Note that when γ = 0, we have the two-parameter Weibull.

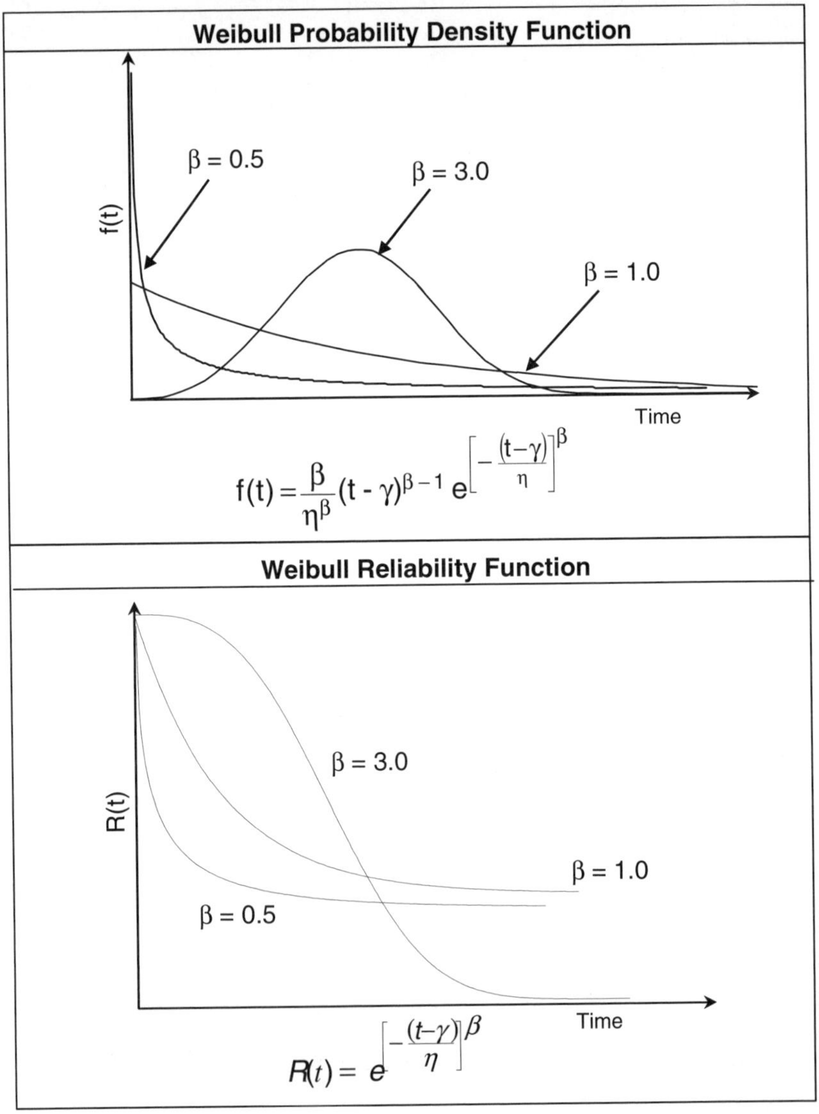

Figure 2.1.2-1. pdf and Reliability Function for the Weibull Distribution

1. *Attributes of the Weibull Distribution.*

   a. The shape parameter, β, describes the shape of the Weibull *probability density function* (*PDF*). (Section 6.3.2 of the RAC *Reliability Toolkit: Commercial Practices Edition* describes the methodology for determining the value for the shape parameter.) The Weibull PDF has no characteristic shape; the shape is determined by the shape parameter and can vary accordingly. When $\beta = 1$, the Weibull function reflects the exponential distribution (when $\gamma = 0$). When $\beta = 3$, the Weibull density function approximates the normal distribution.

   b. The scale parameter, η, is a value called the characteristic life (the life at which point 63.2 percent of the population will have failed).

   c. The location parameter, γ, is the value that represents the failure free period of the equipment. If a system does not have a period where the probability of failure is zero, then $\gamma = 0$ and the Weibull has a two parameter distribution.

2. *Derivation of the Weibull Reliability Function.* The probability density distribution function of the Weibull function is given by Equation 2.7.[2]

$$f(t) = \frac{\beta(t-\gamma)^{\beta-1}}{\eta^{\beta}} e^{-\left(\frac{(t-\gamma)}{\eta}\right)^{\beta}} \tag{2.7}$$

The Reliability, *P(t)*, probability of *t* is derived to be:

$$P(t) = R(t) = 1 - \int_{0}^{t} f(t)\,dt \tag{2.8}$$

Therefore

$$R(t) = e^{-\left(\frac{(t-\gamma)}{\eta}\right)^{\beta}} \tag{2.9}$$

[2] Derived from *The New Weibull Handbook*, Robert Abernethy, North Palm Beach, FL, 1996, page 2-3.

Topic 2.2: Maintainability Measures

Maintainability is an inherent design attribute relating to the ease, accuracy, safety, and economy in the performance of maintenance activities. Maintainability, in its total context, can be measured in terms of a combination of maintenance intervals, task times, personnel labor hour rates, maintenance frequencies, maintenance cost factors, and related logistics support factors. This Section describes those measures most commonly used as constituent to *supportability*.

Topic 2.2.1: Maintenance Categories

Maintenance can be classified into two categories:

1. *Corrective Maintenance*: the unscheduled actions, initiated as a result of failure (or a perceived failure), that are necessary to restore a system to its required level of performance. Corrective maintenance (or repair) actions are triggered by random (unpredictable) malfunctions and failures of the system. Such activities may include troubleshooting, disassembly, repair, removal and replacement, reassembly, alignment and adjustment, and checkout (See Figure 2.2.1-1). This category also includes all software maintenance that is not initially planned, such as *adaptive* maintenance and *perfective* maintenance.
2. *Preventive Maintenance*: the scheduled actions necessary to *retain* a system at a specified level of performance. This category may include periodic inspections, servicing, calibration, condition monitoring, and/or the replacement of designated critical items. *Preventive maintenance*, which may be also described as *scheduled maintenance*, is predictable and may be planned for accomplishment at either prescribed operational or isochronous intervals.[3]

[3] Blanchard, B.S., *Logistics Engineering and Management*, 6th Ed., Pearson Prentice Hall, Upper Saddle River, NJ, 2004, page 58.

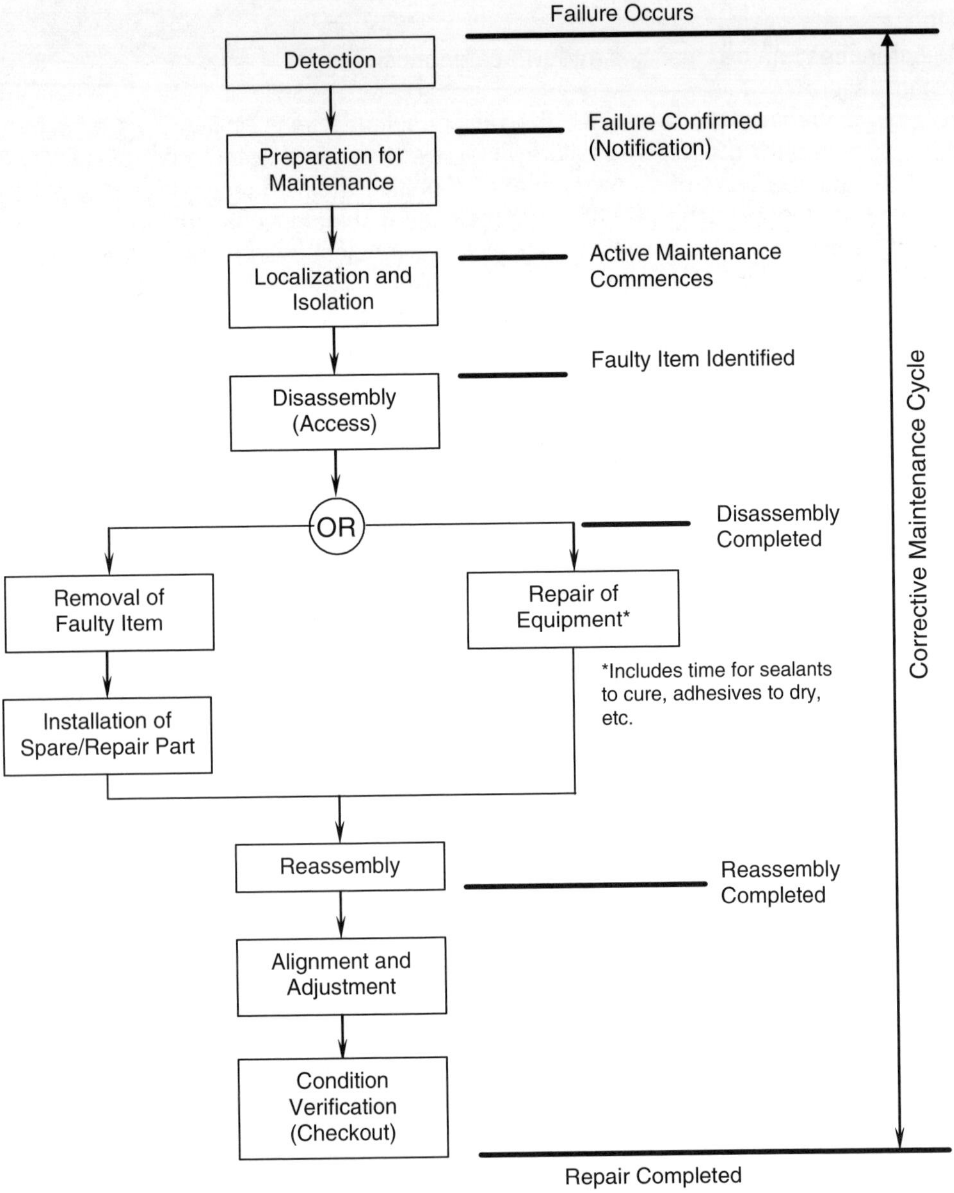

Figure 2.2.1-1. Corrective Maintenance Cycle

Topic 2.2.2: The Metrics of Maintainability

Maintenance constitutes the process of diagnosing and repairing, or preventing system failures. Maintenance time comprises the individual task times associated with the required corrective and preventive maintenance actions for a system (or product). Maintainability is a measure of the ease and rapidity with which a system can be maintained and is measured in terms of the time required to perform maintenance tasks. The more commonly utilized maintainability task time measures are described in the following.

1. *Mean Corrective Maintenance Time (*$\overline{M}_{ct}$*).* Each time a system fails, a series of steps are required to repair or restore the system to its full operational status. Completion of the steps portrayed by Figure 2.2.1-1 defines a corrective maintenance cycle. Mean Corrective Maintenance Time, also known as Mean Time to Repair (MTTR), is a composite value representing the arithmetic average (or mean value) of these individual maintenance cycle times. The following equations describe the derivation of these attributes.[4]

$$\overline{M}_{ct} = \frac{\sum_{i=1}^{n} M_{ct_i}}{n} \tag{2.10}$$

or

$$\overline{M}_{ct} = \frac{\sum_{i=1}^{n} (\lambda_i) M_{ct_i}}{\sum_{i=1}^{n} \lambda_i} \tag{2.11}$$

In the previous two equations, n is the number of task times in the sample, M_{ct_i} is the elapsed time required for the ith individual corrective maintenance action, and λ_i is treated as the corrective maintenance frequency for the individual repair action. The failure rate (λ) is the mathematical counterpart to the preventive maintenance frequency (*fpt*) used in maintainability algorithms.

[4] Blanchard, B.S., *Logistics Engineering and Management*, 6th Ed., Pearson Prentice Hall, Upper Saddle River, NJ, 2004, pages 58-60.

The reciprocal of the MTTR is known as the *repair rate*, as shown in Equation 2.12.

$$\mu = \frac{1}{MTTR} \tag{2.12}$$

where μ = Repair Rate

The probability density function for repair times can usually be expressed by one of the following three common forms.

a. *Normal Distribution*: generally applies to straightforward maintenance tasks, such as basic remove and replacement actions that typically require a fixed amount of time to accomplish and which have little variation. The Normal Distribution for corrective maintenance time values is illustrated by Figure 2.2.2-1.

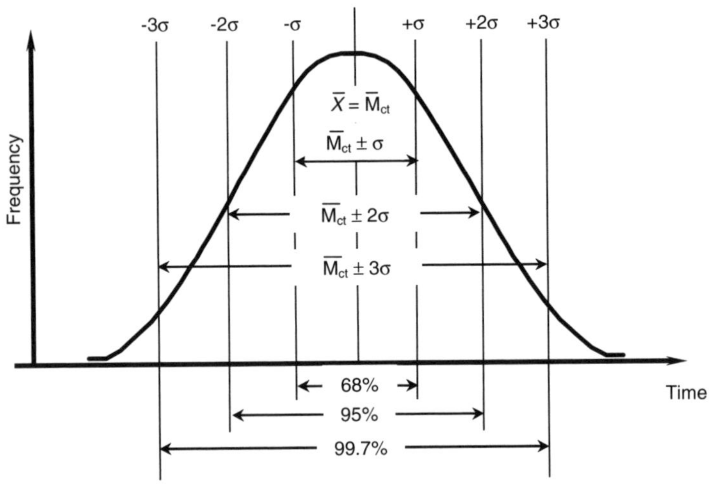

Figure 2.2.2-1. Normal Distribution

b. *Exponential Distribution*: usually applies to equipment with an efficient built-in test (BIT) capability and based on a remove and replace maintenance concept. The maintenance rate is constant. The Exponential Distribution is shown by Figure 2.2.2-2.[5]

5 For further details refer to *Reliability Toolkit: Commercial Practices Edition*, Reliability Analysis Center, 201 Mill Street, Rome, NY, Topic 2.3.

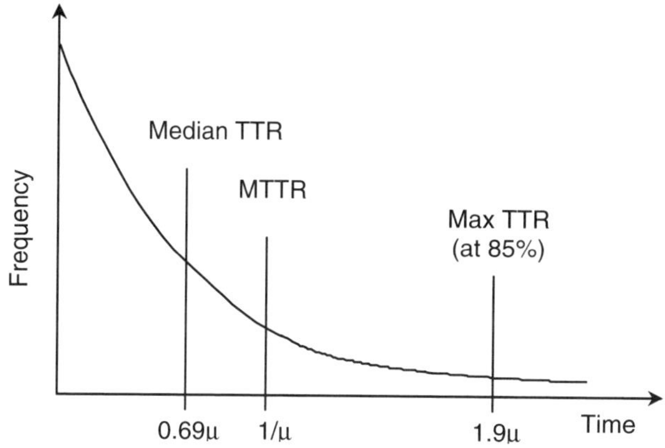

Figure 2.2.2-2. Exponential Distribution

An example of the application of the exponential distribution functions is presented in Appendix C, Exponential Distribution Table.

c. *Log-Normal Distribution*: applies to most maintenance tasks in which the task times and frequencies vary significantly. Experience has indicated that, in the majority of instances, the distribution of maintenance task times for complex systems is log-normal. The Log-Normal Distribution is portrayed by Figure 2.2.2-3.[6]

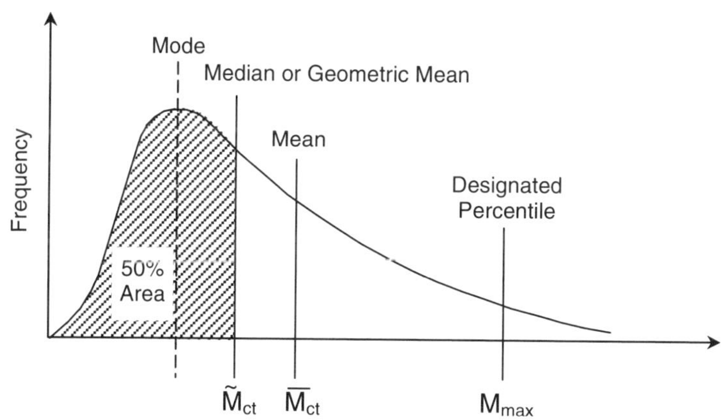

Figure 2.2.2-3. Log-Normal Distribution

[6] Blanchard, B.S., *Logistics Engineering and Management*, 6th Ed., Pearson Prentice Hall, Upper Saddle River, NJ, 2004, pages 64-65.

2. *Mean Preventive Maintenance Time ($\overline{M}_{pt}$).* Preventive (or scheduled) maintenance consists of those actions required to maintain a system at a specified level of performance. Scheduled maintenance includes such functions as periodic inspection, servicing, scheduled replacement of critical items, calibration and overhaul. Mean Preventive Maintenance Time is the average (or mean) elapsed time value for performing preventive or scheduled maintenance on an item:

$$\overline{M}_{pt} = \frac{\Sigma(f_{pt_i})(M_{pt_i})}{\Sigma f_{pt_i}} \qquad (2.13)$$

In the preceding equation, M_{pt_i} is the elapsed time required for the *i*th individual preventive maintenance action and f_{pt_i} is the preventive maintenance frequency for the individual preventive maintenance action. The factor f_{pt} is the mathematical counterpart to the failure rate (λ) used in reliability algorithms.

3. *Mean Active Maintenance Time ($\overline{M}$).* Mean Active Maintenance Time is the average elapsed time to perform both preventive and corrective maintenance.[7] This factor, which consolidates the preventive maintenance time and corrective maintenance time factors, excludes logistics delay time and administrative time elements. The algorithm is shown in Equation 2.14.

$$\overline{M} = \frac{(\lambda)(\overline{M}_{ct}) + (f_{pt})(\overline{M}_{pt})}{\lambda + f_{pt}} \qquad (2.14)$$

where:

λ = Corrective Maintenance (Failure) Rate

f_{pt} = Preventive Maintenance Rate

$\overline{M}_{ct}$ = Mean Corrective Maintenance Time

$\overline{M}_{pt}$ = Mean Preventive Maintenance Time

4. *Median Active Corrective Maintenance Time* ($\tilde{M}_{ct}$). The Median Active Corrective Maintenance Time is determined through mathematical methodology based on the log-normal distribution (i.e., the *common logarithm* as symbolized

[7] Note that active maintenance means that some element of maintenance is being performed by a maintenance person. Inactive time can include delays due to administration or logistics, or other related factors, such as cure times for sealants, paints, adhesives, etc.

by *log*). It is that time value that delineates the repair time values so that 50% are equal to or less than the median and 50% are equal to or greater than the median. The median will normally provide the best average location of the task time data sample. The median for a normal distribution is the same as the mean, as indicated by Figure 2.2.2-1. The median in a log-normal distribution is the same as the geometric mean, MTTRg, as shown in Figure 2.2.2-3. The median is calculated by means of the following formula:

$$\tilde{M}_{ct} = \text{antilog}\frac{\sum_{i=1}^{n}\log M_{ct_i}}{n} = \text{antilog}\frac{\Sigma(\lambda_i)(\log M_{ct_i})}{\Sigma(\lambda_i)} \tag{2.15}$$

In the preceding equation, n is the number of task times in the sample and M_{ct_i} is the elapsed time required for the *i*th individual corrective maintenance action.

5. *Median Active Preventive Maintenance Time* ($\tilde{M}_{pt}$). The Median Active Preventive Maintenance Time is determined through the same methodology, based on the log-normal distribution, as for Median Active Corrective Maintenance Time. It is that time value that delineates the scheduled maintenance time values so that 50% are equal to or less than the median and 50% are equal to or greater than the median. The median for a normal distribution is the same as the mean, as indicated by Figure 2.2.2-1. The median is calculated by means of Equation 2.16.

$$\tilde{M}_{pt} = \text{antilog}\frac{\sum_{i=1}^{n}\log M_{pt_i}}{n} = \text{antilog}\frac{\Sigma(f_{pt_i})(\log M_{pt_i})}{\Sigma(f_{pt_i})} \tag{2.16}$$

In the preceding equation, *n* is the number of task times in the sample, and M_{pt_i} is the elapsed time required for the *i*th individual preventive maintenance action.

6. *Maximum Active Corrective Maintenance Time* (M_{max}). M_{max} can be defined as that value of corrective maintenance time below which a specified percentage of all maintenance actions can expected to be accomplished. M_{max} is related primarily to the log-normal distribution. The 90th or 95th percentile point is commonly designated for the specified value. M_{max} is expressed in the following formula:

$$M_{max} = \text{antilog}[\overline{\log M_{ct}} + Z\sigma_{\log Mcti}] \tag{2.17}$$

where $\overline{\log M_{ct}}$ is the mean of the logarithms of M_{ct_i} and Z is the value corresponding to the specific percentile point at which M_{max} is defined (for example, see Appendix E, Normal Distribution Table; z = + 1.65 for 95%) and $\sigma_{\log M_{ct_i}}$ is given by:[8]

$$\sigma_{\log M_{ct_i}} = \sqrt{\frac{\sum_{i=1}^{N} (\log M_{ct_i})^2 - \frac{\left(\sum_{i=1}^{N} M_{ct_i}\right)^2}{N}}{N - 1}} \qquad (2.18)$$

7. *Logistics Delay Time (LDT).* Logistics Delay Time is that segment of maintenance downtime attributable to elements such as: awaiting delivery and receipt of a spare part after submitting a requisition, waiting for necessary test equipment to perform diagnostics and testing required for a maintenance action, waiting for transportation to provide shipment of the required items, and waiting for use of a facility required for maintenance.

8. *Administrative Delay Time (ADT).* Administrative Delay Time refers to that portion of downtime during which maintenance action is delayed for reasons of an administrative nature. This includes processing and submittal of item requisitions, personnel assignment priorities, labor strikes, organizational constraints, etc. ADT does not include active maintenance time but often constitutes a significant element of total Maintenance Downtime (MDT).

9. *Maintenance Downtime (MDT).* Maintenance Downtime is defined by the total time required (when the system is not operational) to repair and restore a system to full operational status and/or retain a system in operational status. MDT considers Mean Active Maintenance Time ($\overline{M}$), Logistics Delay Time (LDT), and Administrative Delay Time (ADT).[9] The mean or average value is calculated from the elapsed time value for each of the aforementioned functions and the associated frequencies by the same approach used in determining Mean Active Maintenance Time. Figure 2.2.2-4 presents an overview of the elapsed time factors.

---

[8] To illustrate the log-normal distribution methodology, refer to Langford, J.W., *Logistics Principles and Applications*, McGraw-Hill, New York, NY, 1995, pages 508-511.

[9] Maintenance downtime also includes times needed for such factors as cure times for sealants, paints, adhesives, etc.

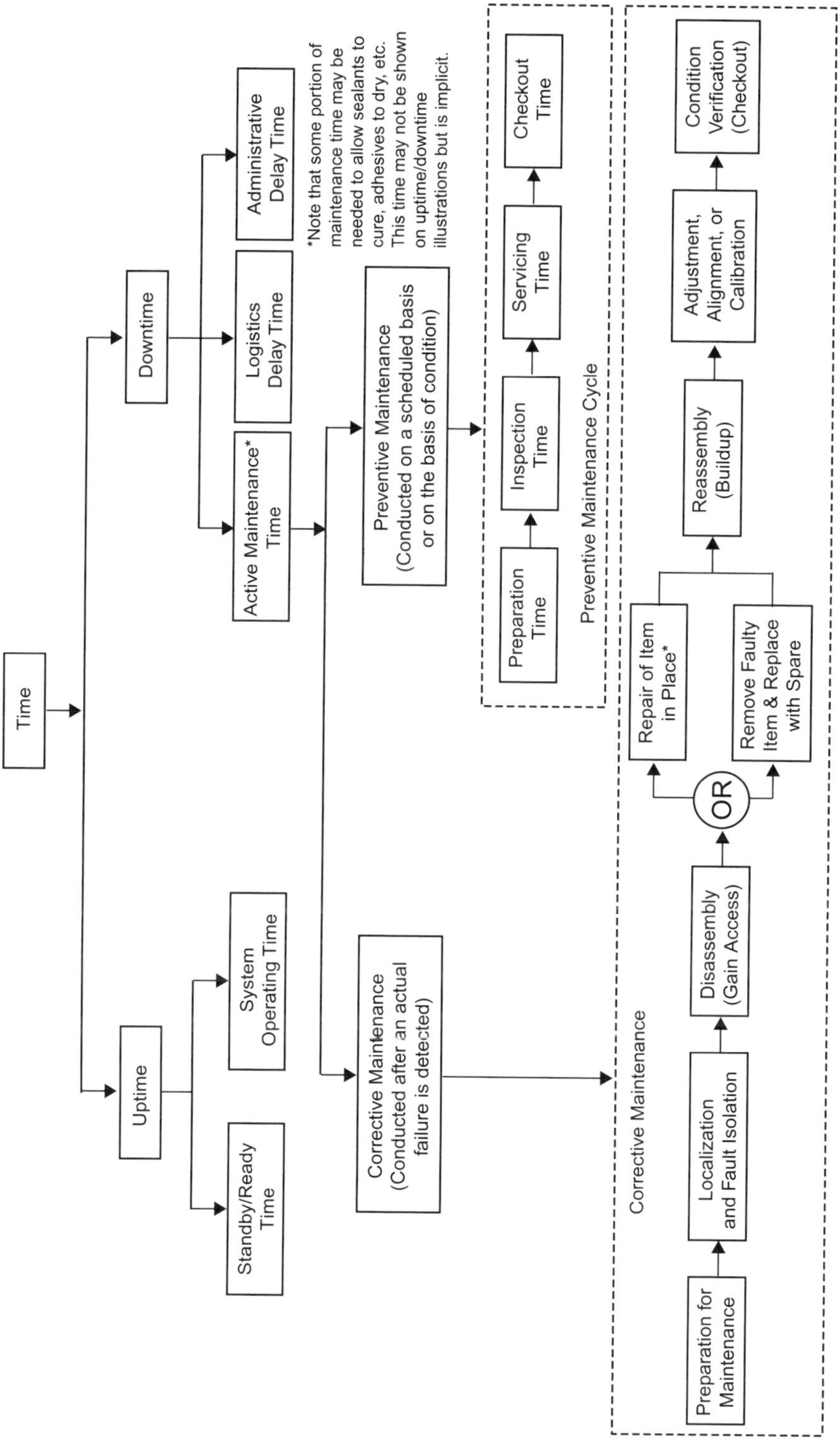

Figure 2.2.2-4. Composite View of Uptime/Downtime Factors

10. *Maintenance Frequency Factors.* Topic 2.1 addresses the measures of reliability, with MTBF and failure rate (λ) being key factors. It is, therefore, obvious that the factors of reliability and maintainability are closely inter-related. The reliability factors, MTBF and λ, provide the basis of determining the frequency of corrective maintenance. Inasmuch as Maintainability concerns those system characteristics pertinent to minimizing corrective maintenance requirements for a system, system reliability, and maintainability requirements must be compatible and mutually supportive.

    Maintainability also deals with those design characteristics that minimize preventive maintenance requirements for the system. Preventive maintenance requirements are sometimes incorporated with the objective of improving system reliability (e.g., reducing failures by specifying replacement of selected components at designated intervals). It must be noted that the introduction of preventive maintenance can prove to be very costly if not carefully controlled. The performance of an excess of preventive maintenance (especially for complex systems) often contributes to degradation of system reliability because failures are frequently induced in the process. It is therefore important to ensure the proper balance between corrective maintenance and preventive at the lowest ultimate overall cost.

    a. *Mean Time Between Maintenance (MTBM).* MTBM is the mean or average interval time between all maintenance actions (corrective and preventive) and can be calculated as shown in Equation 2.19.

$$\text{MTBM} = \frac{1}{\dfrac{1}{\text{MTBM}_u} + \dfrac{1}{\text{MTBM}_s}} \qquad (2.19)$$

where:

MTBM_u is the mean interval of unscheduled (corrective) maintenance and MTBM_s is the mean interval of scheduled (preventive) maintenance.

The reciprocals of MTBM_u and MTBM_s constitute the maintenance rates in terms of maintenance actions per hour of system operation. MTBM_u should be equivalent to MTBF (Mean Time Between Failure), assuming that a combined failure rate is used that includes the consideration of primary inherent failures, dependent failures, manufacturing defects, operator-induced failures, maintenance-induced failures, etc. The reciprocal of MTBM_u, in this case, would define the corrective maintenance frequency (λ) and the reciprocal of MTBM_s would denote the preventive maintenance frequency (f_{pt}). Another way to cite the MTBM algorithm would, therefore, be:

$$\text{MTBM} = \frac{1}{\lambda + f_{pt}} \tag{2.20}$$

The average maintenance interval, MTBM, is a constituent element of maintenance frequency and a major parameter in determining system availability and overall effectiveness.

b. *Mean Time Between Replacement (MTBR).* MTBR is contributory to MTBM and refers to the mean time between component/item replacement. It is a major parameter utilized at the inventory control point (depot) level in determining projected spare parts requirements at the wholesale level so as to ensure sufficient stockage to support anticipated demands.

In many cases, corrective and preventive maintenance actions are accomplished without requiring the replacement of a component part. In other instances, item replacements are required which, in turn, necessitate the availability of a spare part and an impact on the total inventory. MTBR is thusly a significant factor, emanating from both corrective and preventive maintenance activities involving item replacement, and is a key parameter in determining total logistics support requirements.

Topic 2.2.3: Maintainability Human and Financial Resources

1. *Maintenance Labor Hour Factors.* The maintainability factors discussed in Topic 2.2.2 relate to elapsed (mean/average or interval) times. Although mean (average) or interval time values are very important in the performance of maintainability, it is essential to consider the maintenance labor-hours required for the process. Mean or average times can be reduced in many instances by applying additional human resources to accomplish specific tasks; however, such augmentation may prove to be an expensive trade-off, particularly when high skill levels might be required to perform tasks that result in reduction of overall clock time. Maintainability is concerned with ease and economy of the performance of maintenance. The primary objective is to achieve the proper balance of elapsed time, labor time, and personnel skills at a minimum maintenance cost.

   When considering measures of maintainability, it is important not only to address such factors as mean active maintenance times and MDT but to also consider the labor-time element. Some of the additional measures that fall into this category are:

   a. Maintenance labor-hours per system operating hour (MLH/OH).
   b. Maintenance labor-hours per cycle of system operation (MLH/cycle).
   c. Maintenance labor-hours per month (MLH/month).
   d. Maintenance labor-hours per maintenance action (MLH/MA).

   For example, one can determine (as a measure) the Mean Labor Hours for corrective maintenance ($\overline{\text{MLH}}$), which may be expressed as:

$$\overline{\text{MLH}} = \frac{\sum(\lambda_i)(\text{MLH}_i)}{\sum \lambda_i} \tag{2.21}$$

   where λ_i is the failure rate (corrective maintenance frequency) of the *i*th item (failures per hour) and MLH_i is the average maintenance labor hours necessary to accomplish the repair of the *i*th item.

   The same methodology can be applied to determine the mean preventive maintenance labor hours as well as mean total maintenance hours (preventive and corrective maintenance). These factors can be predicted for each level of maintenance and are employed in determining specific support requirements and associated costs at the organizational (retail), intermediate, and depot (wholesale) levels of maintenance.

2. *Maintenance Cost Factors*. For many systems/products, maintenance cost is a major segment of total life-cycle cost. Experience has indicated that maintenance costs are governed significantly by design decisions throughout the early stages of system development. Therefore, It is, essential that total life-cycle cost be considered as a major design parameter commencing with the definition of system requirements.

   Of special interest is *economy* in the performance of maintenance actions. Maintainability is directly concerned with those characteristics of system design that will ultimately result in the performance of maintenance at minimum overall cost. In consideration of maintenance cost, the following are typical figures of merit that may be appropriate as criteria in system design:

   a. Maintenance cost per maintenance action ($/MA)
   b. Maintenance cost per system operating hour ($/OH)
   c. Maintenance cost per month ($/month)
   d. Maintenance cost per mission or mission segment ($/mission)
   e. Ratio of maintenance cost to total life-cycle cost.

   Further discussion and presentation of maintenance cost in the context of life-cycle cost is provided in Section 5, *Supportability Analysis (SA)*.[10]

[10] Blanchard, B.S., *Logistics Engineering and Management*, 6th Ed., Pearson Prentice Hall, Upper Saddle River, NJ, 2004, pages 68-72.

Topic 2.3: Human (Ergonomic) Measures

Human Factors Engineering (HFE) pertains to the human element of the system and the interface(s) between the human being, machine, facilities, as well as the associated software and computer resources. Measuring effectiveness for HFE must consider the anthropometric factors, sensory factors, strength and endurance factors, speed and accuracy factors, body movement (kinesiology) factors, physiological factors (e.g., effects from environmental forces), psychological factors (e.g., human needs, expectations, attitude, motivation), and their interrelationships. The following discusses measurable attributes. Details for analyses related to HFE design are included in Topic 4.2.3.

Topic 2.3.1: Anthropometric Factors

1. *Anthropometry.* Engineering anthropometry concerns the application of human physical dimensions and characteristics with regard to the system requirements for workspace and movement. Figures 2.3.1-1 and 2.3.1-2 portray the human body configurations and pertinent anthropometric data elements keyed to the body dimensions.[11]

   a. *Consideration of Statistical Extremes.* It is standard practice to provide for individuals at one extreme of some anthropometric characteristic in order to accommodate virtually the entire population. Measurement criteria would, therefore, focus on the 95th and 5th percentile values if the accommodation of 100 percent would incur trade-off costs disproportionate to the additional benefits that might be derived.

   b. *Consideration of Adjustable Ranges.* It is important that certain equipment permit adjustments for the range of users who fall within the parameters of the 5th and 95th percentile points.

[11] Derived from Blanchard, B.S. and Fabrycky, W.J., *Systems Engineering and Analysis*, 3rd Ed., Prentice Hall, Upper Saddle River, NJ, 1998, pages 463-464.

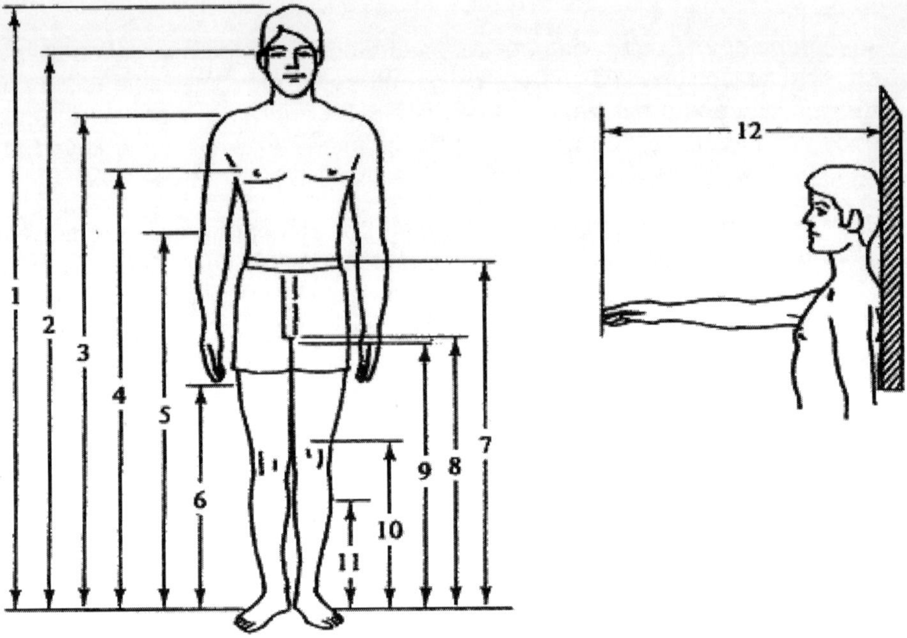

Factors	Percentile values (cm)			
	5th Percentile		95th Percentile	
	Men	Women	Men	Women
Weight (kg)	57.4	46.4	91.6	76.5
Standing body dimensions				
1. Stature	163.8	152.4	185.6	172.2
2. Eye height (standing)	152.1	142.7	173.3	160.1
3. Shoulder (acromiale) height	133.6	123.0	154.2	143.4
4. Chest (nipple) height		110.0		127.3
5. Elbow (radiale) height	104.8		120.0	
6. Fingertip (dactylion) height	61.5		73.2	
7. Waist height	97.5	93.1	115.2	110.1
8. Crotch height	76.3	66.4	91.8	81.4
9. Gluteal furrow height		66.2		79.4
10. Kneecap height	47.5		58.6	
11. Calf height	31.1		40.6	
12. Functional reach	72.7	67.7	90.9	80.4

Figure 2.3.1-1. Anthropometric Data-Standing Body Dimensions

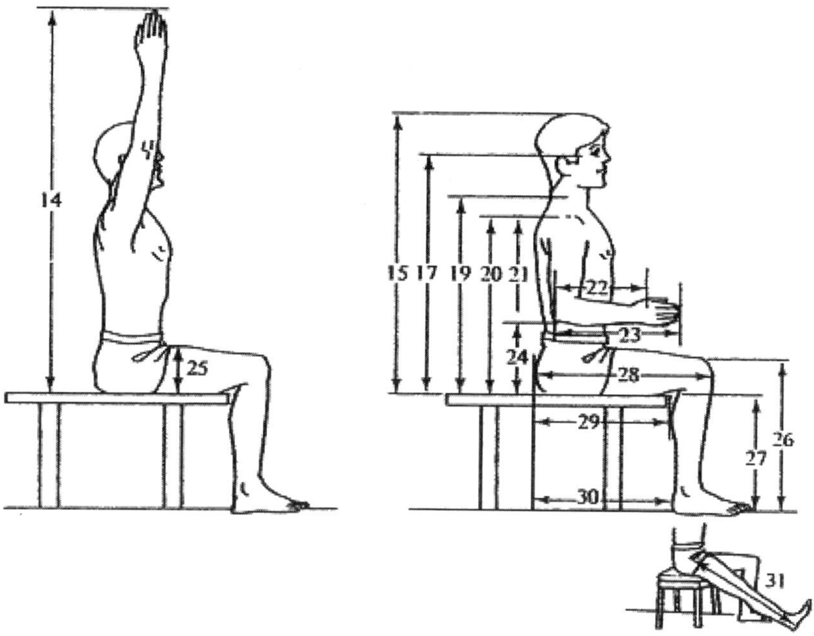

Factors	Percentile values (cm)			
	5th Percentile		95th Percentile	
	Men	Women	Men	Women
Seated body dimensions				
14. Vertical arm reach, sitting	128.6		147.8	
15. Sitting height, erect	84.5	78.4	96.9	90.9
16. Sitting height, relaxed	82.5	76.9	94.8	89.7
17. Eye height, sitting erect	72.8	68.7	84.6	78.8
18. Eye height, sitting relaxed	70.8	67.2	82.5	77.6
19. Mid-shoulder height	57.1	53.7	67.7	62.5
20. Shoulder height, sitting	54.2		65.4	
21. Shoulder-elbow length	33.8	30.2	40.2	36.2
22. Elbow-grip length	32.6		37.9	
23. Elbow-fingertip length	44.3	38.9	51.9	45.7
24. Elbow rest height	17.5	18.7	28.0	26.9
25. Thigh clearance height		10.4		14.6
26. Knee height, sitting	49.7	43.7	58.7	51.6
27. Popliteal height	40.6	38.0	50.0	44.1
28. Buttock-knee length	54.9	52.0	64.3	61.9
29. Buttock-popliteal length	45.8	43.4	54.5	52.6
30. Buttock-heel length	46.7		56.4	
31. Buttock-heel length (diagonal)	103.9		120.4	

Figure 2.3.1-2. Anthropometric Data-Seated Body Dimensions

Topic 2.3.2: Sensory Factors

The five human senses- sight, hearing, touch, smell, and taste describe the stimuli external to the human body. The primary constituent elements of the human factors in human-machine interface are described as the sensory receptors; i.e., eyes (sight), ears (hearing), nasal passages (smell), and epidermal nerves (touch) which serve as the *exterocepters*. These are translated into the *visual, auditory, tactual,* and *olfactory* applications in Human Factors Engineering.[12]

1. *Measures for Visual Applications*. There are measurable variables external to the individual that affect visual discrimination, some of which are:

    a. *Luminance Contrast*: Frequently called brightness contrast.

    b. *Amount of Illumination*: The area or volume to be made visible.

    c. *Time*: The longer the viewing time, the greater the discrimination.

    d. *Luminance Ratio*: The ratio between the luminance of the area of primary visual attention and the surrounding area.

    e. *Glare*: The intensity of illumination.

    f. *Movement*: The movement of a target object and/or the observer.

    g. *Combinations of Variables*: Various combinations of the above which may impact visual performance.

    h. *Visual Acuity*: The ability to differentiate between the detailed features of the various objects observed.

    Of further relevance to the measurable variables are the parameters of the vertical and horizontal visual fields, as shown by Figure 2.3.2-1, and limits of vertical and horizontal color differentiation described by Figure 2.3.2-2.[13]

2. *Measures for Auditory Applications*: Human tolerance for noise is governed by the length of time of exposure and the decibel (dB) level. Prolonged exposure to noise levels exceeding 95 dB is generally unacceptable.

3. *Measures for Tactual Applications*: Sensitivity to pressure reception and changes in temperature are the most significant in tactual applications. The fingertips are the best human indicator of pressure; tactual sensitivity is significantly degraded by low temperatures.

[12] For further discussion see Langford, J.W., *Logistics Principles and Applications*, McGraw-Hill, New York, NY, 1995, p 130, pages 136-143.

[13] Derived from Blanchard, B.S. and Fabrycky, W.J., *Systems Engineering and Analysis*, 3rd Ed, Upper Saddle River, NJ, 1998, pages 469-470.

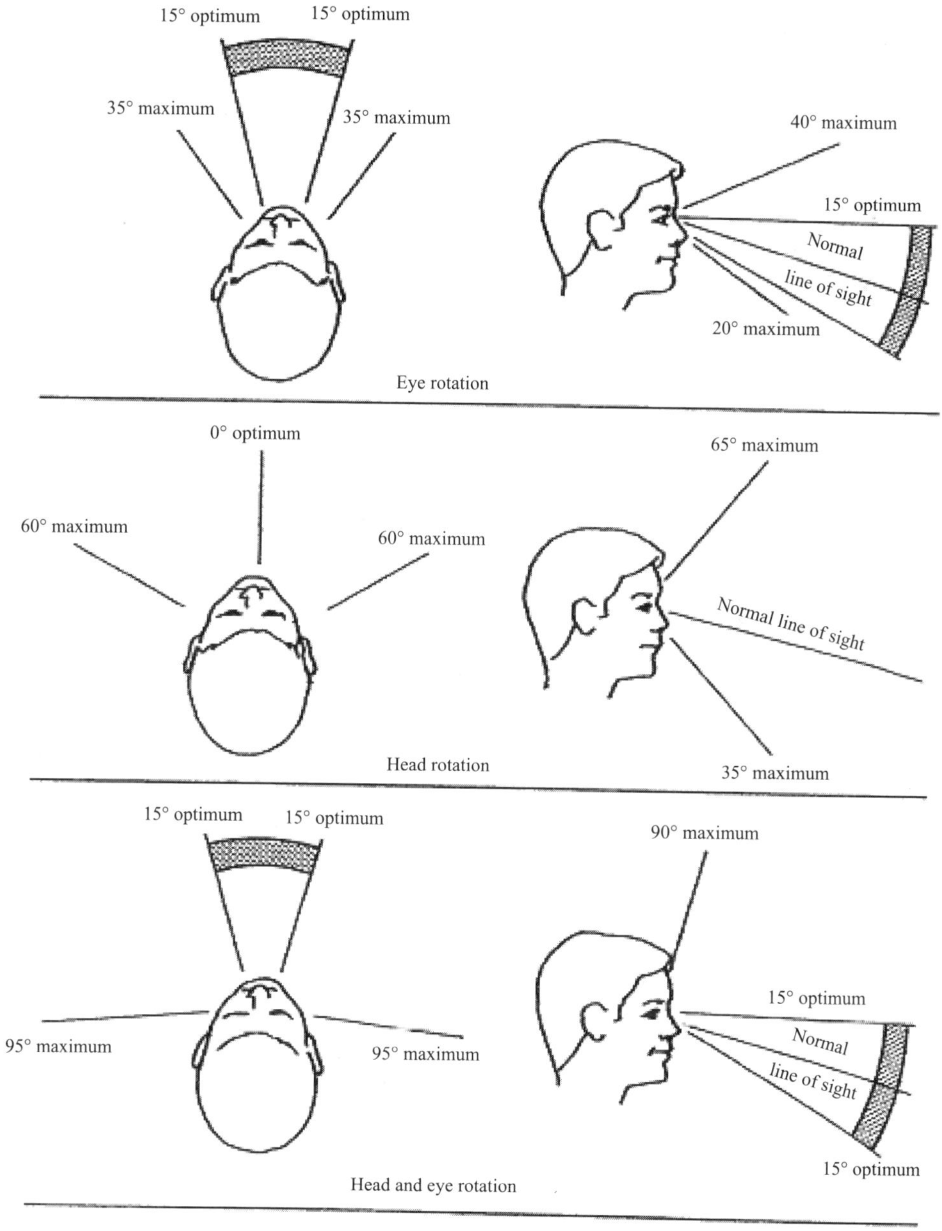

Figure 2.3.2-1. Vertical and Horizontal Visual Field

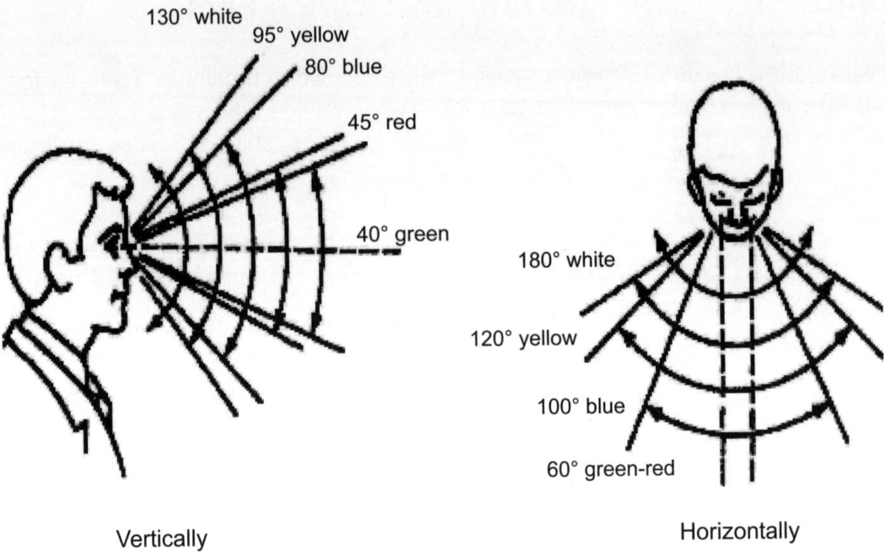

Figure 2.3.2-2. Approximate Limits of Color Differentiation

Topic 2.3.3: Strength and Endurance Factors

Strength is the maximum force muscles can exert isometrically in a single, focused effort; i.e., the muscular capacity to apply force under static conditions. The most frequently observed measurable human characteristics which influence strength and endurance are age, gender, body build, and exercise.[14]

1. *Age*. Strength reaches a maximum by the middle to late twenties and declines gradually thereafter. At the age of 65 strength is normally about 75% of that enjoyed in youth.
2. *Gender.* The strength of women is approximately two-thirds that of men.
3. *Body Build.* Body build is related to strength and endurance. Athletic individuals are generally stronger than others are, but less powerfully built humans may be more efficient.
4. *Exercise*. Exercise can increase strength and endurance to a certain extent. Increases attributable to exercise are usually within 30 to 50 percent above beginning levels.

Figure 2.3.3-1 portrays the results of a study of a sample of human male subjects who made movements in each of a number of directions as shown in Figure 2.3.3-1(a). The results are plotted in Figure 2.3.3-1(b), which shows the maximum strength of the 5th percentile group. The 5th percentile group is typically used for this type of research in order to support the assumption that 95 percent of the male population would have at least that level of strength.[15]

If one contemplates the endurance of humans to maintain a certain muscular force, it can be stated that such ability is related to the magnitude of the force. This is emphasized by Figure 2.3.3-2, which depicts the general pattern of endurance time as a function of force requirements of the task.

[14] Derived from Langford, J.W., *Logistics Principles and Applications*, McGraw-Hill, New York, NY, 1995, page 128.

[15] For further details on HFE, see Sanders, M.S. and E.J. McCormick, *Human Factors in Engineering and Design*, 7th Ed, McGraw-Hill, New York, NY, 1993.

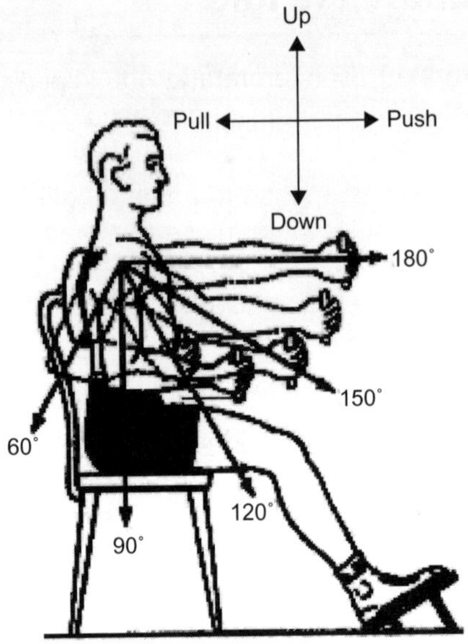

(a)

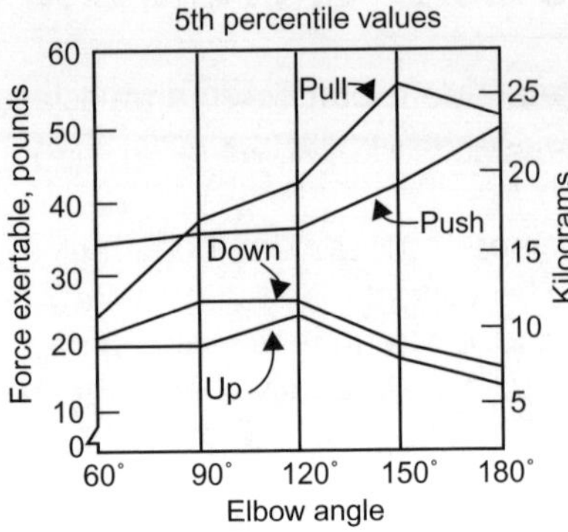

(b)

(a) Side view of subjects being tested for strength in executing push, pull, up, and down movements at each of five arm positions.

(b) Maximum arm strength of the 5th percentile of 55 male subjects.

Figure 2.3.3-1 (a)/(b). Maximum Arm Strength based on 5th Percentile Group

(Source: P.A. Hunsicker, *Arm strength at selected degrees of elbow flexion* (Technical Report 54-548), U.S. Air Force, WADC, 1955)

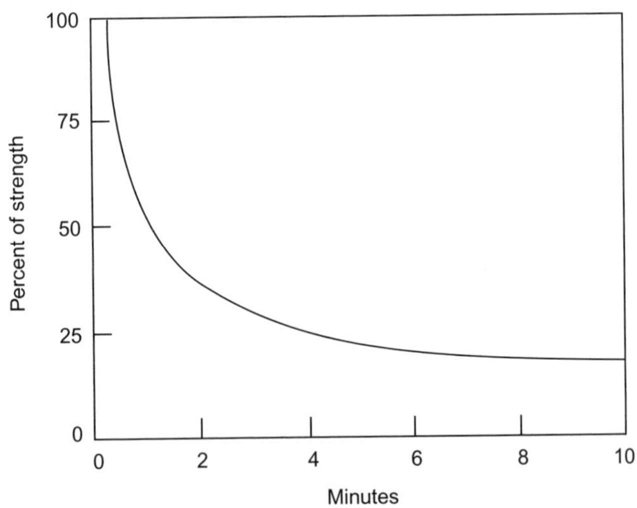

Figure 2.3.3-2. Endurance Time as a Function of Force Requirements

(Source: K.H.E. Kroemer, *Human Factors*, 1970)

Topic 2.3.4: Speed and Accuracy Factors

For measurement purposes, speed is described by interpretation of *stimuli*, while accuracy is defined by how well humans accomplish their functions.[16]

1. *Speed.* Most movements are triggered by response to an external stimulus. The time elapsed in responding to a stimulus determines the reaction time and, consequentially, the speed.
2. *Accuracy.* The accuracy of human performance concerns assessment of human errors: (1) intentional errors, (2) unintentional errors, and (3) omissions.

[16] Extracted from Langford, J.W., *Logistics Principles and Applications*, McGraw-Hill, New York, NY, 1995, pages 129-130.

Topic 2.3.5: Body Movement Factors

Measurement of human body movements features the attributes and dynamics of the human musculoskeletal system-also termed as *kinesiology*. The basic measurable types of movements performed by body members are described in the following.[17]

1. *Flexion*: Bending, or decreasing the angle between the parts of the body.
2. *Extension*: Straightening, or increasing the angle between parts of the body.
3. *Adduction*: Moving toward the midline of the body.
4. *Abduction*: Moving away from the midline of the body
5. *Medial Rotation*: Turning toward the midline of the body.
6. *Lateral Rotation*: Turning away from the midline of the body.
7. *Pronation*: Rotating the forearm so that the palm faces downward.
8. *Supination*: Rotating the forearm so that the palm faces upward.

The basic movements just described are illustrated in Figure 2.3.5-1, along with their corresponding angular values. As with other areas of biomechanics, consideration must be given to individual differences such as effects of physical condition and age.

[17] For further details see Langford, J.W., *Logistics Principles and Applications*, McGraw-Hill, New York, NY, 1995, pages 130-133.

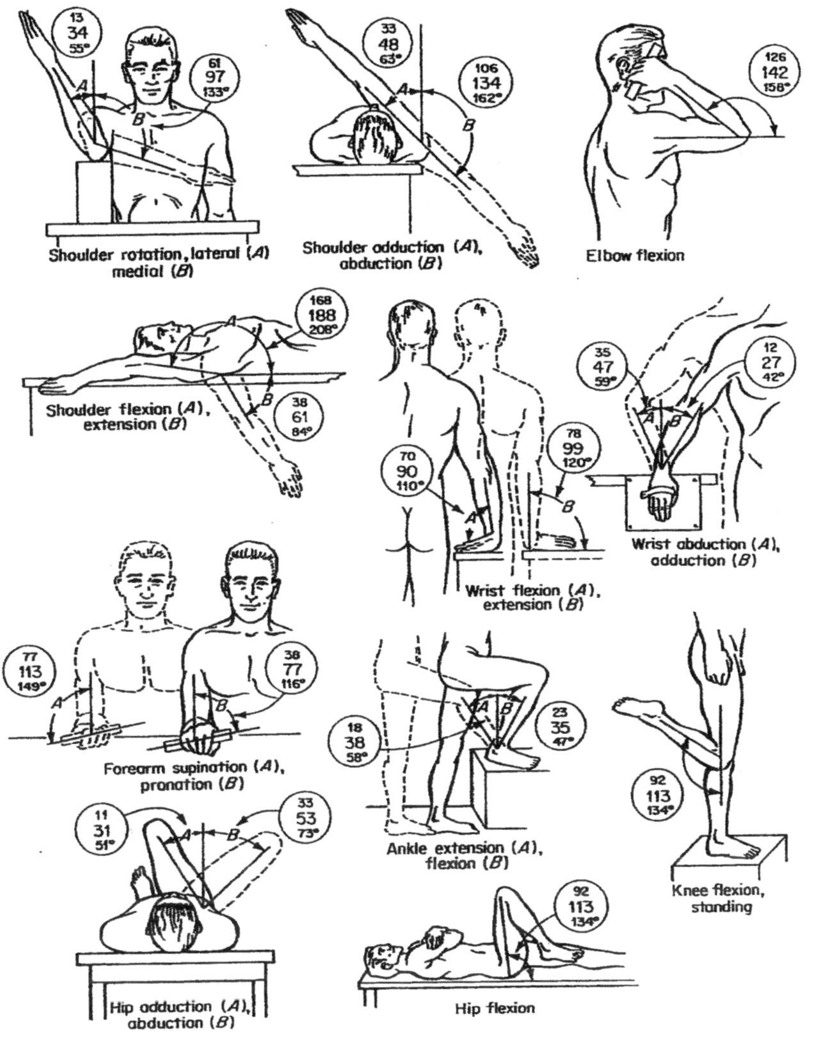

Range of certain movements of the upper and lower extremities, based on a sample of 39 men selected to represent the major physical types in the military services. The three values (in degrees) given for each angle are the 5th percentile, the mean, and the 95th percentile, respectively, of voluntary (not forced) movements. (*Source: Dempster, 1955, as reanalyzed by Baxter, Emanuel, & Truett, 1967.*)

Figure 2.3.5-1. Body Movements

Topic 2.3.6: Physiological Factors

Physiological factors are governed by the immediate environment in which the human must perform. The primary physiological considerations concern the physical space and related facilities which people use, ranging from the immediate environment (e.g., work station), the working environment (e.g., office or factory), to the general environment (e.g., neighborhood, community or city). Equally important are the various aspects of the ambient environment, such as temperature, illumination, atmospheric conditions (including pollution and toxicity), and noise. Pertinent criteria relating to the immediate environment are addressed by previous topics in this section.

Additionally, Table 2.3.6-1 provides the applicable task-related illumination levels; Figure 2.3.6-1 illustrates the Vibration exposure parameters.[18]

Because of the complexities of the relationships among the factors that govern the effects of atmospheric conditions, the Comfort-Health Index (CHI) has been developed which provides data on the sensory, physiological, and health responses usually associated with prolonged exposures at the equivalent temperatures. Table 2.3.6-2 shows the CHI based on both Centigrade and Fahrenheit values.[19]

[18] Derived from Blanchard, B.S. and Fabrycky, W.J., *Systems Engineering and Analysis*, 3rd Ed., Prentice Hall, Upper Saddle River, NJ, 1998, pages 468-472.

[19] For further details see Langford, J.W., *Logistics Principles and Applications*, McGraw-Hill, New York, NY, 1995, pages 154-158.

Table 2.3.6-1. Specific Task Illumination Requirements

(This information (in part) was extracted from MIL-STD-1472, Military Standard, *Human Engineering Design Criteria for Military Systems, Equipment, and Facilities*, Department of Defense, Washington, DC.)

Illumination Levels		
Type of Task	Foot-Candles*	
	Recommended	Minimum
1. Business machine operations	100	50
2. Console surface	50	30
3. Dials, gages, and meters	50	30
4. Factory assembly - general	50	30
5. Factory assembly - precise	300	200
6. Inspection tasks - extra fine	300	200
7. Inspection tasks - rough	50	30
8. Office work, general	70	50
9. Ordinary seeing tasks	50	30
10. Panels	50	30
11. Passageways and stairways	20	10
12. Reading, large print	30	10
13. Reading, small type	70	50
14. Repair work - general	50	30
15. Repair work - instrument	200	100
16. Testing - extra fine	200	100
17. Testing - rough	50	30
18. Transcribing and tabulation	100	50

*As measured at the task object or 30 in. above the floor.

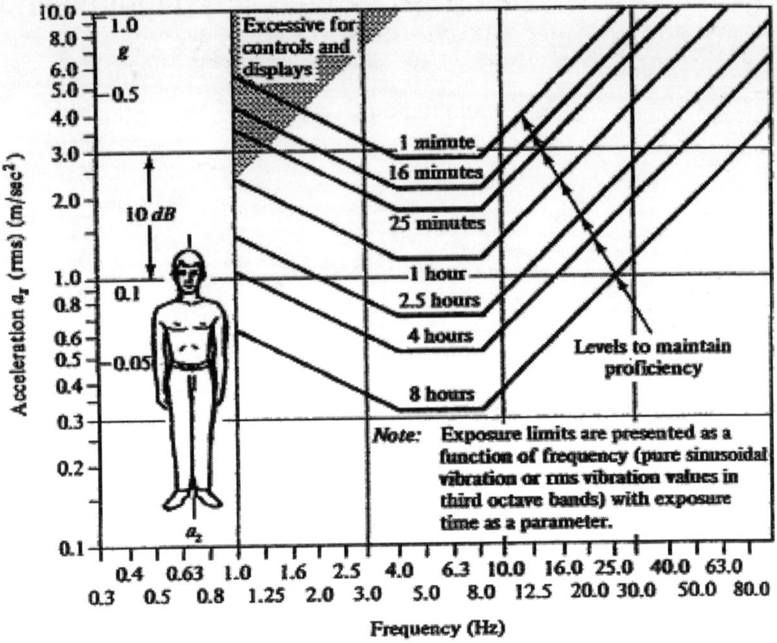

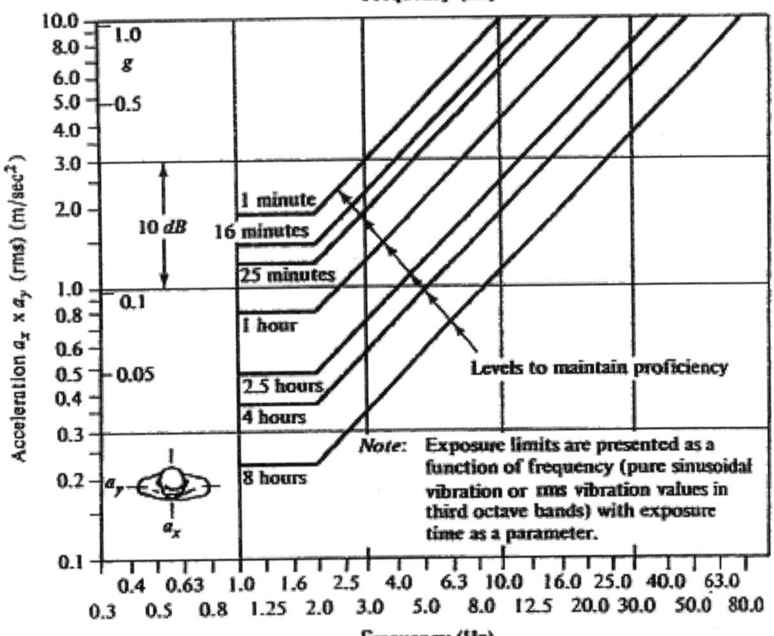

Figure 2.3.6-1. Vibration Exposures Criteria for Longitudinal (Upper Curve) and Transverse (Lower Curve) with Respect to Body Axis

Table 2.3.6-2. Comfort-Health Index

New T_{off} Scale °C	New T_{off} Scale °F	Sensation: Temperature	Sensation: Comfort	Physiology	Health
		Limited tolerance		Body heating	Circulatory collapse
40		Very hot	Very uncomfortable	Failure of regulation Increasing stress caused by sweating and blood flow	↑ Increasing danger of heat strokes, cardiovascular embarrassment
	100	Hot			
35					
	90	Warm	Uncomfortable		
30					
		Slightly warm			
	80				
25		Neutral	Comfortable	Normal regulation by sweating and vascular change Regulation by vascular change	Normal health
	70	Slightly cool			
20					
		Cool	Slightly Uncomfortable	Increasing dry heat loss Urge for more clothing or exercise (behavioral regulation)	
	60				
15		Cold		Vasconstriction	↓ Increasing complaints of dry mucous and skin
10	50	Very cold	Uncomfortable		↓ Muscular pain, impairment of peripheral circulation

Source: ASHRAE, 1977

Topic 2.3.7: Psychological Factors

The human mind, emotion stability, and mental attitude are important elements in formulating the human-machine interface for operation and maintenance of a system. Certain types of systems (for example, the nuclear submarines that spend prolonged mission periods submerged under water) demand critically strict psychological screening of personnel prior to assignment. Among the measurable attributes which constitute psychological criteria are:

- Perceptual ability
- Analytical skills
- Verbal communication skills
- Motivation
- Intelligence
- Persistence
- Innovativeness
- "Team player" qualities
- Dependability
- Work ethic
- Learning proficiency

These human characteristics will be interrelated, with varying emphasis on each attribute, in determining those skills and aptitudes required for the specific system.

Topic 2.4: Safety and Security Measures

For most systems and the users and maintainers of those systems, safety is an important consideration of acceptability. For some systems, safety is a key factor. For example, safety is a key factor in the design, construction, and maintenance of nuclear power plants, commercial aircraft, nuclear weapons, and many other systems. Safety is related to reliability, and many of the analyses associated with one or the other of these two aspects of system performance are beneficial for both. Whereas reliability focuses on preserving functionality, safety focuses on preventing damage or destruction of property and equipment and preventing death or injury of humans.

Security is another issue of great import for many systems and customers. The objectives of designing for security are to detect intruders and unauthorized activities with a high degree of reliability and then to deter and delay them until effective response/engagement can be accomplished. *Physical* security has always and remains an important element of system security. Physical security involves measures to prevent unauthorized access to or use of a system. Today, with the application of world-wide electronic processing and communications in every aspect of our lives, *information* security has become a very critical consideration for data processing and communication systems.

Topic 2.4.1: Safety in the Working Environment

Safety engineering is a companion discipline to and closely interrelated with human factors engineering. The following factors address some of the predominant considerations and factors associated with implementing, managing, and maintaining a safety program in the working environment:

- Ratio of lost labor-hours due to accidents and mishaps to total labor-hours during a designated period.
- Cost per maintenance labor-hour attributable to accidents and mishaps.
- Cost per system operating hour attributable to accidents and mishaps.
- Labor-days per calendar period lost due to accidents and mishaps.
- Number of consecutive days without accidents or mishaps.

Topics 4.2.4 and 5.2.4 describe the approach and analytical methodology for accomplishing a safety hazard and risk analysis based on severity and potential frequency of occurrence of accidents and mishaps that might impact a system.

Topic 2.4.2: Computer and Communication System Security

Unless preventive measures are employed to protect electronic database and communication systems, sensitive organizational activities and information could be subject to compromise or disruption. Typical measures related to this area of security are:

- System hours lost due to downtime caused by disruptions such as "computer virus" attacks, unauthorized access (such as encroachment by "hackers"), electrical power loss, electromagnetic interference, etc., over a prescribed period.
- Cost per operating hour due to computer/communication system downtime.
- Number of operator-induced errors during a prescribed period.
- Number of system-induced errors during a prescribed period.

The aforementioned factors are a typical few of the effectiveness indicators that may be monitored by management. Preventive security devices could include cryptographic encoding systems, "anti-virus" programs, protective "fire-walls", "spam killer" programs, "anti-hacker" programs, and "anti-abuse" programs.

Topic 2.5: Commercial Logistics Measures

The constituent disciplines of commercial logistics are *packaging*, *materials handling*, *warehousing*, *transportation*, *inventory management*, *facility location*, and *logistics information*. These commercial elements are counterparts to the Integrated Logistics Support (ILS) elements of military logistics, as defined in Topic 1.3. The commercial logistics disciplines are inherent to the factors of the primary logistical functions discussed in the subsections that follow.

Topic 2.5.1: Purchasing and Material Flow Factors

Of special interest in the supply chain are the factors associated with *purchasing* and *flow of materials*.

1. *Perfect order fulfillment*, according to the Logistics Management Institute Report LG803R1, reflects the following attributes.[20]
    a. Delivery complete-all items delivered in the quantities ordered.
    b. Delivered on time-according to the customer's definition of on-time delivery.
    c. Documentation complete and accurate-including packing slips, bills of lading, and invoices to support the order.
    d. Delivered in perfect condition, in the correct design configuration needed by the customer and faultlessly installed (as applicable).
2. *Successful delivery percentage* is the number (percentage) of successful deliveries completed relative to the number of orders delivered.
3. *Perfect order percentage* reflects a good quality measure of logistics performance in the supply chain. A logistically "perfect order" meets the following criteria:[21]
    a. Perfectly *entered* (exactly what the customer wants) by means (telephonic or direct entry) the customer desired in a single entry.
    b. Perfectly *fillable* with the exact quantity of each item available for delivery within the customer-specified delivery window.
    c. Perfectly *picked* with the correct quantities of the correct items.
    d. Perfectly *packaged* with the customer-designated packaging and labeling.
    e. Perfectly *shipped* without damage.
    f. Perfectly *delivered* within the customer-designated time window and to the customer-designated location.
    g. Perfectly *communicated* with order status reports available 24 hours a day.
    h. Perfectly *billed* with on-time payment.
    i. Perfectly *documented* with customer-specified documentation means, including paper, fax, EDI, and/or Internet.

The performance metric constitutes a composite based on the percentage of time that each of the nine logistics activities was performed correctly.

[20] LG803R1, *Supply Chain Management: A Recommended Performance Measurement Scorecard*, LMI, McLean, VA, June 1999.

[21] Frazelle, E.H., *Supply Chain Strategy: The Logistics of Supply Chain Management*, McGraw-Hill, New York, NY, 2002, Chapter 3, Section 3.3, page 56.

Topic 2.5.2: Packaging, Handling, and Transportation Factors

Packaging, Handling, and Transportation (PH&T) entail (1) the process of protective and secure packaging design for shipment, (2) timely movement from the shipper to the customer, and (3) retail and bulk package configuration that will efficiently and economically expedite movement within a facility and exploit the capability of standardized pallets, containerization, and handling systems used by the selected transport modes or multi-modal arrangements.

1. *Packaging Measures.* Design of packaging must address the requirements for *mobility* and *transportability.* The following measurement factors, which are typical indicators of successful or inadequate packaging, must be considered (in all cases, the number or percentage is relative to packages shipped):

    a. Number (percentage) of packages damaged en route due to inadequate packaging procedures.

    b. Number (percentage) of packages lost or misrouted due to improper identification of the destination on the address labels.

    c. Number (percentage) of packages with contents sabotaged or pilfered en route due to inadequate security provided by packaging materials.

    In summary, packaging for products that are to be transported must be designed in such a way as to prevent damage, destructive loss, possible degradation, etc.

2. *Transportation Factors.* The function of transportation is the movement of humans and materials from one location to another in the most cost-effective manner possible. The following factors relate to the effectiveness and economy of transportation:

    a. Transportation route, both intra-national and international (e.g., distances, geographical locations, number of nationalities, customs requirements, available labor resources, political, and social factors).

    b. Transport capability and capacity [e.g., type and volume of goods transported, number and frequency of loads, ton-miles per month/year, modes of transportation, legal forms of carriers (*common*, *exempt* or *private*), regulatory requirements, security/safety provisions].

    c. Transportation time (e.g., short-haul versus long-haul, mean delivery time, time per carrier per mode, time per transportation leg).

    d. Transport cost (e.g., cost per shipment, cost of transport per mile/kilometer, cost of packaging and handling).

The basic modes of transportation are *air, highway (motor), pipeline, railroad*, and *water (inland and ocean)*. Combinations of the individual modes are categorized as intermodal, as illustrated by Figure 2.5.2-1.[22]

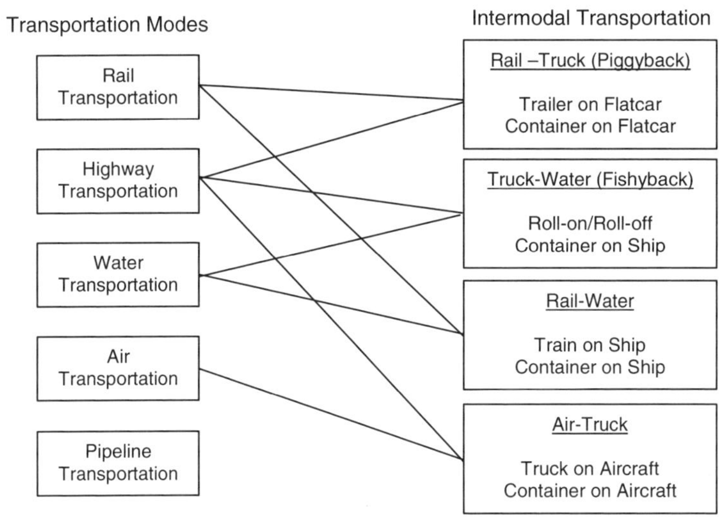

Figure 2.5.2-1. Transportation Modes and Intermodal Interfaces

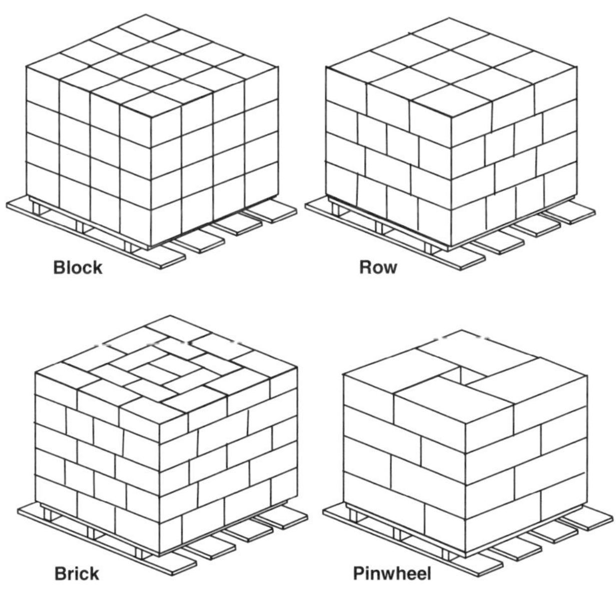

Figure 2.5.2-2. Examples of Possible Pallet Stacking

[22] For a more comprehensive discussion, refer to Langford, J.W., *Logistics Principles and Applications*, McGraw-Hill, New York, NY, 1995, pages 346-354.

3. *Handling Factors.* Materials handling enhances supportability by providing an efficient and responsive flow of material within the context of the facilities for which the system was designed, as well as ensuring efficiency of material movement using the various transportation modes.

   a. *Common Methods of Material Handling.* The overriding consideration for materials handling is pallet utility. Of significance is the packing and stacking of goods on pallets, which in turn are transported from one location to another using various types of lift trucks. The most common pallet size is 40 inches by 48 inches; other sizes used are 48 inches by 48 inches, 48 inches by 36 inches, and 40 inches by 32 inches. The prevalent forms of pallet stacking patterns are illustrated by Figure 2.5.2-2.[23]

   b. As stipulated by the shipper, pallets are integrated and loaded into standard intermodal containers designed for in-transit handling between truck, rail, ship, and air movements (as portrayed by Figure 2.5.2-1). Standard containers are usually 8 feet wide, 8 feet high with lengths of 20 feet, 28 feet, 35 feet, or 40 feet or more.

   c. *Metrics of the Transportation Function.* Effectiveness measures for the transportation process include the following.

      - *Availability* of transportation, or the probability that appropriate capability will be available when required.
      - *Reliability* of transportation, or the probability that the available transportation will complete its mission as planned.
      - *Time* required to transport the product between points.
      - *Maintainability* of a transportation capability, or the probability that after failure, such capability can be restored within a specified time with specified resources.
      - *Cost* of transportation (e.g., cost of a one way trip).
      - *Life-cycle cost* of the transportation capability for a designated time period.

The following measures relate to transportation quality indicators.[24]

$$\text{On-time arrival \%} = \frac{\text{Orders arriving within agreed window}}{\text{Total orders}} \tag{2.22}$$

$$\text{Damage \%} = \frac{\text{Orders arriving without in - transit damage}}{\text{Total orders}} \tag{2.23}$$

---

[23] Refer to Langford, J.W., *Logistics Principles and Applications*, McGraw-Hill, New York, NY, 1995, pages 334-337.

[24] Frazelle, E.H., *Supply Chain Strategy: The Logistics of Supply Chain Management*, McGraw-Hill, New York, NY, 2002, Chapter 3, page 60.

$$\text{Claims-free \%} = \frac{\text{Shipments without claims}}{\text{Total orders}} \qquad (2.24)$$

$$\text{Miles between accidents} = \frac{\text{Total miles driven}}{\text{Number of accidents}} \qquad (2.25)$$

The preceding measures necessarily concern the products being transported.

Topic 2.5.3: Warehousing and Distribution Factors

The major objective of warehousing and distribution is to provide an ideal product flow and acceptable level of service between the producer and the customer by providing warehouses and distribution points at designated locations with sufficient inventory levels to respond to local demand.

1. *Major Functions.* The primary functions are:
   a. *Receiving.* The unloading of inbound goods and products.
   b. *Transfer.* The movements of products into the warehouse for storage, movement to areas for specialized activities such as consolidation, and/or movement to outbound docks for further shipment.
   c. *Order Picking.* The selection of products from storage to fill customers' orders.
   d. *Cross-docking.* The movement of products directly from the inbound dock to the outbound dock without intervening facility activities.
   e. *Shipping*: The packing, loading on an outbound carrier, and shipment of products to the designated receiving points.
2. *Critical Measurement Factors*:
   a. *Time* required to ship a product (from initial notice of a requirement).
   b. *Cost* of each product shipment (from storage to the receiving point).
   c. *Cost* of inventory holding and management for each item in storage.
   d. *Value* of products shipped/*value* of the overall inventory.
   e. *Percentage* of space utilized and *cost* per area of utilization per item.

   The cost effectiveness of the warehousing and distribution can be evaluated by application of the following algorithm:[25]

$$\text{Measure of effectiveness (C)} = a + \frac{b}{V} + c\sqrt{A} \qquad (2.26)$$

where:

C = Cost per dollar's worth of goods distributed.

[25] This measure of effectiveness was provided by Langford, J.W., in a paper entitled "Influence of Logistics Engineering on Commercial Supply Chain Management", *Proceedings of the 18th International Logistics Congress*, Munich, Germany, 2002.

V = Volume of goods handled by the facility per unit of time.

A = Area in square miles covered by the storage facility.

a = Cost per dollar's worth of goods distributed independent of cost of storage and warehousing.

b = Fixed cost of storage facility per unit of time.

c = Cost per mile per dollars worth of goods (vehicle driver's labor, operating costs, etc.) to distribute the product.

3. *Quality Indicators.* The basic quality indicators for a warehouse/distribution center are *inventory accuracy*, *picking accuracy*, *shipping accuracy*, and *warehouse damage percentage.* These measures can be quantified as follows:[26]

$$\text{Inventory accuracy} = \frac{\text{Number of warehouse locations without discrepancies}}{\text{Total number of warehouse locations}} \tag{2.27}$$

$$\text{Picking accuracy} = \frac{\text{Number of items picked without errors}}{\text{Total number of items picked}} \tag{2.28}$$

$$\text{Shipping accuracy} = \frac{\text{Number of items shipped without errors}}{\text{Total number of items shipped}} \tag{2.29}$$

$$\text{Warehouse damage percentage} = \frac{\text{Value (\$) of warehouse damages per year}}{\text{Value (\$) of shipments per year}} \tag{2.30}$$

[26] Frazelle, E.H., *Supply Chain Strategy: The Logistics of Supply Chain Management*, McGraw-Hill, New York, NY, 2002, Chapter 3, Section 3.3, pages 60-62.

Topic 2.5.4: Supply Chain Factors

The supply chain encompasses those functions associated with the forward flow of activities depicted in Figures 1.3-1, 1.3-2, and 1.3-3. The measurements pertaining to the supply chain entail technical, economic, and quality perspectives.

1. *Technical Perspective.* This is the top level overview that may be stated by the following criteria:
   a. *Capability.* The ability to accomplish all the functions required (e.g., purchasing, materials handling, transportation, warehousing).
   b. *Availability.* The ability to respond to any, or all, of the requirements at any time, when needed.
   c. *Quality.* The process responsiveness as indicated in terms of the objectives illustrated by Figure 2.5.4-1.
2. *Cost Perspective.* To complete the proper balance implied by Figure 2.5.4-1, it is necessary to evaluate the issue of cost and, more precisely, the life-cycle cost (LCC) associated with the infrastructure portrayed by Figure 1.3-1 and its impact on the LCC for the system overall. The total cost of the infrastructure may be broken down as shown in the following equation.[27]

Total Logistics Cost (TLC)

$$\text{TLC} = \begin{bmatrix}\text{Cost of}\\ \text{Development}\\ \text{of Logistics}\end{bmatrix} + \begin{bmatrix}\text{Cost of}\\ \text{Producting}\\ \text{Logistics Elements}\end{bmatrix} + \begin{bmatrix}\text{Cost of}\\ \text{Operational}\\ \text{Logistics}\end{bmatrix} + \begin{bmatrix}\text{Cost of}\\ \text{Maintenance and}\\ \text{Support (Sustaining)}\end{bmatrix}$$

$$\Downarrow \qquad (2.31)$$

$$\begin{bmatrix}\text{Cost of Material}\\ \text{Acquisition}\\ \text{(Purchasing)}\end{bmatrix} + \begin{bmatrix}\text{Cost of}\\ \text{Material Processing}\\ \text{(Material Handling)}\end{bmatrix} + \begin{bmatrix}\text{Cost of}\\ \text{Distribution}\\ \text{(Transportation)}\end{bmatrix} + \begin{bmatrix}\text{Cost of}\\ \text{Customer}\\ \text{Service}\end{bmatrix}$$

[27] Based on Blanchard, B.S., *Logistics Engineering and Management*, 6th Ed., Pearson Prentice Hall, Upper Saddle River, NJ, 2004, pages 89-90.

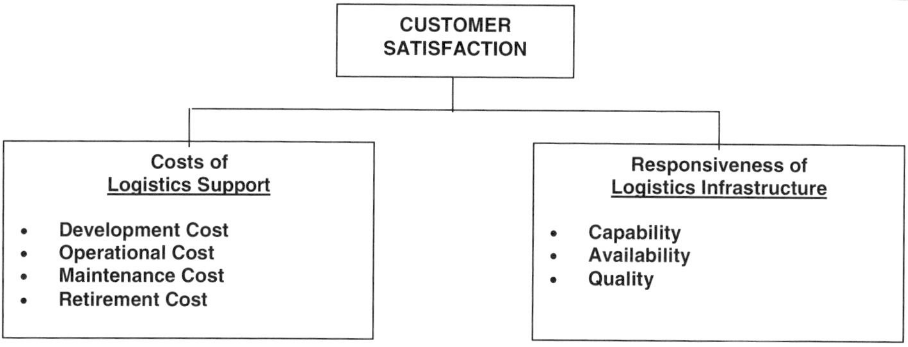

Objectives:

- **The right product:** being able to deliver a quality product in response to a customer need.
- **The right location:** delivering the product where the customer wants it.
- **The right time:** delivering the product when the customer wants it.
- **The right service:** providing fast, efficient, and worldwide support when required.
- **The right information:** providing the customer with the information needed (as perceived by the customer) in real time, 24 hours a day, seven days a week.
- **The right value:** ensuring that the customer believes that he/she has received a great product, with great service, at the right price.

On Every Occasion When Logistic Support is Required!

Figure 2.5.4-1. Supply Chain Goals and Objectives

3. *Quality Perspective*. The term *quality* is related to the degree to which the system is able to attain and maintain a status reflecting complete customer satisfaction. This entails not only system performance and reliability but also the responsiveness of the logistics and maintenance support infrastructure in meeting customer requirements. The following discusses specific quality measurement criteria:

   a. *Response Time*. The time to react in response to an identified need.

   b. *Total Processing Time*. From the identification of a need for a product or service to delivery to the point of consumption or completion of the service.

   c. *Total Processing Cost*. Total cost for processing an item through the logistics and maintenance support infrastructure (e.g., packaging, handling, documentation).

   d. *Management of Obsolescence*. The process time and cost for removing an obsolete item from the inventory.

   e. *Product Defect Rate*. The defect rate in terms of products delivered or services provided per designated period of time.

Topic 2.6: Maintenance and Support Measures

This topic concerns the basic measures pertaining to system maintenance and support as portrayed in Figure 2.6-1. The focus is, therefore, on the metrics associated with organization; spares, repair parts and related inventories; test and support equipment; transportation and handling; maintenance facilities; computer resources; and technical data.

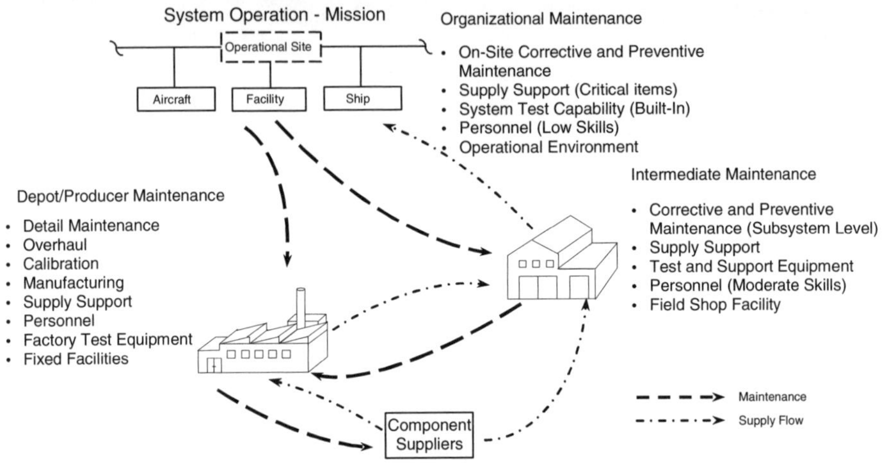

Figure 2.6-1. System Support Infrastructure

Topic 2.6.1: Organizational Factors

The organizational measures associated with the maintenance organization are essentially the same that are typified by any organization. Especially relevant to logistics and system support are the following:[28]

1. *Labor Time*. This factor encompasses both unscheduled and maintenance.
   a. Maintenance labor-hours per system operating hour (MLH/OH).
   b. Maintenance labor-hours per mission cycle (or segment of a mission).
   c. Maintenance labor-hours per month (MLH/month).
   d. Maintenance labor-hours per maintenance action (MLH/MA).
2. *Indirect Labor Time*. Labor time required to support system maintenance activities (overhead).
3. *Personnel Attrition Rate*. Rate (in percentage) of turnover in personnel.
4. *Personnel Training Rate*. Worker-days of formal training per year of system operation and support.
5. *Work Order Process Rate*. Number of maintenance work orders processed per unit of time and average time required for work order processing.
6. *Mean Administrative Delay Time*. Average time from receipt of an item for maintenance to the point when active maintenance actually begins.
7. *Economic Factors*. These entail cost performance criteria, as typified by:
   a. Logistics cost per month.
   b. Cost per maintenance action.
   c. Logistics cost per person in organization.

[28] Based on Blanchard, B.S., *Logistics Engineering and Management*, 6th Ed., Pearson Prentice Hall, Upper Saddle River, NJ, 2004, pages 101-102.

Topic 2.6.2: Spares, Repair Parts, and Related Inventory Factors

1. *Probability of Success with Spares Availability Considerations.* This concerns the probability of system success by having a spare available in time *t* (assuming that failures occur randomly and are exponentially distributed). The computation determining system success probability with *n* spare parts as backup is stated as:

$$P = e^{-\lambda t} + (\lambda t e^{-\lambda t}) + \frac{(\lambda t)^2 e^{-\lambda t}}{2!} + \frac{(\lambda t)^3 e^{-\lambda t}}{3!} + \ldots + \frac{(\lambda t)^n e^{-\lambda t}}{n!} \qquad (2.32)$$

The objective is to determine the probability that x failures will occur if an item is placed in operation for t hours and each failure is corrected with an item replacement as it occurs. With n items in the system, the number of failures will be $n\lambda t$ and the general Poisson expression becomes:

$$P = \sum_{x=0}^{x=n} \frac{(n\lambda t)^x e^{-n\lambda t}}{x!} \qquad (2.33)$$

where:

P = Probability of success (spare available)
n = Number of "back-up spares" available for system
t = System operating cycle
λ = Failure rate

2. Spare-Part Quantity Determination. Spare part quantity determination is a function of probability of having a spare part available in inventory when required to respond to a demand. Equation 2.34, derived from the Poisson distribution, is useful for spare part quantity determination.

$$P = \sum_{n=0}^{n=s} \frac{(R)[\{-\ln R\}^n]}{n!} \qquad (2.34)$$

where:

P = Probability having a peculiar spare part available when required
S = Number of spares carried in inventory
R = Composite component reliability; $R = e^{-K\lambda t}$
K = Quantity of spare parts used

lnR = Natural logarithm of R
n = Designator for quantitative sequence of individual spares; e.g., n = 0, n = 1, n = 2,.....n = s.

Figure 2.6.2-1 shows a spares nomograph which greatly simplifies determination of spares and which is an alternative to the formula.

The scalar structure of the spares nomograph is:

Scale (1): K = Number of parts

Scale (2): λ = Component failure rate (expressed as failures per 1,000 operating hours)

Scale (3): Index line for plot purposes

Scale (4): T = Reorder cycle (expressed as hours or calendar units)

Scale (5): KλT = Composite reliability, a product of scalar elements 1, 2, and 4 (hours)

Scale (6): P = Probability of part availability, also noted as confidence Value

Scale (7): S = Number of spares required to be on hand to assure availability at specified confidence level P

To illustrate use of the spares nomograph, assume the following:

K = 20
λ = 0.1 (Failures per 1,000 hours)
T = 3 months, based on 24 hour per day system operation (2,160 hours)
P = 0.95, desired probability (or confidence) of having a spare available when desired

Sequential plot steps:

Step 1 Connect the scale (2) point where λ = 0.1 to the point at scale (1) where K = 20 with a straight line, thereby intersecting the index line, scale (3).

Step 2 Plot a straight line from the point on scale (4) where T = 3 months or 2,160 hours, through the point where the index line intersects the line previously plotted during Step 1, to scale (5). This line will join scale (5) at the point where KλT is 4.32.

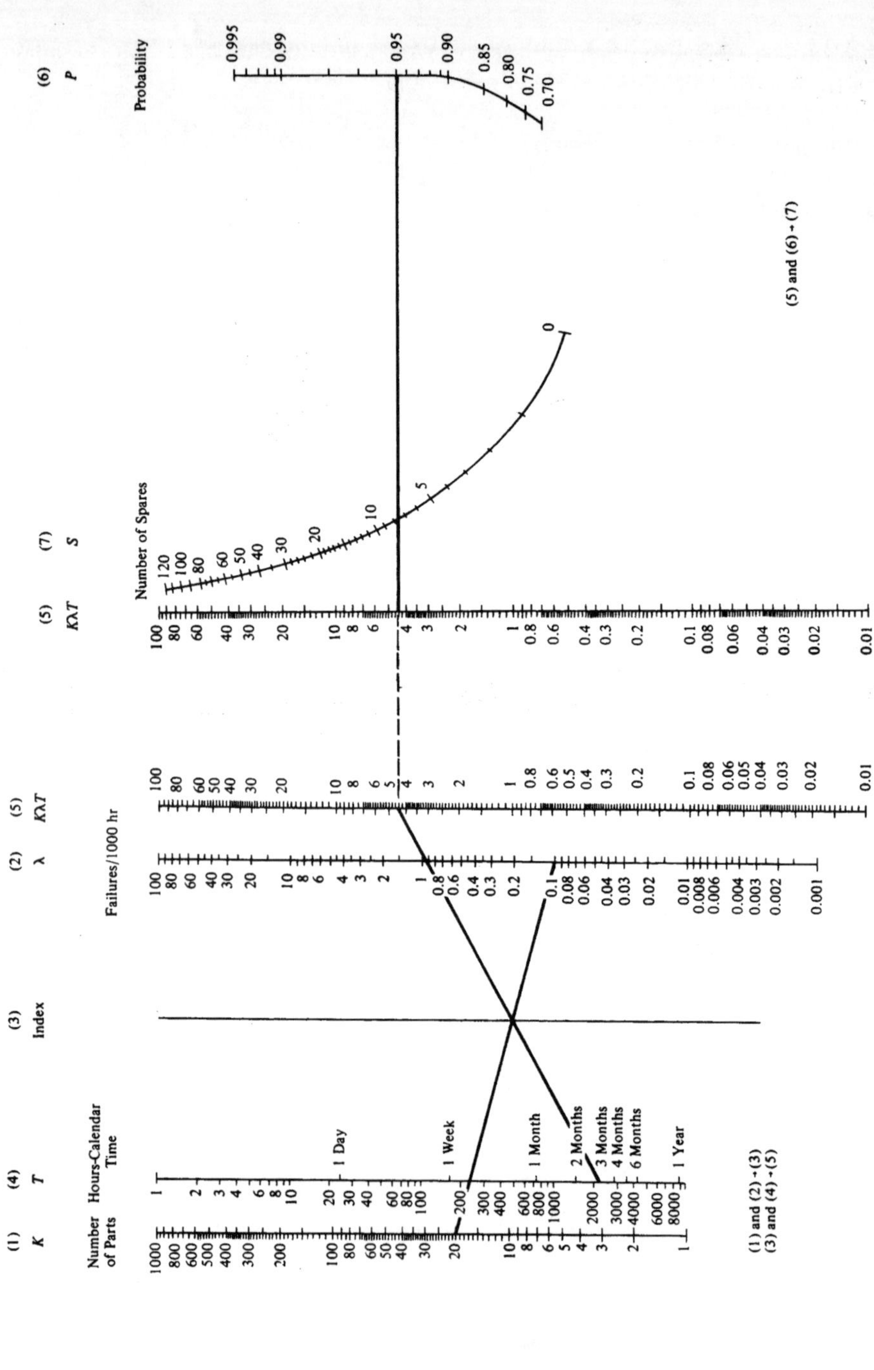

Figure 2.6.2-1 (a). Spare Part Requirement Nomograph

Figure 2.6.2-1 (b). Spare Part Requirement Nomograph

Step 3 Plot a straight line from this point, 4.32 on scale (5), to the 0.95 probability point P on scale (6). This line will intersect scale (7) at the point where S = 8.

Step 4 The value of S determined by the intersection of the Step 3 plot line with scale (7) is the required spares level. For this case, the required level is eight (8) spare parts.

Compared to using the complex algorithm, the utility to the analyst of the spares nomograph is obvious.[29]

3. *Inventory Considerations.* This perspective concerns the inventory cycle, cost of inventory, and economic order quantity (EOQ).

   a. *Inventory Cycle.* The inventory cycle is based on the following elements:

      - *Operating Level.* Quantity of items required to support normal demands during the interval between item orders and deliveries.
      - *Safety Stock.* Additional stock required as a cushion for unexpected demands.
      - *Reorder Cycle.* Interval of time between successive orders.
      - *Procurement Lead Time.* Time span between date of item order to receipt of the shipment in inventory.
      - *Pipeline.* Distance between supplier and customer, measured in days of supply (based on daily demand rate on the inventory).
      - *Order Point.* Point in time when orders are issued for replacement items.

Figure 2.6.2-2(a) portrays an idealistic view of the inventory cycle. As demands are not always constant and the reorder cycle might change with time, Figure 2.6.2-2(b) presents a more realistic scenario.

   b. *Inventory Cost.* The total cost of inventory is reflected by the following expression:

$$\text{Total Cost (TC)} = C_i D + \frac{C_h Q}{2} + \frac{C_p D}{Q} \qquad (2.35)$$

where:

C_i = Acquisition cost of inventory item

C_p = Average item ordering cost in dollars per order

[29] The sparing model is discussed in Langford, J.W., *Logistics Principles and Applications*, McGraw-Hill, New York, NY, 1995, pages 387-389.

C_h = Cost of holding an item in inventory (percent of item price multiplied by the item purchase price)

D = Annual Demand for the item (assumed to be constant)

Q = Most economical quantity.

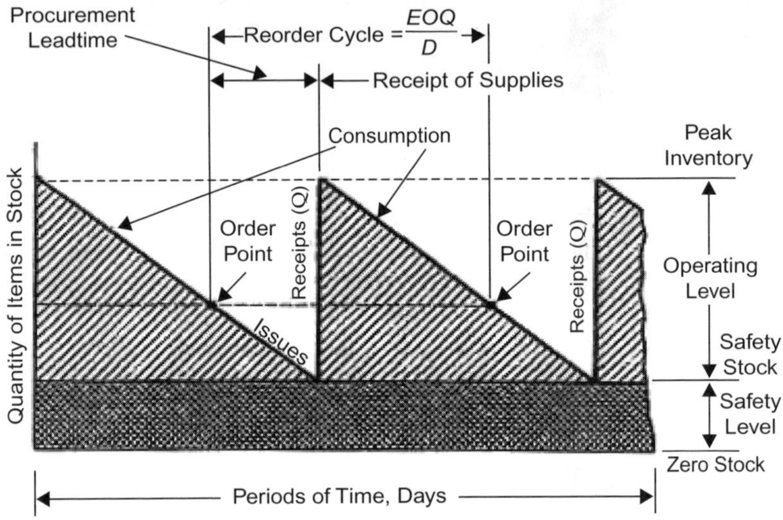

Figure 2.6.2-2 (a). Theoretical Inventory Cycle

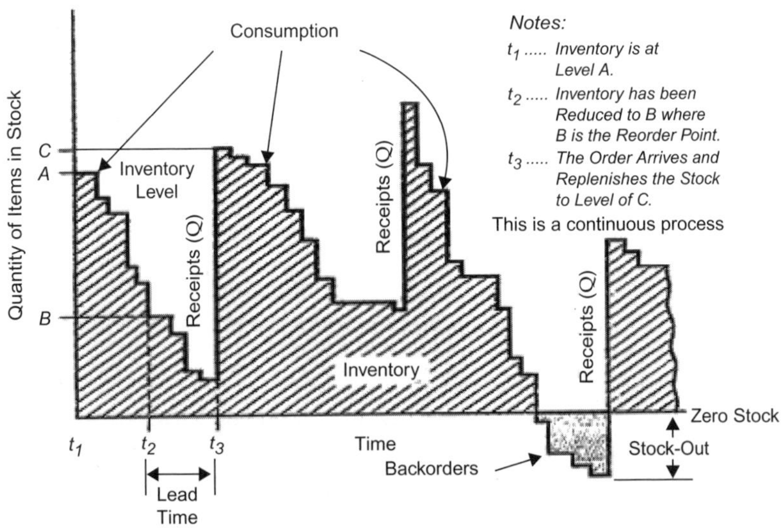

Figure 2.6.2-2 (b). Representation of Actual Inventory Cycle

The inventory cost elements are graphically depicted in Figure 2.6.2-3.

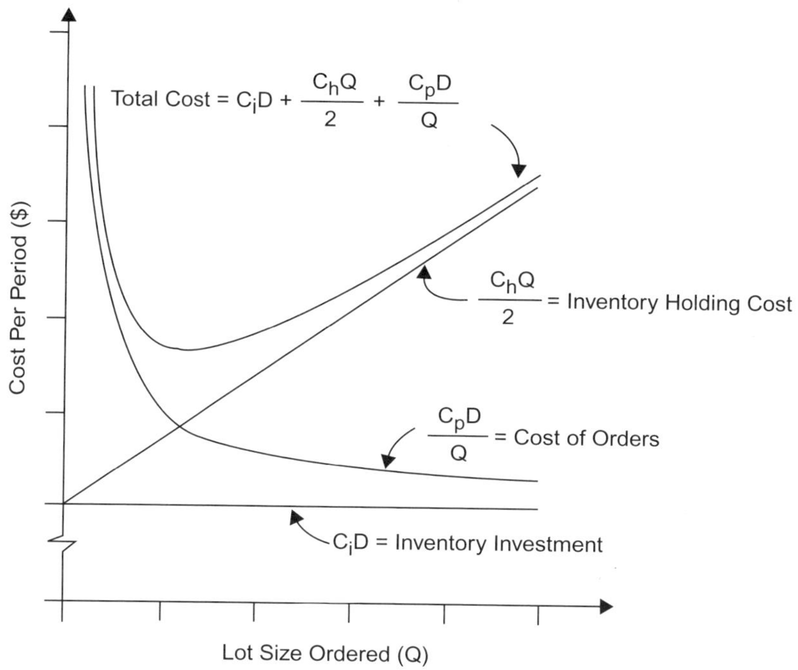

Figure 2.6.2-3. Inventory Cost Considerations

c. Economic Order Quantity. The Inventory Total Cost (TC) formula is the basis for the economic order quantity (EOQ), mathematically derived through the process of Calculus. Determining the first derivative of TC with respect to Q results in the following formula for EOQ:[30]

$$Q * = EOQ = \sqrt{2C_pD/C_h} \tag{2.36}$$

*The EOQ concept shown in Equation 2.36 is a simplified representation of the major consideration in an inventory. There are many variations, and additional material should be evaluated before deciding on a specific application.

[30] The calculus methodology is discussed in Langford, J.W., *Logistics Principles and Applications*, McGraw-Hill, New York, NY, 1995, pages 384-385.

Topic 2.6.3: Test and Support Equipment Factors

The purpose of *test and support equipment* is to sustain, in a cost effective manner, the maintainability parameters of the system, typically the mean preventive maintenance time ($\overline{M}_{pt}$) and mean corrective maintenance time ($\overline{M}_{ct}$). The utilization of test and support equipment involves consideration of the ambiguity group, use of maintenance assistance modules (MAMs), diagnostics effectiveness, and economic effectiveness.[31]

1. *Ambiguity Group.* The term *ambiguity group* refers to the level of testing to which a test unit can localize a group of components or submodules which includes the cause of system failure. Within that grouping, the testing design scheme may employ the option of using individual testing devices to isolate the specific malfunctioning unit or a set of MAMs for that purpose.

2. *Maintenance Assistance Module (MAMs).* MAMs represent a maintenance practice by which each component of a set of specially marked modules or components identical to those in the ambiguity group is substituted for the corresponding module in the ambiguity group until the faulty module is identified. The use of MAMs, or Built-in Test (BIT) diagnostic units, is determined by cost-effectiveness trade-off analyses. In many cases, MAMs may be the preferred economic alternative.

3. *Diagnostics Effectiveness Measures.* The effectiveness of diagnostic equipment can be measured in terms of *utility, reliability,* and *economic criteria.*

   a. *Diagnostic Utility:* The ratio of the failures the diagnostic device is able to detect to the expected total types of failures.

   $$\text{Diagnostic Utility} = \frac{\text{No. of malfunctions detectable}}{\text{No. of potential failures}} \tag{2.37}$$

   b. *Reliability.* The reliability of the testing device in detecting malfunctions (as designed) within the ambiguity group.

   $$\text{Probability of success of detection} = \frac{\text{No. of failures detected}}{\text{No. of failures experienced}} \tag{2.38}$$

   c. *System-level Reliability.* The probability of success in detecting a malfunction at system level, based on system-level fault indicators. This is computed by determining the mathematical product of the reliability values of

[31] These considerations apply primarily to complex electronic test equipment.

the individual testing devices used for measuring ambiguity groups as defined within the system.[32]

d. *Economic Criteria*. Typical economic factors are:

- Cost per individual test program.
- Cost per product tested.
- Cost per maintenance action on test equipment.

[32] See Langford, J.W., *Logistics Principles and Applications*, McGraw-Hill, New York, NY, 1995, pages 445-448.

Topic 2.6.4: Transportation and Handling Factors

Although the material in Topic 2.5.2 addresses some of the major forms of transportation, the focus in this section is on the functions conveyed by Figure 2.6-1, *System Support Infrastructure.* Especially significant are the transportation requirements related to the system maintenance functions. These would, by example, include the following:

1. Transportation of components removed from the system at the organizational/retail maintenance level and returned to intermediate level maintenance or directly to the depot/manufacturer/supplier.
2. Transportation of faulty components removed at the intermediate maintenance level and returned to the depot/manufacturer/supplier.
3. Transportation of the replenishment spare/repair parts, maintenance personnel, and/or support equipment to the locations where maintenance activities are being performed.
4. Transportation of obsolete items for recycling and/or disposal.

The prime objective is to minimize system maintenance downtime (MDT), or, more specifically, logistics delay time (LDT) and administrative delay time (ADT) wherever corrective and/or preventive maintenance are being performed. The critical measures should include *transportation availability*, *transportation time*, and *transportation cost*.

Topic 2.6.5: Maintenance Facility Factors

Facilities are required to support activities involved in performing active maintenance tasks, providing warehousing for spare/repair parts and related inventories, and housing for administrative functions. The primary measures for facilities include the following:

1. Item process time or turnaround time (TAT); i.e., elapsed time required to process an item for maintenance and returning it to full operational status.
2. Facility utilization; i.e., the ratio of the time utilized to the time available for use, percent utilization in terms of space occupancy.
3. Energy utilization in the performance of maintenance; i.e., energy unit consumption per maintenance action, cost of energy consumption per increment of time or per maintenance action.
4. Total facility cost for system operation and support; i.e., total cost per month, cost per maintenance action.

Although specific quantitative measures associated with facilities may vary among individual systems, the aforementioned factors are considered relevant in most cases.[33]

[33] Based on Blanchard, B.S., *Logistics Engineering and Management*, 6th Ed., Pearson Prentice Hall, Upper Saddle River, NJ, 2004, page 115.

Topic 2.6.6: Computer Resources and Maintenance Software Factors

Computer software may be evaluated in terms of language levels or complexity, program length as defined by number of course code lines, cost per maintenance subroutine, or something of a comparable nature. As with equipment, reliability, maintainability, and quality are significant considerations in software development. There has been extensive discussion on how to measure software reliability without complete agreement as to the specific measures or levels of acceptable performance. It has been established that there are distinct differences between the features of software reliability and hardware reliability. This section addresses the unique attributes of software reliability and typical software evaluation models.

1. *Hardware Reliability vs. Software Reliability.* An accepted definition of software reliability is *the probability of failure-free operation of a software component or system in a specified environment during a prescribed period of time.* A failure is defined as an *unacceptable departure of program operation from program requirements.* Table 2.6.6-1 details the variances between the constituent elements of hardware reliability and software reliability.[34]
2. *Computer/Software Evaluation Models.* Some of the models utilized for evaluation of software are described in the subsections that follow.
    a. *Poisson Model (Time related).* The Poisson model is applicable when it assumed that errors can exist randomly and that their emergence is a function of the time the program is run.
    b. *The Musa Model.* The Musa model uses program execution time as the independent variable.[35]
    c. *The Littlewood Models.* The Littlewood models attempt to take account of the fact that different program errors have different probabilities of causing failure.

Additional evaluation models are described in Topic 6.3.7, Reliability Prediction of Software, of the RAC *Reliability Toolkit: Commercial Practices Edition.*

[34] Details discussed in O'Connor, Patrick D.T., *Practical Reliability Engineering*, 2nd Ed., John Wiley and Sons Ltd., Chichester, England, 1985, pages 210-219.

[35] A complete description of the features and example of the Musa model are provided in Langford, J.W., *Logistics Principles and Applications*, McGraw-Hill, New York, NY, 1995, pages 461-464.

Table 2.6.6-1. Hardware Reliability vs. Software Reliability

Hardware	Software
1. Failures can be caused by deficiencies in design, production, installation, use, and maintenance.	1. Failures are primarily due to design errors, with production (copying), use, and maintenance (excluding corrections) having negligible effect.
2. Failures can be due to wear or other energy-related phenomena. Sometimes warning is available before failure occurs.	2. There is no wearout phenomenon. Software failures occur with no warning.
3. Maintenance can be performed to retain the design level of reliability or to restore it after failure.	3. The only repair is by redesign (reprogramming), which, if it removes the errors and introduces no others, will increase reliability.
4. Reliability can depend upon wearout phenomena, stresses, manufacturing quality, or burn-in; i.e., failure rate can be decreasing, increasing, or constant with respect to operating time.	4. Reliability is less dependent on outside factors. Reliability improvement over time may be affected, but the improvement is not related to operating time. Rather, it is a function of the effort put into detecting and correcting failures.
5. Failure can be related to the passage of operating or storage time.	5. Reliability is not related to time in the same way as it is for hardware. Failures occur when a program step or path, which is in error, is executed.
6. Reliability is related to environmental factors.	6. The external environment only affects reliability to the extent it might affect program inputs.
7. Reliability can be predicted in theory from knowledge of design, environmental, and use factors.	7. Reliability cannot be predicted on any physical basis because it entirely depends on human factors in design. Some *a priori* approaches have been proposed.
8. Reliability can sometimes be improved by redundancy.	8. Reliability cannot be improved by redundancy if the parallel program paths are identical because if one path fails, the other(s) will have the same error and also fail. It is possible to provide redundancy by having parallel paths, each with having a different program that was written and checked by different teams.
9. Failure can occur to components of a system in a pattern that to some extent can be predicted based on the stresses on the components and other factors. Reliability-critical lists and Pareto analysis to prioritize failures are useful techniques.	9. Failures are not usually predictable from analyzing separate statements. Errors are likely to exist randomly throughout the program and any statement may or may not be in error. Reliability-critical list and Pareto analysis are not appropriate.

Topic 2.6.7: Technical Data and Information System Measures

The information age has had a major impact on logistics with the advent of computers and associated software, new electronic information networks, and electronic commerce.

1. *Objectives of Technical Data and Information (TD&I) Systems.* The objectives of TD&I include:
    a. Simplifying generation and processing of technical data through better packaging, eliminating redundancies, reducing processing time, and making information more accessible to requiring organizations.
    b. Providing an expeditious means for introduction of design changes and for better configuration management and control.
    c. Providing a mechanism for greater asset visibility through traceability of items in transit and location of items in inventories.
    d. Enabling faster, timely, accurate, and more reliable communications among multiple logistical locations.
2. *Metrics for TD&I Systems.* The major factors for measuring TD&I system effectiveness include the following:[36]
    a. *Logistics Response Time.* Time elapsed from the point when a system support requirement is first identified until that requirement has been satisfied (e.g., time for item provisioning and procurement, time for item shipment from inventory to the user, time required to acquire necessary maintenance personnel or test equipment).
    b. *Data Access Time.* Time required locating and acquiring the required data/information.
    c. *Item Location Time.* Time required locating a given asset whether in use, in transit, or in inventory (asset visibility).
    d. *Information Processing Time.* Time required for the processing of messages.
    e. *Change Implementation Time.* Time required processing and implementing a design change.
    f. *Cost.* Cost of transmission per bit of data, the cost per data access action, etc.

[36] See also Blanchard, B.S., *Logistics Engineering and Management*, 6th Ed., Pearson Prentice Hall, Upper Saddle River, NJ, 2004, pages 116-117.

Topic 2.7: System Retirement and Material Recycling/Disposal Factors

Activities during the system retirement and material recycling and disposal phase include:

- Processing of obsolete items out of the inventory.
- Recycling of obsolete material for other uses.
- Decomposition and disposal of material that cannot be used for other purposes.

The key measurement factors during the system retirement and disposal phase pertain to the *time to process an item out of inventory*. Economic measures include the:

- Cost of processing an item out of the inventory.
- Cost of recycling an item for other uses.
- Cost of decomposing and disposing of an item that has no other application.

All material recycling and disposal processes must necessarily consider environmental impact. A feature peculiar to items with military applications is the process and cost of demilitarization.[37]

[37] Derived from Blanchard, B.S., *Logistics Engineering and Management*, 6th Ed., Pearson Prentice Hall, Upper Saddle River, NJ, 2004, page 117.

Topic 2.8: System Measures of Effectiveness (MOEs)

Figure 2.8-1 portrays the balance between technical factors which can be combined to result in a measure of system effectiveness, and the economic factors which can be combined to represent the total life-cycle cost (LCC) for a system. The integration of these two perspectives results in the overall MOE for the system.

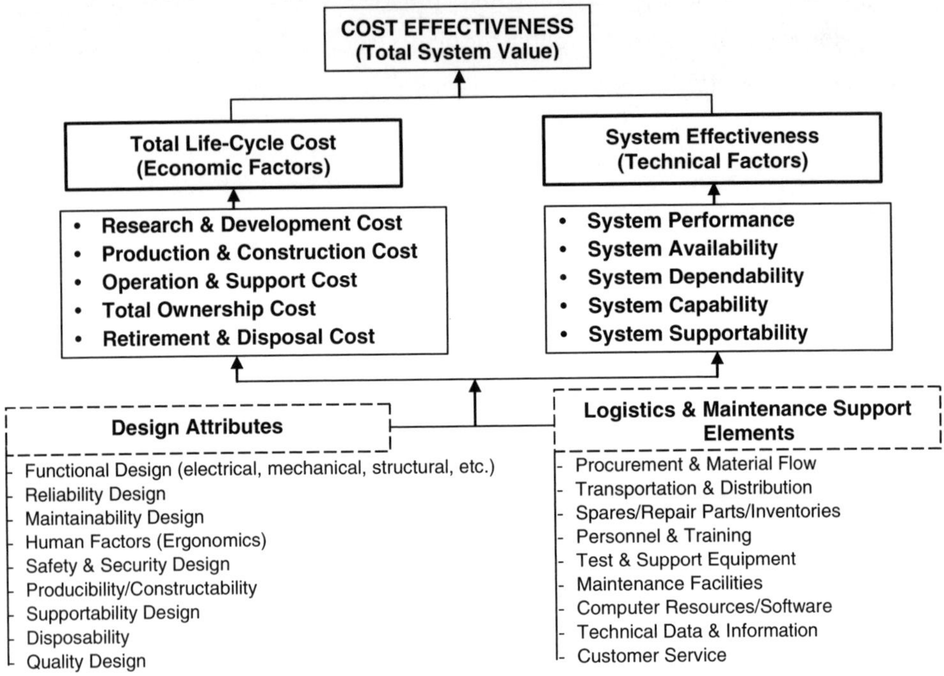

Figure 2.8-1. Basic Ingredients of Cost Effectiveness

Topic 2.8.1: Availability Factors

The term *Availability* is used in a variety of contexts. In simple expression, it is determined by the ratio of the uptime relative to the uptime plus downtime:

$$\text{Availability} = \frac{\text{Uptime}}{\text{Uptime + Downtime}} \tag{2.39}$$

It is frequently used to express system readiness. In some cases, it will connote the probability that a system will not only be ready when required but will complete its mission in a satisfactory manner. *Availability* can also be used as a probability that all is satisfactory at any given point in the accomplishment of a mission. Topic 2.8.2 discusses the perspective of production plant loading where the Availability parameter is used to express available production capacity. This section focuses on *Availability* as related to logistics readiness of a system, considering the three most common figures of merit (FOMs). The specific FOMs are based on the measures of reliability and maintainability (e.g., MTBF, MTBM, MDT, $\overline{M}$, $\overline{M}_{ct}$). In the system readiness context, the three basic parameters of Availability are *Inherent Availability (A_i)*, *Achieved Availability (A_a)*, and *Operational Availability (A_o)*.

1. *Inherent Availability (A_i).* Inherent (or design) Availability is the probability that a system, when used under stated conditions and in an ideal support environment (readily available tools, spare parts, maintenance personnel, etc.), will be ready for commitment to system mission operations at any point in time. It excludes preventive maintenance actions and logistics delay times. The expression for Inherent Availability is:

$$A_i = \frac{\text{MTBF}}{\text{MTBF} + \overline{M}_{ct}} \tag{2.40}$$

   where:

   A_i = Inherent Availability
   MTBF = Mean Time Between Failure
   $\overline{M}_{ct}$ = Mean Corrective Maintenance Time

2. *Achieved Availability (A_a).* Achieved Availability is the probability that a system, when used under an ideal logistics environment (e.g., readily available tools, spare parts, maintenance personnel), will be ready for commitment to system operations at any point in time. This definition is similar to the definition for A_i

except that preventive maintenance is included. As is the case for A_i, logistics delay time is excluded. The equation for Achieved Availability is:

$$A_a = \frac{MTBM}{MTBM + \overline{M}} \tag{2.41}$$

where:

A_a	=	Achieved Availability
MTBM	=	Mean Time Between Maintenance
$\overline{M}$	=	Mean Active Maintenance Downtime

3. *Operational Availability (A_o).* Operational Availability is the probability that a System, when in an *actual* operating environment, will be ready for commitment to system mission operations at any point in time. Operational Availability is expressed as:

$$A_o = \frac{MTBM}{MTBM + MDT} \tag{2.42}$$

where:

A_o	=	Operational Availability
MTBM	=	Mean Time Between Maintenance
MDT	=	Maintenance Downtime

Although this is the conventionally accepted A_o algorithm, there have been variations of the basic equation developed among users to reflect operational availability based on specific mission requirements and utilization of data bases reflecting unique management perspectives pertinent to their missions. All applications of A_o are based on the equational concept of *uptime* divided by *uptime plus downtime.*[38]

[38] The U.S. Navy OPNAVINST 3000.12 stipulates three types of operational availability (A_o) factors: *Continuous Availability*; *Intermittent Availability*; and *Single Shot Availability*.

Topic 2.8.2: System Effectiveness (SE)

As measures of system effectiveness can vary among individual systems, it is necessary to determine the specific system performance factors that relate to the operational requirements. Some combination of these relevant factors may be used to derive an overall measure of system performance. One approach to determining system effectiveness (SE) is to derive a mathematical function based on *performance (P), availability (A),* and *dependability (D),* or

$$SE = (P)(A)(D) \tag{2.43}$$

where P constitutes the appropriate combination of system operational factors (i.e., range, weight, accuracy, size, speed); A is the probability that the system will be ready and operable when activated; and D refers to the probability that a system will successfully complete its mission, given that the system is operable at the start of its mission. One can employ different terms, but it is important to clearly define each term as it relates to the mission scenarios.

A more specific example of a current application of MOE is described by the concept of *total productive maintenance (TPM)* used in the commercial manufacturing plant (i.e., the factory). The objectives of TPM are to:

1. Maximize the overall effectiveness of the equipment and production process by maximizing availability of the production process through improved reliability and maintainability, with the goal of minimizing downtime. This element entails *maintenance prevention (MP)* and *maintainability improvement (MI)* which considers the incorporation of reliability, maintainability, and supportability characteristics in design.
2. Establish a life-cycle approach in the performance of *preventive maintenance (PM)* through application of *reliability centered maintenance (RCM)* methodology.
3. Involve all operating groups, from the top down and throughout the production organization, in the initial planning for as well as subsequent implementation of a maintenance program.
4. Initiate a program based on the promotion of maintenance through *motivational management* and the development of autonomous small-group activities (i.e., the team approach).
5. Include *human factors engineering* as a critical element of the overall system.

2.8.2: SYSTEM EFFECTIVENESS (SE)

The measure of TPM can be expressed in terms of *overall equipment effectiveness (OEE)* as a function of *availability (A), performance rate (P)* and *quality rate (Q)*:

$$\text{OEE} = (\text{availability})(\text{performance rate})(\text{quality rate}) \tag{2.44}$$

where:

$$\text{availability (A)} = \frac{\text{loading time - downtime}}{\text{loading time}} \tag{2.45}$$

loading time = total time (days, months or minutes) available for manufacture of product

downtime = time down due to stoppages

and

$$\text{performance rate (P)} = \frac{\text{(output) (actual cycle time)}}{\text{(loading time - downtime)}} \times \frac{\text{(ideal cycle time)}}{\text{(actual cycle time)}} \tag{2.46}$$

where:

output = number of products produced

ideal cycle time = designed cycle time (e.g., minutes) per product

actual cycle time = actual time (e.g., minutes) experienced per product

and

$$\text{quality rate (Q)} = \frac{\text{(input) - (quality defects + startup defects + rework)}}{\text{input}} \tag{2.47}$$

where:

input = total number of products processed

defects = total number of defects based on established accept/reject criteria

Essentially, the manufacturing plant is treated as a *system*, and the OEE factor is a measure of system effectiveness.

Topic 2.8.3: Economic and Life-Cycle Cost (LCC) Factors

Planning for logistics resources to ensure system life cycle support and making decisions based on economic considerations are driven by the methodologies governing the time value of money. The following will focus on development of system life-cycle cost stream profiles; economic evaluation of alternative systems; analyzing projected cash flow, pay-back points and profitability of proposed equipment production alternatives; and projections based on current and future costs so as to enable future decisions at designated points in time.

1. *LCC Cost Stream.* One side of the spectrum in Figure 2.8-1 concerns economic factors, with life-cycle cost (LCC) being the major area of emphasis. Typical categories of total life-cycle-cost are portrayed by the example in Figure 1.6.12-1. The significant output of the cost data is establishment of a *cost stream*, which tabulates the costs projected on an annual basis throughout the number of years in the system life cycle. The annualized costs are initially actual ("then year") costs, which summarize the estimated costs of all LCC categories for each projected year. The total actual cost for each projected year is then calculated in terms of *present value cost,* which represents that "then year" cost in terms of current equivalent costs, giving consideration to the *discount factor* applied for each year. The present value cost for each projected year is then tabulated to construct a present value cost stream, thereby providing a picture of the system costs for the projected year on a current year cost basis. Summation of all the annual present value costs provides a total cost profile of the system on a current year basis. For illustration of the *cost stream* approach, based on Figure 1.6.12-1, a life-cycle cost summary is constructed to portray the four major cost categories of Research, Design, and Development (C_R); Production and/or Construction (C_P); Operation and Support (C_O); and Retirement/Material Recycling/Disposal (C_D). The LCC summaries of two hypothetical systems (System A and System B) are shown at Figure 2.8.3-1 and Figure 2.8.3-2.

   The *Total Actual Cost* (C) summarizes the total of the four major cost categories ($C_R + C_P + C_O + C_D$). The *Total Present Value Cost* ($C_{5\%}$) shows the current equivalent cost by applying the discount factor to the total actual cost projected for a designated year (n) in the system life cycle. The formula for determining present value cost for each *cost stream* year is:

$$\text{Present value (PV)} = F\left[\frac{1}{(1+i)^n}\right] \tag{2.48}$$

where:

F = Future year cost (specific year in the cost stream)
i = Discount rate (interest rate)
n = The specific year in the cost stream (i.e., n = 5 connotes the fifth year)

As an example, based on Figure 2.8.3-1, determine the present value of the System A cost projected to be a total actual of $220,000 during the fifth year, assuming a discount factor of five percent:

F = $220,000
i = 0.05 (5%)
n = 5

$$\text{Present value (PV)} = F\left[\frac{1}{(1+i)^n}\right] = \$220{,}000\ (0.78353) = \$172{,}377$$

The life-cycle cost summary for System B, depicted by Figure 2.8.3-2, is based on the same methodology as just described. Topic 5.2.1 provides further details as well as an example of life-cycle cost analysis.[39]

2. *System Life-Cycle Cost Profile of Alternative Systems.* The present value approach is an important analytical tool in comparing life-cycle costs of competing alternatives for approach to development, production, operation, and phase-out of a system. In the evaluation of alternative systems, each configuration is represented by a unique life-cycle cost profile, because cost-generating activities will vary from one case to the next. Reliability and maintainability factors will vary and the specific logistics support requirements will be different for each alternative. Figure 2.8.3-3 reflects the cost profiles for System A and System B as determined by cost data shown in Figure 2.8.3-1 and Figure 2.8.3-2.

   The comparison of possible courses of action requires that the costs associated with the alternatives under evaluation be related on an equivalent basis. The point in time when the decision is to be made is usually established as the *present time* or *now.* For each system profile, the costs for each year in the life cycle are discounted to the present value. The plot points for the profiles for System A and System B depicted in Figure 2.8.3-3 correspond to the annualized present-value factors shown in Figure 2.8.3-1 (for System A) and Figure 2.8.3-2 (for System B).

[39] Details and applications of LCC are provided in Langford, J.W., *Logistics Principles and Applications*, McGraw-Hill, New York, NY, 1995, pages 263-290.

Projected Life = 10 years **Discount Rate = 5%**

Life-Cycle Activity	Activity Cost Code	Cost for Year (n) in System Life Cycle										
		1	2	3	4	5	6	7	8	9	10	Total
Research, Design, and Development	C_R	\$100,000	\$125,000	\$150,000	-	-	-	-	-	-	-	\$375,000
Production and/or Construction	C_P	-	\$25,000	\$175,000	\$180,000	\$160,000	-	-	-	-	-	\$540,000
Operation and Support	C_O	-	-	\$30,000	\$60,000	\$60,000	\$80,000	\$80,000	\$80,000	\$50,000	\$10,000	\$450,000
Retirement/Material Recycling/Disposal	C^D	-	-	-	-	-	-	-	-	\$20,000	\$35,000	\$55,000
Total Actual Cost (Budgetary)	C	\$100,000	\$150,000	\$355,000	\$240,000	\$220,000	\$80,000	\$80,000	\$80,000	\$70,000	\$45,000	\$1,420,000
Discount Factor (0.05)	-	0.95238	0.90703	0.86384	0.82270	0.78353	0.74622	0.71068	0.67684	0.64461	0.61391	-
Total Present Value (Current Equivalent Cost)	$C_{5\%}$	\$95,238	\$136,055	\$306,663	\$197,448	\$172,377	\$59,698	\$56,854	\$54,147	\$45,123	\$27,626	\$1,151,229

Figure 2.8.3-1. Life Cycle Cost Summary - System A

Projected Life = 10 years **Discount Rate = 5%**

Life-Cycle Activity	Activity Cost Code	Cost for Year (n) in System Life Cycle										
		1	2	3	4	5	6	7	8	9	10	Total
Research, Design, and Development	C_R	$120,000	$130,000	$170,000	-	-	-	-	-	-	-	$420,000
Production and/or Construction	C_P	-	$50,000	$200,000	$200,000	$150,000	-	-	-	-	-	$600,000
Operation and Support	C_O	-	-	$25,000	$30,000	$30,000	$60,000	$60,000	$60,000	$30,000	$10,000	$305,000
Retirement/Material Recycling/Disposal	C^D	-	-	-	-	-	-	-	-	$15,000	$10,000	$25,000
Total Actual Cost (Budgetary)	C	$120,000	$180,000	$395,000	$230,000	$180,000	$60,000	$60,000	$60,000	$45,000	$20,000	$1,350,000
Discount Factor (0.05)	-	0.95238	0.90703	0.86384	0.82270	0.78353	0.74622	0.71068	0.67684	0.64461	0.61391	-
Total Present Value (Current Equivalent Cost)	$C_{5\%}$	$114,286	$163,265	$341,217	$189,221	$141,035	$44,773	$42,641	$40,610	$29,007	$12,278	$1,118,333

Figure 2.8.3-2. Life Cycle Cost Summary - System B

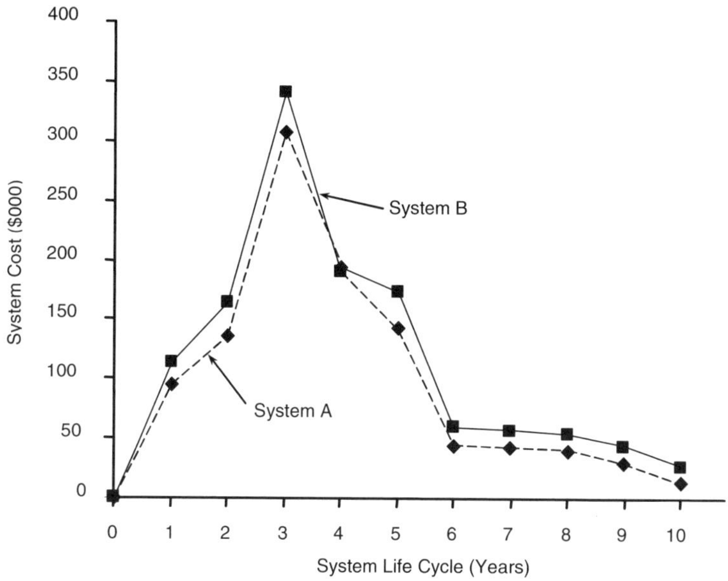

Figure 2.8.3-3. System Life Cycle Cost Profile of Alternatives - System A and System B

3. *Analysis of Production Alternatives.* An example of the application of discounting and evaluation of alternatives is provided by the hypothetical case of a manufacturing firm considering the possibility of introducing a new product into the market that is expected to meet sales projections over a designated period of time (i.e., at least seven years). The firm must invest in capital equipment and start up costs in order to manufacture the product. Based on a survey of potential sources, two equipment alternatives (Equipment A and Equipment B) are considered to be potential selections. A seven-year life cycle and an interest rate of 5% are assumed for both alternatives.

   Table 2.8.3-1 depicts the projected cash flow for Equipment A, including a schedule of both anticipated disbursements (expenses) and benefits (in terms of revenues).

   For Equipment A, an initial capital investment of $20,000 is required. The cumulative cash flow indicates net benefits of $69,000 ($102,000-$33,000), and the cumulative present value shows net benefits of $48,328 ($80,482-$32,154).

   Figure 2.8.3-4 shows the cash flow and payback curve (defined by cumulated receipts less disbursements) and indicates an estimated payback point at three years and 8.7 months.

Table 2.8.3-1. Net Present Value Comparison - Equipment A

Year n	Cash Flow		Discount Factor (5%)	Present Value	
	Benefits	Costs		Benefits	Cost
0	-	$20,000	1.00000	-	$20,000
1	-	$8,000	0.95238	-	$7,619
2	$8,000	$5,000	0.90703	$7,256	$4,535
3	$12,000	-	0.86384	$10,366	-
4	$18,000	-	0.82270	$14,809	-
5	$25,000	-	0.78353	$19,558	-
6	$21,000	-	0.74622	$15,671	-
7	$18,000	-	0.71068	$12,792	-
Total	$102,000	$33,000	-	$80,482	$32,154

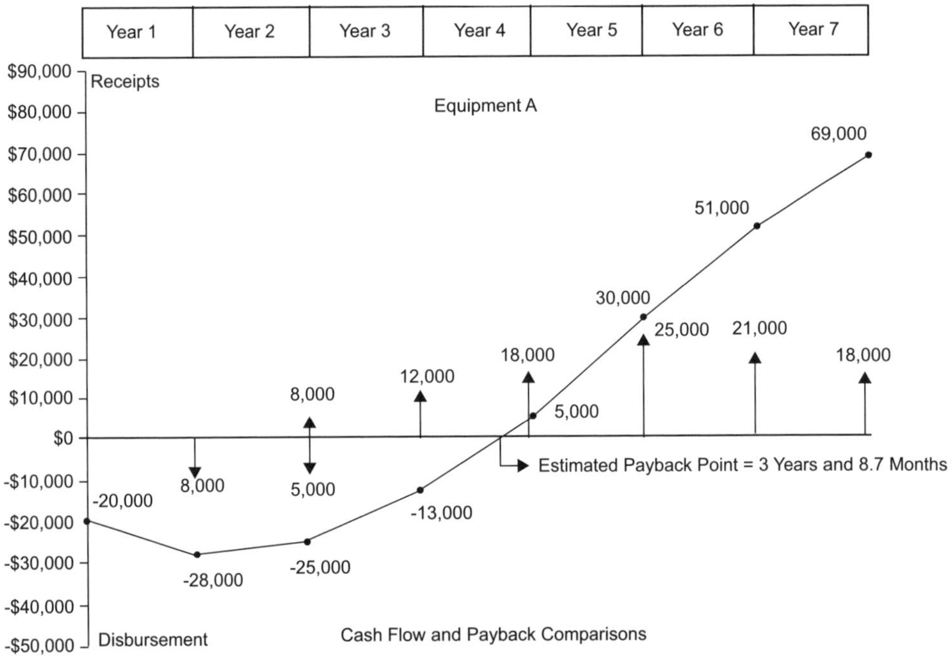

Figure 2.8.3-4. Cash Flow and Payback - Equipment A

Equipment B is analyzed with the same methodology as that applied for Equipment A. Table 2.8.3-2 depicts the projected cash flow for Equipment B, including a schedule of both anticipated disbursements (expenses) and benefits (in terms of revenues).

Table 2.8.3-2. Net Present Value Comparison - Equipment B

Year n	Cash Flow		Discount Factor (5%)	Present Value	
	Benefits	Costs		Benefits	Cost
0	-	$25,000	1.00000	-	$25,000
1	-	$10,000	0.95238	-	$9,524
2	$10,000	$6,000	0.90703	$9,070	$5,442
3	$15,000	-	0.86384	$12,958	-
4	$20,000	-	0.82270	$16,454	-
5	$28,000	-	0.78353	$21,939	-
6	$25,000	-	0.74622	$18,656	-
7	$21,000	-	0.71068	$14,924	-
Total	$119,000	$41,000	-	$94,001	$39,966

For Equipment B, an initial capital investment of $25,000 is required. The cumulative cash flow indicates net benefits of $78,000 and the cumulative present value notes net benefits of $54,035.

The cash flow and payback curve at Figure 2.8.3-5 indicates an estimated payback point at three years and 9.6 months.

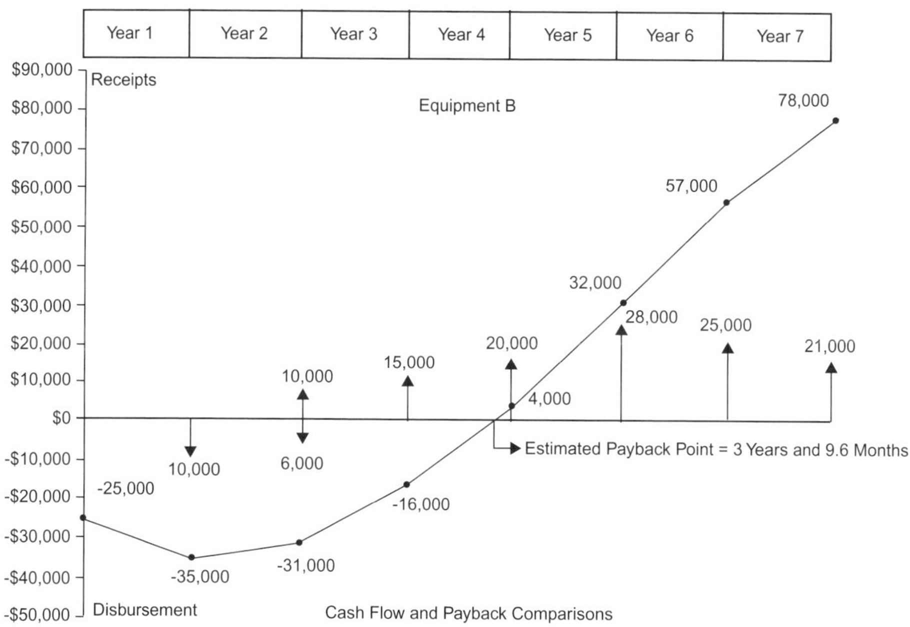

Figure 2.8.3-5. Cash Flow and Payback - Equipment B

Based on comparison of the net cash flow and present value benefits for the two alternatives, System B is the most logical selection.

4. *Present Value-Future Projection.* It may be desirable to project financial analysis to some future point in time, at which point a major investment decision would be made. Such cases would involve, in addition to the present value method, application of the future-value formula:

$$\text{Future Value (F)} = P[(1 + i)^n] \qquad (2.49)$$

where:

P = Present Value
i = Discount rate
n = The specific year in the cost stream

As an example, assume that \$10,000 is invested today in some activity and the annual interest rate is 5%, the compound rate at the end of the 3rd year is computed as:

$$F = 10{,}000\,[(1.05)^3] = (10{,}000)(1.157625) = \$11{,}576$$

With the application of present value and future value algorithms, the analyst can relate to the life cycle and convert revenues and costs to any future decision point. Figure 2.8.3-6 illustrates this relationship, where the projected decision point is at the end of the second year (year 2).

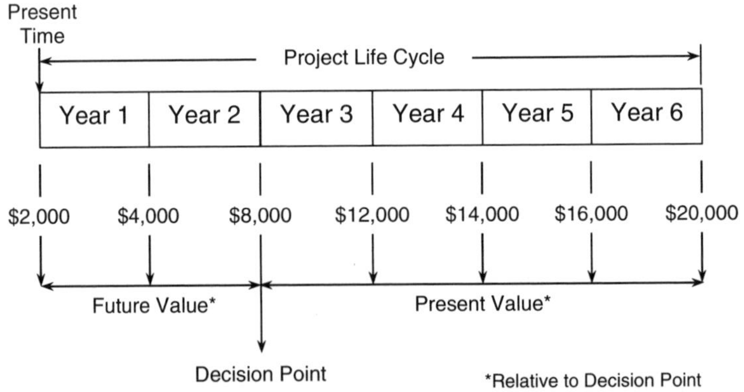

Figure 2.8.3-6. Present Value–Future Value Projection

As shown by Figure 2.8.3-6, the discounted value of the six-year cost stream at 5% is:

Year	Apply	Calculation
1 and 2	Future value	2,000 (1.1025) + 4,000 (1.05) + 8,000
3, 4, 5, & 6 (After decision point)	Present value	12,000 (0.95238) + 14,000 (0.90703) + 16,000 (0.86384) + 20,000 (0.82270)

The total of the annualized values for the six year cost stream is:

$$P_{(decision\ point)} = 2{,}205 + 4{,}200 + 8{,}000 + 11{,}429 + 12{,}698 + 13{,}821 + 16{,}454 = \$68{,}807$$

In summary, revenues and costs may be treated differently based on the immediate considerations. Money has time value, and discounting is appropriate in the comparison of alternative cost profiles. For economic analyses, discounting is employed to relate all revenues and costs to a specific decision point, whether now or in the future.

Topic 5.2.1 provides additional discussion and an example of LCC analysis[40].

[40] Langford, J.W., *Logistics Principles and Applications*, McGraw-Hill, New York, NY, 1995, pages 263-290.

Topic 2.8.4: Cost Effectiveness (CE)

Cost effectiveness (CE) is typically expressed in terms of the metrics of system effectiveness correlated with the metrics of LCC. Establishing the relationship of technical factors and economic factors for a system may constitute a CE *figure of merit (FOM)*. Specific methodologies and applications may vary, but some examples of FOMs are:[41]

$$\text{Cost Effectiveness FOM} = \frac{\text{availability}}{\text{life-cycle cost}} \tag{2.50}$$

or

$$\text{Cost Effectiveness FOM} = \frac{\text{performance}}{\text{life-cycle cost}} \tag{2.51}$$

or

$$\text{Cost Effectiveness FOM} = \frac{\text{logistics effectiveness}}{\text{life-cycle cost}} \tag{2.52}$$

or

$$\text{Cost Effectiveness FOM} = \frac{\text{overall equipment effectiveness}}{\text{life-cycle cost}} \tag{2.53}$$

or

$$\text{Cost Effectiveness FOM} = \frac{\text{facility space}}{\text{life-cycle cost}} \tag{2.54}$$

Other CE options include an array of logistics-related FOMs such as *cost per operating hour, cost per ton-mile, cost of fuel consumption per mile,* etc. These types of factors are contributory to development of estimates for LCC for the system. Such estimates are derived from blending of the equipment attributes and FOMs, based on interaction between the technical and cost analysis expertise. This is a critical process that will govern the proper identification and collection of the required data. As an example, Figure 2.8.3-7 illustrates the relationship between Reliability (MTBF) and total LCC, in which the objective is to design a system to meet the specified MTBF and budgetary constraints-yet being cost effective.

[41] Based on Blanchard, B.S., *Logistics Engineering and Management*, 6th Ed., Pearson Prentice Hall, Upper Saddle River, NJ, 2004, pages 87-89.

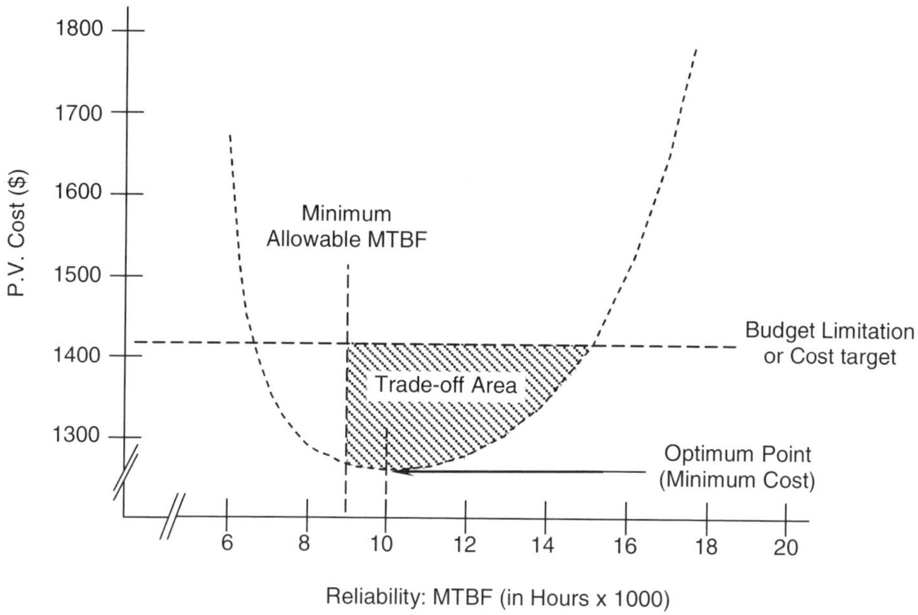

Figure 2.8.3-7. Reliability (MTBF) vs. Present Value Cost ($)

The same methodology can be applied to trade-off analyses involving all pertinent logistics attributes relative to cost. It is underscored that the selected data utilized will significantly affect the validity of the analysis.

For More Information:

1. Blanchard, B.S., *Logistics Engineering and Management*, 6th Ed., Pearson Prentice Hall, Upper Saddle River, NJ, 2004.
2. Blanchard, B.S., *System Engineering Management*, 3rd Ed., John Wiley & Sons, Hoboken, NJ, 2004.
3. Blanchard, B.S. and W.J. Fabrycky, *Systems Engineering and Analysis*, 3rd Ed., Prentice Hall, Upper Saddle River, NJ, 1998.
4. Langford, J.W., *Logistics Principles and Applications*, McGraw-Hill, New York, NY, 1995.
5. MIL-STD-882D, *Department of Defense Standard Practice for System Safety*, Department of Defense, Washington, DC, 19 January 2000.
6. MIL-HDBK-470, *Designing and Developing Maintainable Products and Systems*, Department of Defense, Washington, DC, 4 August 1977.
7. MIL-HDBK-217F (Notice 2), *Reliability Prediction for Electronic Equipment*, Department of Defense, Washington, DC, 28 February 1995.

8. "Designing and Assessing Supportability in DOD Weapon Systems: A Guide to Increased Reliability and Reduced Logistics Footprint," Prepared by the Office of Secretary of Defense, October 24, 2003.
9. O'Connor, P.D.T., D. Newton, and R. Bromley, *Practical Reliability Engineering*, 3rd Ed., John Wiley and Sons, Hoboken, NJ, 1996.
10. Knezevic, J., *Reliability, Maintainability and Supportability: A Probabilistic Approach*, McGraw-Hill Book Company Europe, Maidenhead, England, 1993.
11. "Reliability Toolkit: Commercial Practices Edition," Reliability Analysis Center, 201 Mill Street, Rome, NY.
12. "Maintainability Toolkit: A Practical Guide for Designing and Developing Maintainable Products and Systems," Reliability Analysis Center, 201 Mill Street, Rome, NY.
13. "Quality Toolkit: A Practical Guide for Developing and Implementing a Quality Program for Products and Services," Reliability Analysis Center, 201 Mill Street, Rome, NY.

Section 3: Supportability in the System Life Cycle

Insight

Critical in the design and development of new systems (and in the re-engineering of existing systems) is the need to address supportability from the beginning! In the early identification of the need for a new system and in the accomplishment of a feasibility analysis, logistics and the maintenance support infrastructure must be considered as a major element of the prospective system and must be viewed in the context of the entire system life cycle. In the definition of system operational requirements and the maintenance concept, the specific requirements for supportability and the maintenance support infrastructure begin to take shape. From this, the identification and prioritization of the appropriate technical performance measures (TPMs) and performance-based logistics (PBL) factors lead to the specification of specific supportability criteria as an input to the overall system design. Accomplishment of the functional analysis leads to partitioning (functional packaging) and the identification of the specific elements of the system and, through the subsequent allocation process, specific design-to requirements are established for the various sub-systems and lower-level components. Logistics and the *design for supportability* must be "inherent" within this overall top-down, front-end process if the objectives described in Section 1 are to be realized.

The best way to accomplish these objectives is through implementation of the *system engineering process*, described in Topic 1.6.1. Referring to Figure 1.6.1-1, there is a top-down/bottom-up approach, viewing the system as a *whole* (with all of its elements to include logistics and the maintenance support infrastructure), over its entire *life cycle* (refer to Figure 1.7-1), providing an *integrated* and *interdisciplinary* approach in its design and development (with the design for supportability being one of the prime design objectives), and so on. The objective in this section is to provide some emphasis with regard to the first two blocks in Figure 1.6.1-1; i.e., *system requirements* and *functional analysis and requirements allocation.*

Topic 3.1: The System Life Cycle

For the purposes of discussion, and as a "baseline" for reference, the system life-cycle model illustrated in Figure 3.1-1 (see Figure 1.7-1) is assumed. Referring to the figure, the shaded area provides the emphasis throughout Section 3 of this Toolkit.

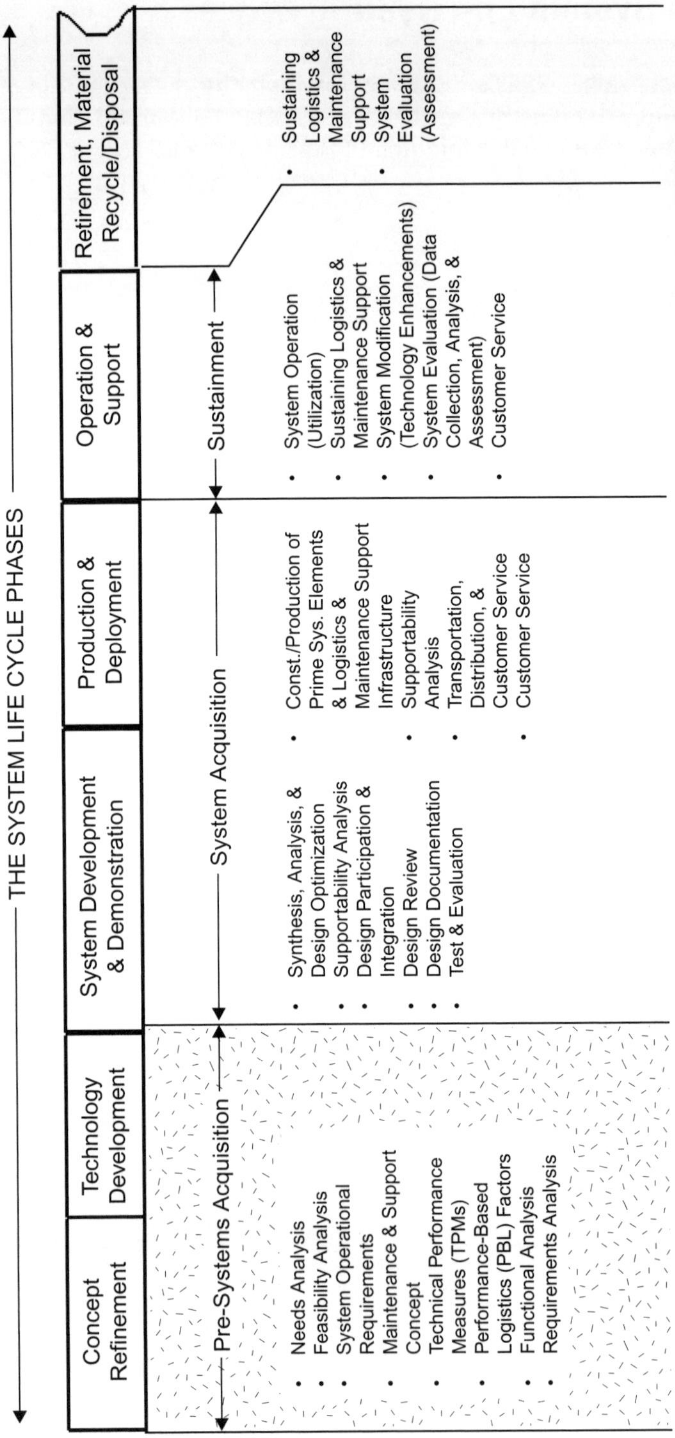

Figure 3.1-1. The System Life Cycle - Shaded Area is Pre-Systems Acquisition

Topic 3.2: The System Engineering Process

While application of the system engineering process is life-cycle oriented (i.e., the process is applicable in each and every phase of the system life cycle), the critical steps to ensure that the *design for supportability* objectives are realized are as indicated by the shaded area in Figure 3.1-1. These steps are highlighted by the topics to follow.

Topic 3.2.1: Problem ("Need") Identification and Feasibility Analysis

The system engineering processes commences with the identification of a want or a desire for something and is based on a real (or perceived) deficiency. For example, the current system capability is not adequate in terms of meeting certain performance goals, is not available when needed, cannot be properly supported, or is too costly in terms of operation. As a result, a new system requirement is defined along with the priority for introduction, the date when the new system capability is required for customer use, and the anticipated resources necessary for acquiring the new system. To ensure a good start, a complete *statement of the problem* should be presented in specific qualitative and quantitative terms, in enough detail to justify progressing to the next step.

The requirement for identifying the *need* (as a starting point) may seem to be rather basic, or self-evident; however, one often finds that a design effort is initiated as a result of a personal interest or other subjective basis, without the requirements for such first having been adequately defined. In the software area, in particular, there is often a tendency to accomplish a lot of coding before identifying the need and the functional requirements for such. Additionally, there are instances when the responsible design engineer(s) sincerely believes that he/she knows what the customer needs, without having involved the customer in the process. Often, the "design-it-now-fix-it-later" philosophy prevails which, in turn, can turn out to be rather costly.

Defining the problem is sometimes the most difficult part of the process, particularly if one is in a rush to get going; yet, the number of false starts and the ultimate risks can be significant unless a good foundation is established from the beginning. A complete description of the need, expressed in quantitatively stated performance parameters where possible, is essential. It is at this point when the basic requirements for logistics and the maintenance support infrastructure are first identified and begin to take shape. The question is -- *what type of a logistics and maintenance support infrastructure is envisioned?* Although highly "conceptual" at this point, one needs to start asking the appropriate questions in any event.

Through the needs analysis, the basic functions that the system must perform are identified (i.e., primary functions, secondary functions), along with the geographical location(s) where these functions are to be performed and the anticipated period of performance. In essence, one must define the "WHAT" requirements (versus the "HOWs"). In response, there may be a number of different *technological* approaches considered to be feasible in solving the problem at hand (i.e., correcting the deficiency).

A *feasibility analysis* is accomplished with the objective of evaluating the different technological approaches that may be considered in responding to the specified functional requirements. For instance, in the design of a communication system, should one incorporate a fiber-optics, cellular, or conventional hard-wired approach? In designing an aircraft, to what extent should one incorporate composite materials? When designing an automobile, should one apply very high-speed integrated electronic circuitry in certain control applications, or should one select a more conventional electro-mechanical approach?

At this point, it is necessary to (1) identify the various possible design approaches that can be pursued to meet the requirements; (2) evaluate the most likely candidates in terms of performance, effectiveness, logistics requirements, and life-cycle economic criteria; and (3) recommend a preferred approach for application. The objective here is to select an overall *technical* approach, and NOT to select specific hardware, software, and related system components. There may be many different alternatives; however, the number of possibilities must be narrowed down to a few feasible options, consistent with the state-of-the-art and the availability of resources.

It is at this early stage in the life cycle (i.e., the Concept Refinement Phase) when major decisions are made relative to adopting a specific design approach. Not only will the results of a feasibility analysis have a great impact on the operating characteristics of the system but on the production and maintenance support requirements as well. The selection (and application) of a given technology has reliability and maintainability implications, may significantly affect the requirements for spare parts and test equipment, may impact transportation and handling requirements, may have an impact on manufacturing methods, and will certainly affect life-cycle cost (LCC). Thus, it is at this stage when the overall requirements for *supportability* must be initially addressed.

Topic 3.2.2: System Operational Requirements

Once a system requirement (i.e., the need) and a technical approach have been identified, it is necessary to expand on the anticipated operational requirements. At this point, the following questions must be asked: *what is the specific mission and associated operational scenarios that must be accomplished? Where (geographically) are these scenarios to be accomplished and for how long? What are the anticipated quantities of equipment, software, people, etc., required and where are they to be located? How is the system to be utilized in terms of on-off cycles, hours of operation per designated time period, etc.? What are the expected operational effectiveness goals for the system? What are the expected environmental conditions to which the system will be subjected throughout its operational life?*

Table 3.2.2-1 provides a summary of the key objectives in defining the operational requirements for the system. If one is to design and develop a system to meet a given customer requirement, it is important that the various responsible members of the design team know the mission objectives and how the system will be utilized in accomplishing these objectives. Of particular interest is the anticipated geographical deployment and type of operational scenarios that are expected to be accomplished. Referring to Figure 3.2.2-1, a few examples are presented. While one certainly cannot cover all of the future areas of operation, some initial assumptions as to operational scenarios, anticipated utilization, the stresses that the system is expected to see, etc., must be made. The question is -- *How can one design without having a pretty fair idea as to how the system is to be utilized?* This question is particularly relevant when determining the design requirements for reliability, maintainability, and supportability. Thus, it is appropriate to identify a few of the more rigorous operational profiles and to design with these in mind.

Table 3.2.2-1. System Operational Requirements

1. Mission definition	Identification of the prime mission of the system and alternative or secondary missions. This may be defined through a series of typical scenarios or operational profiles, and associated system utilization requirements. These scenarios reflect the *dynamics* of the system operating characteristics.
2. Operational deployment (distribution)	Identification of the geographical location(s) where the system and its elements are to be located; i.e., quantity of equipment, software, personnel, facilities, etc., to be distributed and the time duration for distribution.
3. Performance and physical parameters	Definition of the operating characteristics or functions of the system; e.g., speed, acceleration, throughput, accuracy, output, size, weight, process time, etc. These factors must be directly related to the applicable mission/operational scenario(s).
4. Effectiveness factors	Definition of cost-effectiveness, operational availability, dependability, supportability (PBL factors), MTBM, MDT, MLH/OH, facility utilization, readiness rate, and related requirements. These factors must be directly related to each applicable mission/operational scenario.
5. Operational life cycle (horizon)	Anticipated time that the system will be in operational use. This represents the baseline for determining the total inventory profile.
6. Operational environment	Definition of the environment in which the system is expected to operate (e.g., temperature, vibration, humidity, arctic or tropics, mountainous or flat terrain, airborne, shipboard, ground fixed or mobile, etc.). This reflects what the system will experience as it accomplishes its mission.

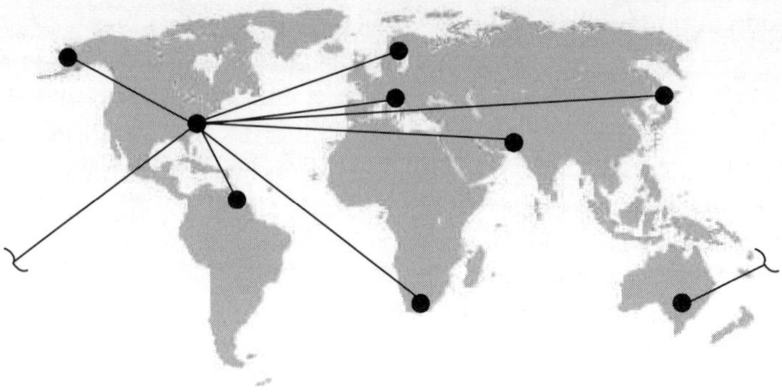

SYSTEM OPERATIONAL REQUIREMENTS (Geographical Distribution)

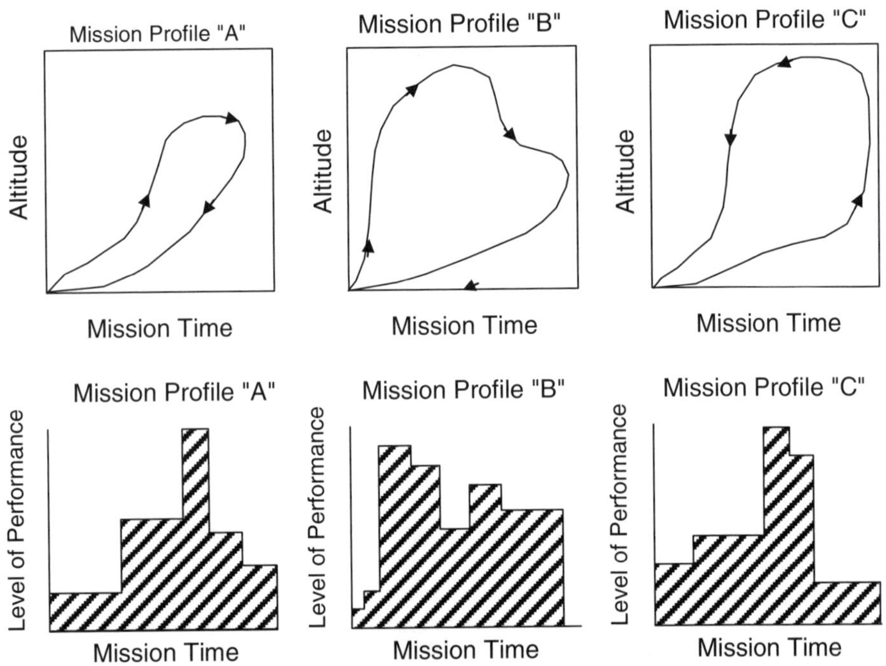

TYPICAL SYSTEM OPERATIONAL PROFILES

Figure 3.2.2-1. System Operational Requirements – Geographic Distribution and Operational Profiles

Topic 3.2.3: System Maintenance and Support Concept

The system *maintenance concept*, developed during the Concept Refinement Phase, constitutes a "before-the-fact" series of illustrations and statements pertaining to how the system (and its elements) should be designed for supportability. The objective is to address the following questions: *Where will the anticipated maintenance and support activities for the system be accomplished? To what level(s) of maintenance should the system and its elements be designed; i.e., should the system be designed to be supported at one, two, three, or four levels of maintenance? To what depth (in the design of the system or component) should maintenance be accomplished? Who (what organizations) will be responsible for maintenance at each level? What are the effectiveness requirements (design-to PBL factors) for the logistics and maintenance support infrastructure? What are the expected environmental conditions to which the system will be subjected during the performance of maintenance?*

Table 3.2.3-1 provides a summary of the key objectives in defining the maintenance concept for the system. This concept definition constitutes an early attempt to describe the infrastructure that will be required for the sustaining support of the system throughout its planned life cycle, as well as the specific "design-to" criteria for the various elements of logistics (see Figure 1.3-5). As the system development process progresses, and through the iterative and on-going supportability analysis (SA) effort, the maintenance concept evolves into a detailed *maintenance plan.* The maintenance plan, which is based on the results of the SA, is an "after-the-fact" description of the way in which the system will ultimately be maintained and supported, and the resources required in accomplishing such.

With regard to the "levels" of maintenance, corrective and preventive maintenance may be performed on the system itself (or an element thereof) at the operational site where the system is utilized by the customer, in an intermediate shop near the customer's operational site, or at a remote depot, the producer's manufacturing plant, a supplier's factory, or at a third-party logistics facility. Maintenance level pertains to the division of functions and tasks for each area where maintenance is performed. Anticipated frequency of maintenance, task complexity, personnel skill-level requirements, special facility needs, safety and security issues, and so on, dictate to a great extent the specific functions to be accomplished at each level. Depending on the nature and mission of the system, there may be two, three, or four levels of maintenance. Figure 3.2.3-1 provides an illustration of the relationships and "flow" between the various levels. For the purposes of further discussion, maintenance may be classified as *organizational*, *intermediate*, and *producer/supplier/depot.*

Table 3.2.3-1. System Maintenance and Support Requirements

1. Levels of maintenance and support	Identification of the expected levels of maintenance, i.e., organizational (on-equipment at the operational site), intermediate (fixed shop or mobile facility), and depot (manufacturer, organic, supplier, third party) maintenance shop (remote facility).
2. Repair policy	Extent (or depth) to which maintenance will be accomplished; i.e., whether an item is to be non-repairable, partially repairable, or fully repairable at a given level of maintenance.
3. Maintenance responsibility	Description of the organization(s) responsible for the accomplishment of maintenance and support at each applicable level; i.e., customer, system operator (user), producer, third party, or a combination thereof.
4. Maintenance support elements	Description of the basic input design "criteria" for the various elements of logistics and maintenance support; i.e., spares/repair parts and associated inventories, test and support equipment, personnel, transportation and handling capability, facilities, computer resources, and data and information (see Figure 1.3-5).
5. Effectiveness requirements	Identification of the performance-based logistics (PBL) requirements; i.e., operational availability (A_o), maintenance downtime (MDT), logistics effectiveness, logistics responsiveness, transportation reliability, test equipment availability, facility utilization, logistics data processing time, spares availability, life cycle cost (LCC), etc. (see Figure 1.5-1).
6. Maintenance environment	Definition of the environment that the system (and its elements) is likely to experience during the accomplishment of maintenance and support activities at different levels. This includes the anticipated environmental requirements during transportation and storage.

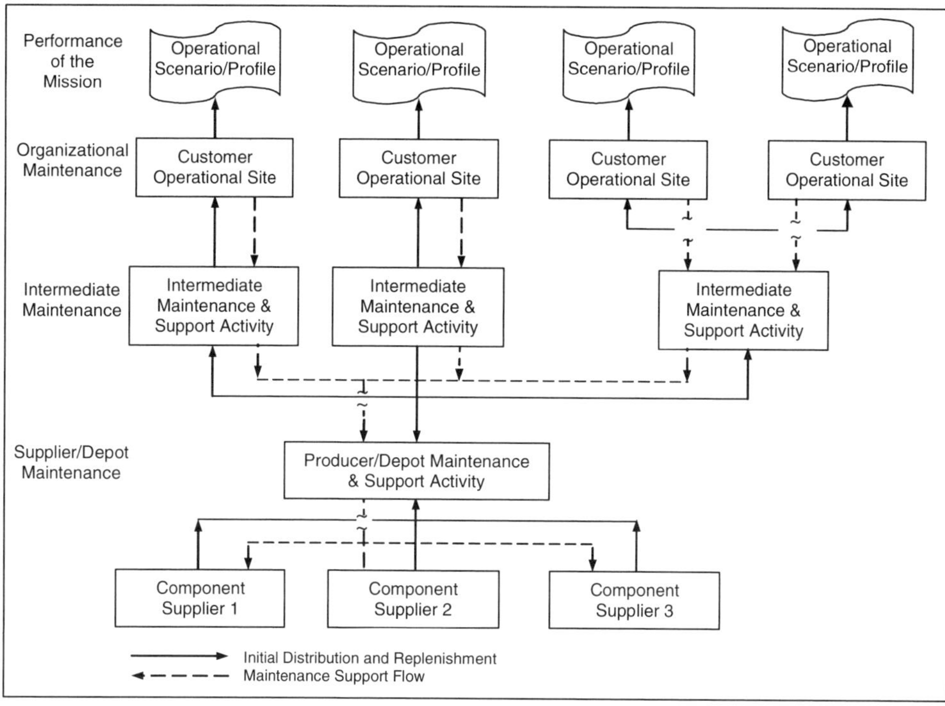

Figure 3.2.3-1. System Maintenance and Support Flow (Reference Figure 1.3-4)

1. *Organizational maintenance*: Organizational maintenance is performed at the operational site (e.g., airplane, vehicle, manufacturing production line, ship, communication facility). Generally, it includes tasks performed by the using organization on its own equipment/software. Organizational-level personnel are generally involved with the operation and use of equipment/software, and have minimum time available for detailed system maintenance. Maintenance at this level normally is limited to periodic checks of equipment performance, visual inspections, cleaning of system elements, verification of software, some servicing, external adjustments, and the removal and replacement of some components. Personnel assigned to this level generally do not repair the removed components but forward them to the intermediate level. From the maintenance standpoint, the least skilled personnel are assigned to this function, and the design of equipment/software must take this fact into consideration; i.e., the design for simplicity, design for ease in diagnostics, design for ease in removal and replacement, etc.

2. *Intermediate maintenance*: Intermediate maintenance tasks are performed by mobile, semi-mobile, and/or fixed specialized organizations and installations. At this level, end items may be repaired by the removal and replacement of major

modules, assemblies, or piece parts. Scheduled maintenance requiring equipment disassembly may also be accomplished. Available maintenance personnel are usually more skilled and better equipped than those at the organizational level and are responsible for performing more detailed maintenance tasks.

Mobile or semi-mobile units are often assigned to provide close support to deployed operational systems. These units may include vans, trucks, or portable shelters containing some test and support equipment and spares/repair parts. The objective is to provide on-site maintenance, beyond that accomplished by organizational-level personnel, to facilitate the return of the system to its complete operational status on an expedited basis. A mobile unit may be used to support more than one operational site. A good example is the maintenance vehicle (and its crew) that is deployed from a fixed maintenance shop to an airplane, parked at a commercial airline terminal gate, that needs extended maintenance.

Fixed installations (permanent shops) are generally established to support both the organizational-level tasks and the mobile or semi-mobile units. Maintenance that cannot be performed by the lower levels, due to limited personnel skills, support equipment, and facilities, are performed here. High personnel skills, additional test and support equipment, more spares, and better facilities often enable equipment repair to the module and piece part level. Fixed shops are usually located within specified geographical areas, and rapid maintenance turnaround times are not as imperative here as at the lower levels of maintenance.

3. *Producer, supplier, or depot maintenance*: This constitutes the highest level of maintenance, and supports accomplishment of the tasks above and beyond the capabilities at the intermediate level. Physically, this may include a specialized repair center providing high-level support for a number of different systems in the inventory, and it may constitute a regional depot, a producer's manufacturing plant, a supplier's plant, or a combination thereof. This facility is fixed and mobility is not a problem. Complex and bulky equipment, large quantities of spares, environmental control provisions, safety/security provisions, calibration capabilities, special clean rooms, and so on, are available. The anticipated high volume potential often fosters the use of assembly-line techniques, which in turn, permits the use of relatively unskilled labor for a large portion of the workload, with a concentration of highly skilled specialists in certain key areas such as troubleshooting and fault diagnostics, calibration, and quality control. This level of maintenance includes the complete overhauling and/or rebuilding of equipment and the performance of highly complex tasks.

The three levels of maintenance just discussed are covered in Figure 3.2.3-1. However, the "flow" of various items may vary somewhat depending on the anticipated frequency of maintenance, the criticality of the item relative to its impact on the accomplishment of the system mission, the anticipated complexity of maintenance, product warranty requirements, safety and security issues, and so on. While, in many instances, all three levels may be involved in a single repair action (i.e., chain of events), there are numerous instances where a faulty repairable item will be sent directly back to the depot or to the original equipment manufacturer (OEM), involving only two levels of maintenance. If, for example, the frequency of maintenance for a given repairable is low and the establishment of a repair capability at the intermediate level is costly, then it may be feasible to establish a two-level maintenance approach. If there are proprietary items requiring corrective maintenance, then the faulty items may be shipped directly back to the responsible supplier. If, on the other hand, there are mission-critical items where the anticipated failure rate is relatively high, then it may be appropriate to accomplish a designated amount of corrective maintenance at the organizational level where the turnaround time is relatively short. In any event, there may be any number of variations to the "model" illustrated in Figure 3.2.3-1. However, it is important at this early stage in the system life cycle to develop an overall "concept" as an input to the design process.

Figure 3.2.3-2 (an extension of the flow process in Figure 3.2.3-1) represents an example of a proposed "repair policy," showing the levels of maintenance for a given system, the proposed flow of items between one level and another, and selected PBL-related metrics applicable in design and covering major criteria for each of the three levels shown. Referring to the figure, the system shall be designed such that the reliability MTBF shall be equal to or greater than 1,800 hours, the turnaround time (TAT) at the intermediate level shall not exceed 4 hours, the mean corrective maintenance time ($\overline{M}_{ct}$) at the depot level shall be 3 hours or less, and so on. These factors, derived from higher-level factors (see Figure 1.5-1), are then specified as "design-to" requirements, applicable both to the design of the prime elements of the system and the design of the maintenance support infrastructure.

As a final point, it is important that one commence with an early "conceptualization" of the projected activities such as those shown in Figures 3.2.3-1 and 3.2.3-2 during the Concept Refinement Phase. In the past, this level of definition has primarily been "downstream" and as a result of the supportability analysis (SA) and the development of a detailed maintenance plan. However, if one is to *design for supportability*, it is critical that these issues be addressed from the beginning, evolving from a broad conceptualization into a more refined approach as the system development process progresses.[1]

[1] The maintenance concept and its development is discussed in detail in Blanchard, B.S., *Logistics Engineering and Management*, 6th Ed., Pearson Prentice Hall, Upper Saddle River, NJ, pages 139-145, 2004.

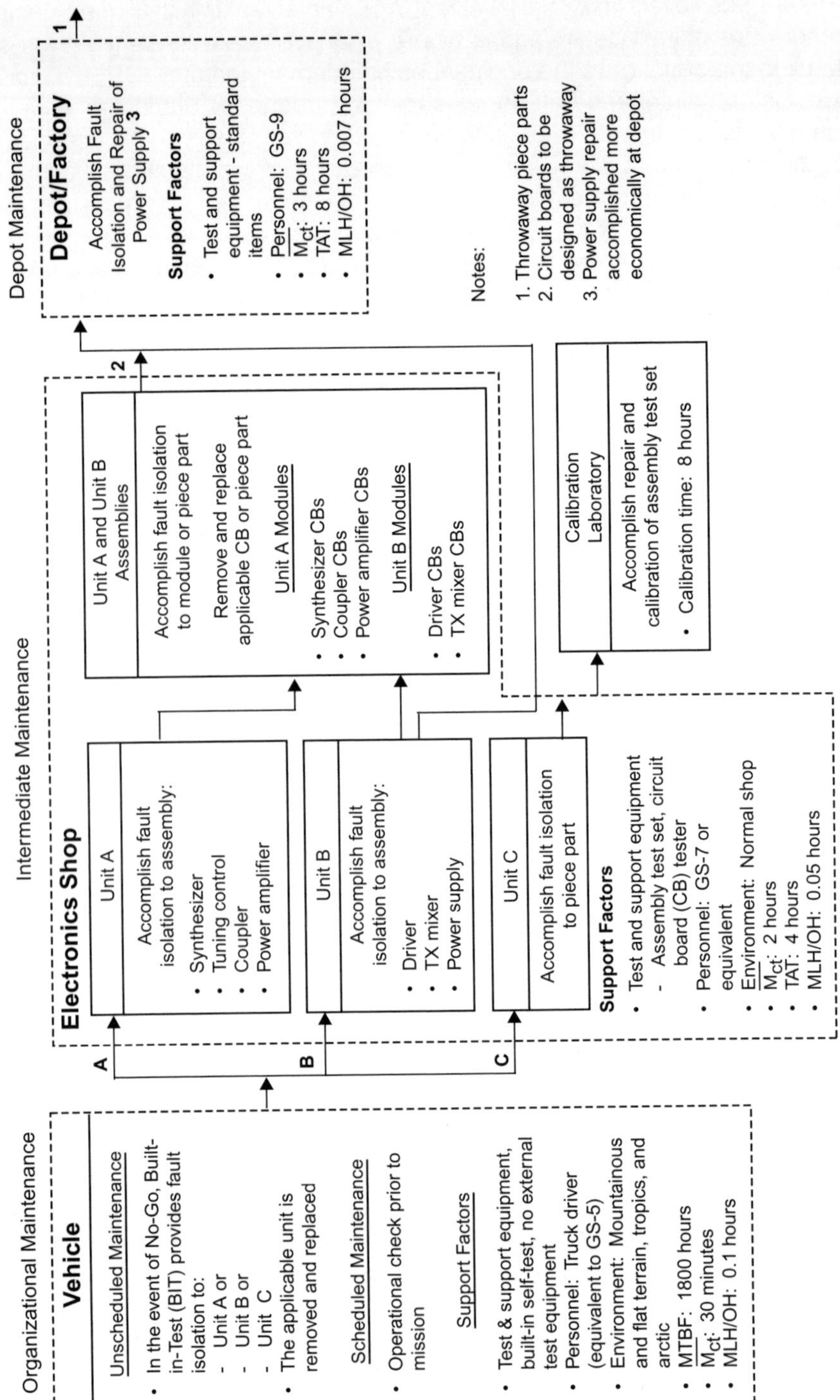

Figure 3.2.3-2. System XYZ Maintenance Flow (Repair Policy - Example)

Topic 3.2.4: Identification and Prioritization of Technical Performance Measures (TPMs)

Through the development of operational requirements and the maintenance concept for the system, specific *performance-related* factors are identified and applied with the objective of ensuring that the system will be designed to satisfactorily accomplish its mission. These factors, identified as *technical performance measures (TPMs)*, may be applied as design-to criteria for the prime mission-related elements of the system and for those elements that are necessary to provide sustaining maintenance and support of the system throughout its life cycle.[2]

A number of different metrics may be applicable in defining the design requirements for a system, and priorities need to be established in order to determine the relative degree of importance in the event that design trade-offs (i.e., compromises) are necessary. For example, in the design of a vehicle is *speed* more important than *size*? For a manufacturing plant, is *production quantity* more important than *product quality*? In a communication system, is *range* more important than *reliability* or *clarity of message*? For a computer capability, is *capacity* more important than *processing speed*? In equipment packaging, is *packaging density* more important than providing *accessibility for diagnostics and maintenance*? Are *human factors* more important than *cost*?

All of these issues are important and there will likely be a set of minimum requirements in each area. However, there may be any number of different design options, and the responsible designer needs to understand which are more important than others, as well as the interrelationships that exist among them. If compromises in design have to be made, *which area(s) is more critical and where additional emphasis is required in arriving at a recommended design solution?* Further, the designer needs to ensure that the ultimate system configuration incorporates the necessary attributes (or characteristics) that will be responsive to the TPM requirements, and are consistent with the established priorities.

A basic approach in establishing specific design priorities is shown in Figure 3.2.4-1. It is essential that good communications between the customer and the responsible design team be realized from the beginning. In one or more early design review meetings, the TPMs, derived from the development of system operational requirements and the maintenance concept, are reviewed and evaluated in terms of relative degree(s) of importance. The most critical factors are identified and, thus, lead to areas where special design emphasis may be required. This first step (i.e., block 1, Figure 3.2.4-1), representing the *voice of the customer (VOC)*, identifies the WHATs, and the results may take a form such as shown in Table 3.2.4-1. Referring to the table,

[2] TPMs refer to critical "performance" measures for the system as an entity. Included within are the performance-based logistics (PBL) factors discussed under Topic 1.5 (refer to Figure 1.5-1).

the most critical TPMs are Operational Availability (A_o), unit life-cycle cost (LCC), logistics response time, system velocity, and so on.

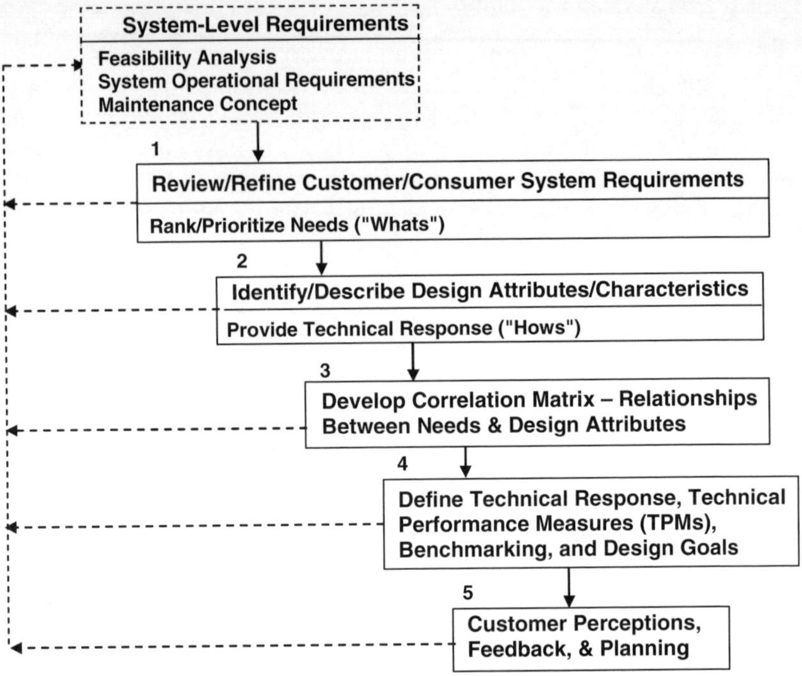

Figure 3.2.4-1. Basic Steps in the Technical Performance Measure (TPM) Prioritization Process

Table 3.2.4-1. Prioritization of Technical Performance Measures (TPMs)

TPM	Quantitative Requirement (Metric)	Current Benchmark		Relative Importance (Customer Desires)
		Metric	System	
Operational Availability (A_o)	98% (min)	98.5%	H	26%
Velocity (mph)	125 (min)	100	B	11%
Weight (pounds)	125K (max)	150K	H	9%
Size (feet)	Length: 125 Width: 12 Height: 10 (max)	Length: 136 Width: 15 Height: 12	B	6%
Human factors (error rate/yr)	Less than 1%	2%	D	5%
Maintainability (MTBM)	300 hours (min)	275 hours	H	6%
Logistics Response Time	2 hours (max)	6 hours	H	12%
Information Process Time	0.5 hours (max)	2 hours	B	5%
Life Cycle Cost ($/Unit)	$1.5 million (max)	$3.3 million	B	20%
				100%

Referring to Figure 3.2.4-1 (block 2), the next step is to identify the attributes, or characteristics, that need to be included and inherent within the ultimate system design configuration to comply with the requirements in Table 3.2.4-1. By providing a good technical response, we begin to define the HOW requirements (in response to the WHATs). In other words, given that Operational Availability (A_o) is a key requirement, *what specific characteristics need to be built into the design in order to ensure that a 98% A_o for the system will be attained?*

An excellent tool that can be applied to aid in the establishment and prioritization of TPMs, as well as for the identification of the appropriate technical responses, is the *quality function deployment (QFD)* model. Implementation of the QFD model requires a team approach to help assure that the "voice of the customer" is reflected in the ultimate design. The purpose is to establish the necessary requirements and to translate these requirements into technical solutions.

Customer requirements and preferences are defined, weighted based on perceived degree of importance, and attributes are described. The QFD method provides the design team with a complete understanding of customer desires, forces the customer to prioritize these desires, and enables a comparison of one design approach against another. Each customer attribute is then satisfied by a technical solution.[3]

The QFD process involves constructing one or more matrices, the first of which is often referred to as the *House of Quality (HOQ).* A modified version of the HOQ is presented in Figure 3.2.4-2. Commencing on the left side of the structure, the customer needs are identified and those needs are ranked in terms of priority, the levels of importance being specified quantitatively. This side reflects the *WHATS* that must be addressed. A team, with representation from both customer and various responsible design organizations, determines the priorities through an iterative process of review, evaluation, revision, re-evaluation, and so on. The top part of the HOQ identifies the designer's proposed *technical* response relative to the attributes (characteristics) that must be incorporated into the design in order to respond to the needs (i.e., the voice of the customer). This part constitutes the *HOWS*, and there should be at least one technical solution for each identified customer need. The interrelationships among attributes (or technical correlations) are identified, as well as possible areas of conflict. The center part of the HOQ conveys the strength or impact of the proposed technical response on the identified requirement. The bottom part

[3] Four good references to the QFD method are: (1) Akao, Y., *Quality Function Deployment: Integrating Customer Requirements into Product Design*, Productivity Press, Portland, OR, 1990; (2) Cohen, L, *Quality Function Deployment: How to Make QFD Work for You*, Addison-Wesley, Reading, MA, 1995; (3) Revelle, J.B., J.W. Moran, and C. Cox, *The QFD Handbook*, John Wiley & Sons, Hoboken, NJ, 1997; and (4) *Quality Toolkit: A Practical Guide for Developing and Implementing a Quality Program for Products and Services*, Reliability Analysis Center (RAC), 201 Mill Street, Rome, NY 13440-6916, pages 30-32, 2002.

allows for a comparison between possible alternatives, and the information on the right side of the HOQ is used for planning purposes.[4]

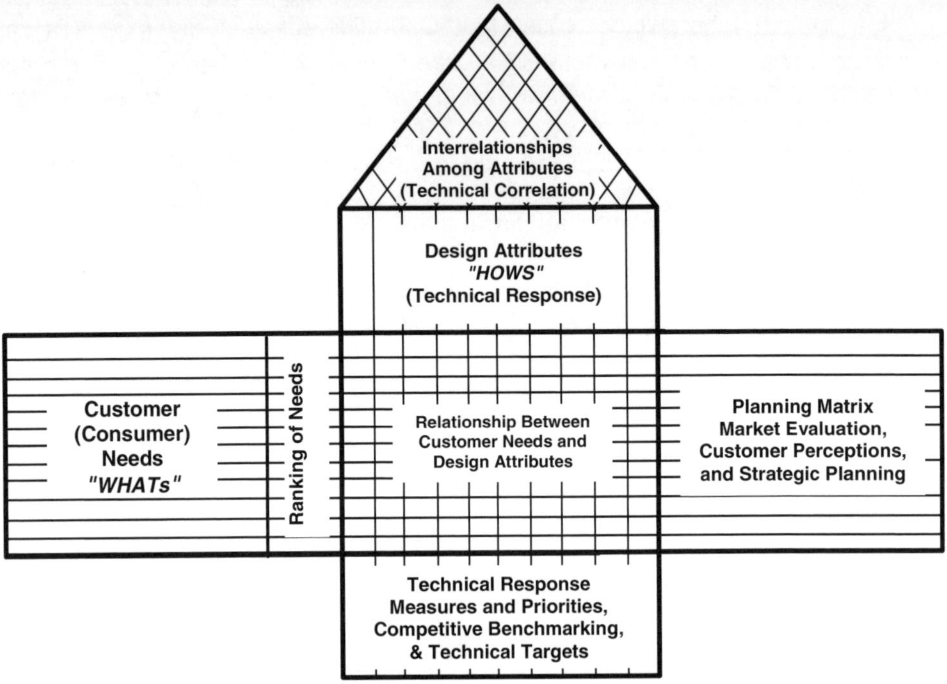

Figure 3.2.4-2. Modified House of Quality (HOQ)

The QFD method is used to facilitate the translation of a prioritized set of subjective customer requirements into a set of system-level requirements during conceptual design (i.e., the Concept Refinement Phase). A similar approach may be used to subsequently translate system-level requirements into a more detailed set of requirements at each stage in the design and development process. In Figure 3.2.4-3, the *HOWS* from one house become the *WHATS* for a succeeding house. Requirements may be developed for the system, subsystems, components, manufacturing process, and for the maintenance and support infrastructure. The objective is to ensure the required justification and traceability of requirements from the top down; i.e., from the system and down to the maintenance and support infrastructure.

[4] Hauser, J.R., and D. Clausing, "The House of Quality," *Harvard Business Review*, May-June 1988, pages 63-73.

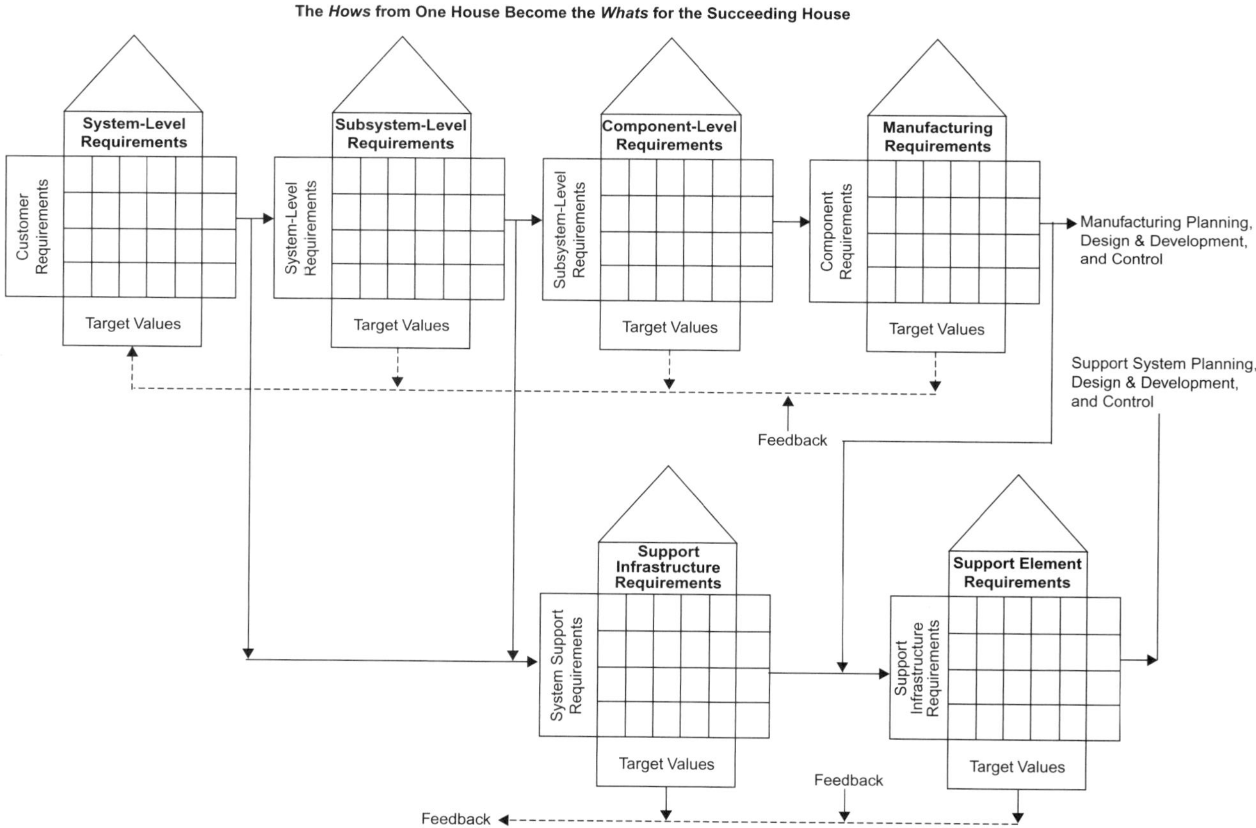

Figure 3.2.4-3. Family of Houses

(Source: Blanchard, E.S., Logistics Engineering and Management, 6th Edition, Pearson Prentice-Hall, NJ, 2004)

Topic 3.2.5: Development of Performance-Based Logistics (PBL) Requirements

Referring to Topic 1.5 (Section 1), the objective in developing PBL requirements is to emphasize the importance of considering logistics and supportability factors in the same context with other system *performance* requirements, and to specify these factors early in the design process (during the Concept Refinement Phase) as "design-to" criteria -- an *input* to design and not as an "after-the-fact" requirement. In Figure 1.5-1, there are many different factors, presented in a hierarchical structure, which could be considered within the overall PBL context. However, the objective is to identify and select those factors that are directly applicable and "tailored" for a given system, and to apply these at the system level and, through functional analysis and allocation, down to the various lower-level elements as appropriate.

Referring to the top-down/bottom-up system development process in Figure 1.6.1-1 (block 1), an early requirement includes the identification and prioritization of *technical performance measures (TPMs)*, as described under Topic 3.2.4. These measures represent ***performance*** requirements for a given system, and these requirements must not only cover the more conventional factors dealing with system capacity, output, speed, size, weight, etc., but must include those measures that have a direct impact on logistics and system support as well. For example, the factors identified in Table 3.2.4-1 (Topic 3.2.4) reflect the results of a QFD analysis, indicating specific design-to requirements and their respective degrees of importance. In identifying those factors that would have the greatest impact on supportability, one would likely select Operational Availability (A_o), Maintainability (MTBM), Logistics Response Time, Information Process Time, and Life-Cycle Cost (LCC). While the degree of impact may vary somewhat, these factors could be classified as PBL requirements included within the overall TPM spectrum. Then, with these PBL requirements specified at the system level, lower level requirements can then be developed and presented as shown in Figure 1.5-1. At this early point in the life cycle, it is essential that the design requirements for logistics and the maintenance and support infrastructure be identified and specified, in meaningful quantitative terms, as an input to design; i.e., the *design for system supportability.*[5]

[5] A complete review and understanding of the measures described in Section 2 of this Toolkit is essential if one is to select the appropriate metrics and apply these as a design requirement for the system and its elements. Basically, the various factors described in Section 2 may be included as PBL requirements (at different levels of the system hierarchy), as long as the application of such is meaningful.

Topic 3.2.6: Functional Analysis and Allocation

An essential task in early conceptual and preliminary design (i.e., Concept Refinement and Technology Development Phases) is the development of a *functional* description of the system, which in turn, serves as a basis for the identification of the resources necessary for the system to accomplish its objective(s). A *function* refers to a specific or discrete action (or series of actions) that is necessary to achieve a given objective; that is, an operation that the system must perform to accomplish its mission, a logistics activity that is required for the transportation of material, or a maintenance action that is necessary to restore the system to operational use. Such actions will ultimately be accomplished through the use of equipment, software, people, facilities, data, or combinations thereof. However, at this point, the objective is to specify the *WHATS* and **not** the *HOWS*; that is, *what* needs to be accomplished versus *how* it is to be done. The functional analysis is an iterative process of breaking requirements down from the system-level, to the sub-system, and as far down in the hierarchical structure as necessary to identify input design criteria and/or constraints for the various elements of the system.[6]

Referring to Figures 1.6.1-1 and 1.7-1, the functional analysis may be initiated in the early stages of conceptual design (i.e., the Concept Refinement Phase) as part of the problem definition and needs analysis task, and when functions that the system must perform in order to fulfill the needs of the customer are first identified. These *operating* functions are then expanded and formalized through the development of system operational requirements (see Topic 3.2.2). Primary *logistics and maintenance support* functions for the system, which evolve from operational requirements, are identified as part of the maintenance concept development process (see Topic 3.2.3). Subsequently, these functions must be expanded to include *all* of the activities from the initial identification of need to retirement of the system (refer to Figure 1.6.1-1).

The accomplishment of a functional analysis can be facilitated through the use of functional flow block diagrams (FFBDs), as illustrated in Figure 3.2.6-1. Block diagrams are developed primarily for the purposes of structuring system requirements into functional terms. They are developed to illustrate basic system organization and to identify functional interfaces. The functional analysis (and the generation of functional flow block diagrams) is intended to facilitate the completion of system design and development in a comprehensive and logical manner.

Referring to Figure 3.2.6-1, top-level requirements are identified, expanded to the second level (i.e., sub-functions), and down to the depth required for the purposes of

[6] In applying the principles of system engineering, one should not identify and/or initiate the purchase of one piece of equipment, or module of software, or data item, or element of logistic support without first having justified the need for such through the functional analysis. On many projects, items are often purchased based on what is perceived as a "requirement" but later turn out not to be necessary. This practice can turn out to be quite costly.

acquiring good design "definition." For example, this process may be applied to each block (and all of the activities) identified in Figure 1.3-2; i.e., a breakdown of design and development functions in block 2, production/construction functions in block 4, distribution and warehouse functions in block 6, operational functions in block 7, intermediate-level maintenance functions in block 8, and so on. This, in turn, may lead to the development of functions at a lower level. The objective is to ensure the "traceability" of requirements from the top-down.

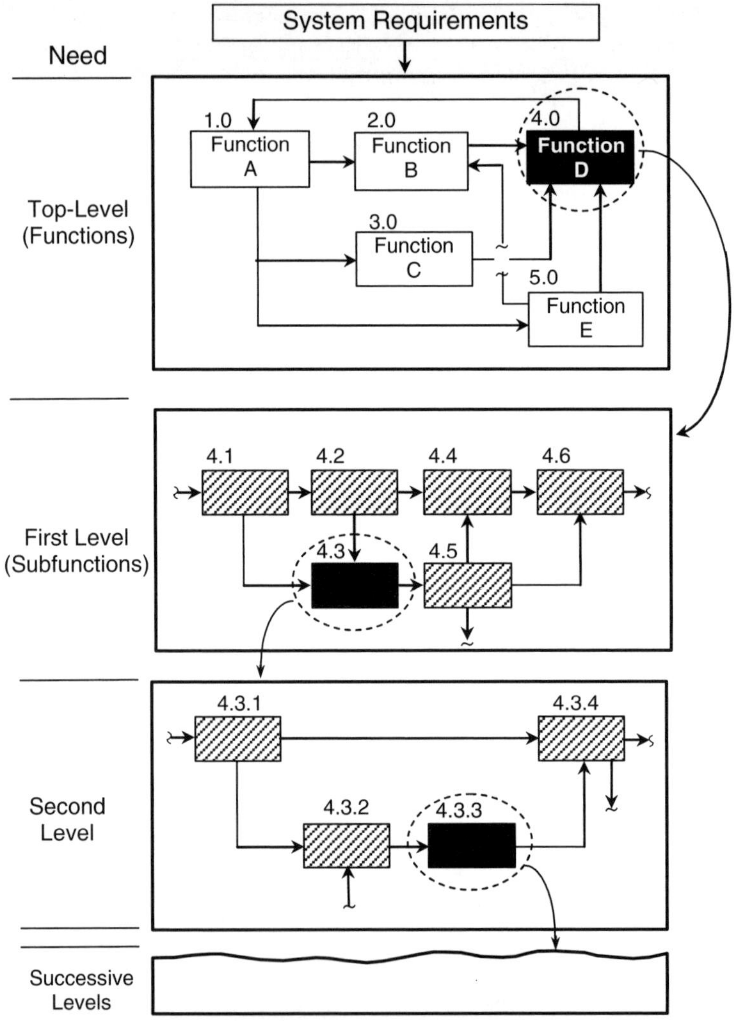

Figure 3.2.6-1. System Functional Indenture Levels

In accomplishing the functional analysis, one should commence with a description of *operational functions*, which covers all of the "forward flow" activities identified in Figure 1.3-2 (i.e., requirements, design and development, production/construction, and system utilization). The logistics and supply chain functions, illustrated in Figure 1.3-3, are included in this flow. Associated with each functional block are the applicable TPM/PBL factors, described in Topics 3.2.4 and 3.2.5, indicating what is necessary from a "performance" perspective in order to fulfill mission requirements for the system in a satisfactory manner; i.e., the quantitative metrics pertaining to each applicable function. An abbreviated example of *operational* functions is shown in Figure 3.2.6-2; i.e., blocks 1.0 – 6.0 and 4.0 – 4.4. For each of these functional blocks there is a *GO/NO-GO* option, with a "GO" leading into the next operational function. In the event of a "NO-GO," this will lead into a series of *maintenance and support* functions; i.e., blocks 15.1 – 15.3 and 20.1 – 20.3. A more detailed example of a maintenance functional flow diagram is shown in Figure 3.2.6-3.

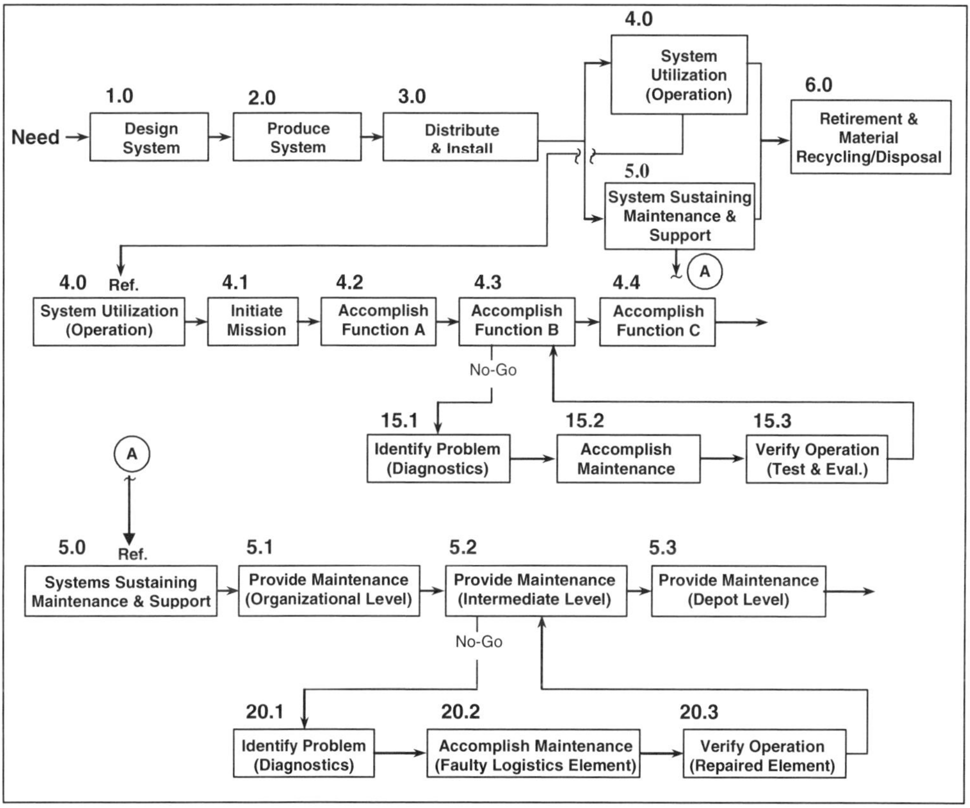

Figure 3.2.6-2. Functional Flow Block Diagram (Abbreviated)

For each of the blocks in a functional flow diagram, the designer/analyst should be able to specify the *input* requirements, expected *outputs*, external *controls* and/or *constraints*, and the projected *mechanisms* (resources) necessary to accomplish the specific function in question.

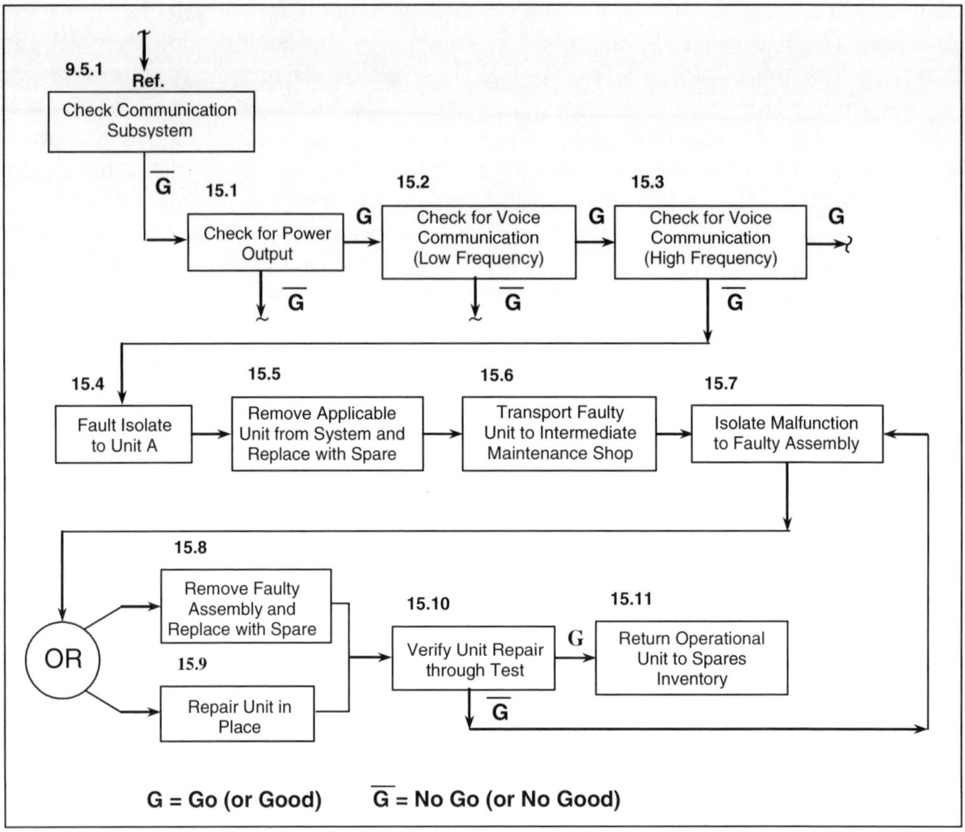

Figure 3.2.6-3. Maintenance Functional Flow Diagram

Referring to Figure 3.2.6-4, the functional block represents some specific "requirement" (in terms of the *WHATs*), there are expected measures/metrics associated with the function being evaluated, and often there are controls/constraints within which the function must be accomplished. Given such, there may be a number of different alternative approaches that should be considered in response to the functional requirement, trade-off studies are conducted, alternatives are evaluated against criteria developed from established technical measures (i.e., TPM/PBL factors), and a preferred approach is recommended. It is at this point that one begins to identify the specific requirements for hardware, software, people, facilities, data, or various combinations thereof (i.e., the *HOWs*). The process reflected in Figure 3.2.6-4 should be applied in determining the resource requirements for each individual function, included in the overall functional description of the system. These resources are then combined, integrated, and lead to the identification of the various lower-level elements of the system.

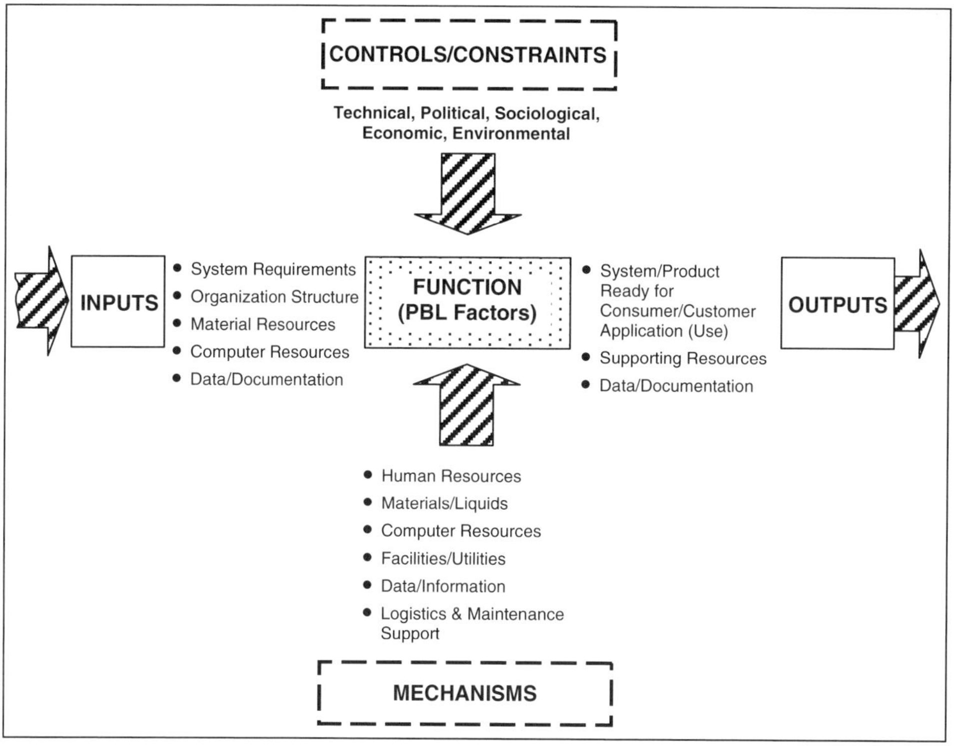

Figure 3.2.6-4. Identification of Resource Requirements

Having a top-level definition through the functional analysis, a breakdown of the system into its elements (components) is accomplished by *partitioning*. This step involves a breakout of the system into sub-systems and lower-level entities. The challenge is to identify and group closely-related functions into packages, employing a common resource (e.g., equipment, software) to accomplish multiple functions to the extent possible. Although it may be relatively easy to identify individual functional requirements and associated resources on an independent basis, this process may end up with a configuration that could be rather costly when it comes to system packaging, weight, size, and so on. The questions are -- *what hardware or software can be selected that will perform multiple functions? How can new functions be added without adding any new physical elements to the system structure?*

The partitioning of the system into elements is evolutionary in nature. Common functions may be grouped or combined in such a way as to provide a system packaging scheme, with the following objectives in mind:

1. System elements may be grouped by geographical location, a common environment, or by similar types of equipment/software.

2. Individual system "packages" should be as independent as possible with a minimum of "interaction effects" with other packages. A design objective is to be able to remove and replace a given package without having to remove and replace other packages in the process, or requiring an extensive amount of alignment and adjustment as a result.

3. In breaking down a system into sub-systems, select a configuration in which the communication between the different sub-systems is minimized. In other words, whereas the sub-system *internal* complexity may be high, the *external* complexity should be low. Breaking down the system into packages requiring high rates of information exchange between the packages should be avoided.

An overall design objective is to break down the system into elements such that only a very few (if any) critical events can influence or change the inner workings of the various packages that make up the system architecture. This facilitates an *open-architecture* approach in design.

As a result of partitioning, a system may be broken down into its elements (components) such as shown in Figure 3.2.6-5. System functions are identified, broken down into sub-functions, and grouped into four equipment units, different modules of software, elements of data/information, the various elements of the maintenance and support infrastructure, and so on. Having identified the major system elements, the next step is to *allocate* or *apportion* the requirements specified for the system down to the level desired to provide a meaningful input to design. For example, given an Operational Availability (A_o) of 0.998 at the system level, *what specific design-to requirement(s) should be specified for Unit "A," Unit "B," Unit "C," and Unit "D?"* This step involves a top-down distribution of the quantitative and qualitative criteria developed in Topic 3.2.4. Given the prioritized technical performance measures (TPMs) in Table 3.2.4-1, *what requirements should be specified for each of the levels (elements) shown in Figure 3.2.6-5 that will, when combined, meet the overall requirements initially specified for the system?*

In summary, the functional analysis is first implemented (at a top-level) during the early stages of conceptual design (i.e., Concept Refinement Phase) and is expanded as system design definition progresses (i.e., Technology Development Phase). Accomplishing a functional analysis helps to ensure:

1. That all facets of system design and development, production, operation, support, and phase-out (retirement) are covered; that is, all significant activities within the system life cycle.

2. That all elements of the system are fully recognized and defined; that is, prime equipment, software, facilities, personnel, data, and the elements of logistics and maintenance support.

3. That a means is provided for relating system packaging concepts and support requirements to specific system functions; that is, satisfying the requirements of good *functional* design.
4. That the proper sequences of activity and design relationships are established, along with critical design interfaces.

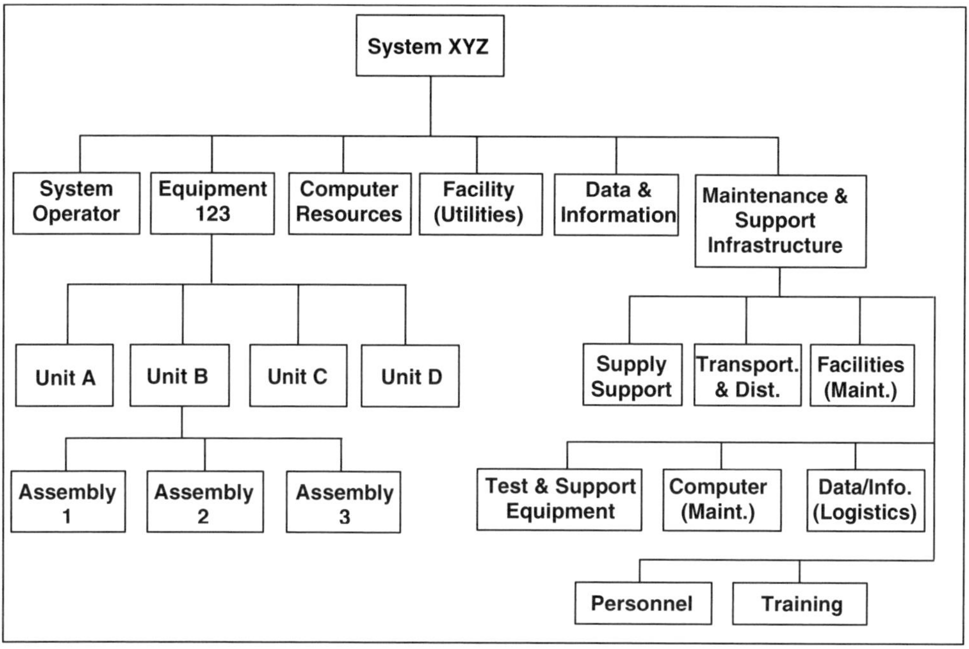

Figure 3.2.6-5. Hierarchy of System Elements

The functional analysis constitutes a critical step in the early system design and development effort, and it forms a *baseline* for many activities that are accomplished subsequently. For instance, it serves as a basis in the development of the following:

1. Reliability and maintainability models (block diagrams)
2. Failure mode, effects, and criticality analysis (FMECA)
3. Reliability-centered maintenance (RCM) analysis
4. System safety/hazard analysis
5. Operator task analysis (OTA)
6. Maintenance task analysis (MTA)
7. Level of repair analysis (LORA)
8. Supportability analysis (SA)

Topic 3.2.7: Allocation of Logistics and Supportability Requirements

Referring to Figure 3.2.6-5, the proposed system configuration has been broken down into its various elements; i.e., people, equipment, software and computer resources, facilities, data/information, and the maintenance support infrastructure. Additionally, the prime equipment has been broken down into Unit "A," Unit "B," etc., and the maintenance support infrastructure has been broken down to include the functional entities of supply support, transportation and distribution, test and support equipment, and so on. The question is -- *given the requirements for the system, what specific design-to requirements should be specified for Unit "A," Unit "B," supply support, test and support equipment, and so on?*

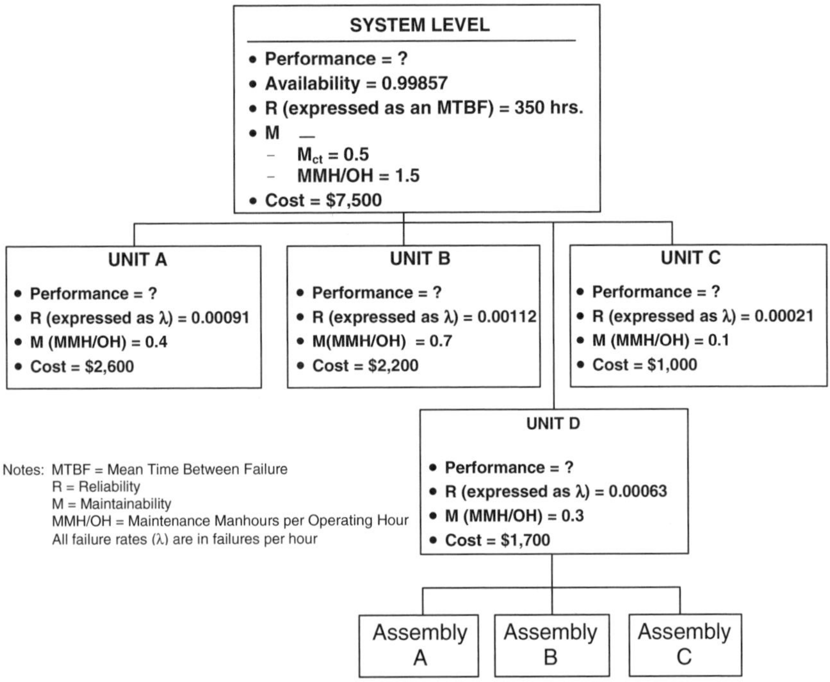

Figure 3.2.7-1. Allocation of Requirements

Top-level requirements must be *allocated* or *apportioned* downward to the level necessary for providing a good and meaningful input to the design. For example, Figure 3.2.7-1 illustrates a breakout of "Equipment 123" (see Figure 3.2.6-5) into four basic "units" that make up the equipment; i.e., Unit "A," Unit "B," Unit "C," and Unit "D." The requirements at the system level (e.g., Reliability MTBF of 350 hours) have been allocated to each of the units; thus, the reliability design goal for Unit "A" is 0.00091 failures per hour, the design goal for Unit "B" is 0.00112 failures per hour, and so on. At the same time, the reliabilities for the four units, when combined, must be equivalent to the reliability at the system level (i.e., 350 hours). In other words, there must be a top-down/bottom-up "traceability" of requirements.[7]

Employing the same basic approach, one also needs to establish the design-to requirements for the critical elements of the logistics and maintenance support infrastructure (refer to Figure 3.2.6-5). Such requirements, which must evolve directly from the PBL factors established earlier (see Topics 1.5 and 3.2.5), may take a variety of forms depending on the type and complexity of system and the criticality of its mission. An example of a few design-to goals are noted here:

1. The response time for the logistics and maintenance support infrastructure shall not exceed 4 hours.
2. The probability of spares availability at the organizational level of maintenance shall be at least 95%.
3. The transportation time between the location where organizational maintenance is accomplished and the intermediate maintenance shop shall not exceed 6 hours.
4. Test equipment utilization in the intermediate maintenance shop shall be at least 80%, and the reliability of the test equipment shall be 95% or better.
5. Personnel skill levels at the organizational level of maintenance shall be equivalent to grade *x* or below.
6. The maintenance facility at the intermediate level shall be designed for a minimum of 75% utilization.
7. The turnaround time (TAT) in the intermediate maintenance shop shall be 2 days (or less), and 10 days (or less) in the depot maintenance facility.
8. The time for processing logistics information shall not exceed 10 minutes.
9. The process time for removing an obsolete item from the operational inventory shall not exceed 12 hours, and the cost per item processed shall be *y* dollars or less.

[7] The requirements allocation process is discussed in detail in Blanchard, B.S., *Logistics Engineering and Management*, 6th Ed., Pearson Prentice Hall, Upper Saddle River, NJ, pages 162-172, 2004. Reliability allocation is also covered in "Reliability Toolkit: Commercial Practices Edition," Reliability Analysis Center (RAC), 201 Mill Street, Rome, NY 13440, pages 51-53. Maintainability allocation is also covered in "Maintainability Toolkit: A Practical Guide for Designing and Developing Maintainable Products and Systems," Reliability Analysis Center (RAC), 201 Mill Street, Rome, NY 13440, pages 71, 168-174.

3.2.7: ALLOCATION OF LOGISTICS AND SUPPORTABILITY REQUIREMENTS

Referring to Figure 3.2.6-5, specific design goals should be established for each of the system elements shown. Further, these goals, when combined, must be directly supportive of the overall requirements for the system.[8]

[8] In the past, very little (if any) attention has been directed to the establishment of early "before-the-fact" design-to requirements for logistics and the maintenance support infrastructure. Yet, if the goals pertaining to the *design for supportability* are to be realized, then we need to start addressing these requirements not only in the acquisition of new systems but in the re-engineering of existing systems. In any event, it is critical that some serious attention be given to logistics and system support early in conceptual design (i.e., the Concept Refinement Phase).

Topic 3.2.8: The Specification of Supportability Requirements for System Design

Having identified the specific "input" requirements for design of the various elements of the system, it is important that such requirements be properly stated in an applicable *specification(s)*. While the classifications and formats of specifications may actually vary from one program to the next, the following generic classifications have been assumed for reference throughout this Toolkit:[9]

1. *System specification (Type A)*: includes the technical, performance, operational, and support characteristics for the system as an entity. Operational requirements, the maintenance concept, the results from feasibility analysis, the applicable system-level TPMs and PBL factors, and a description of system requirements in functional terms are presented. All system-level design requirements are included.
2. *Development specification (Type B)*: includes the technical requirements for any element or component below the system level where research, design, and development are accomplished. This may cover a new equipment item, assembly, computer program, facility, critical item of support, and so on. Each development specification must include the performance, effectiveness, and supportability characteristics that are required in the evolving of design from the system level and down.
3. *Product specification (Type C)*: includes the technical requirements for any item below the top system level that is currently in the inventory and can be procured as a "commercial off-the-shelf (COTS)" item. This specification may cover standard system components (equipment, assemblies, units, cables), a specific element of software, a spare part, a maintenance tool, and so on.
4. *Process specification (Type D)*: includes the technical requirements that cover a "process" associated with any element or component of the system (e.g., purchasing process, machining process in the factory, material handling process, packaging and transportation process, information flow process).
5. *Material specification (Type E)*: includes the technical requirements that pertain to raw materials, mixtures (e.g., paint, chemical compounds), and/or semi-fabricated materials (e.g., electrical cable, piping) that are used in the manufacture and assembly of the product.

Associated with most programs (with varying forms of contractual arrangements) is a hierarchy of specifications, such as illustrated in Figure 3.2.8-1. At the top is the

[9] MIL-STD-490, "Specification Practices," Department of Defense, Washington, DC.

System (Type A) Specification, which should include all system-level requirements for logistics and the *design for supportability*. These requirements are then allocated downward to the various elements or components of the system, and the design-to requirements for these are included in the appropriate lower-level specifications. It is essential that the requirements for logistics and supportability be "traceable" from the top-down.

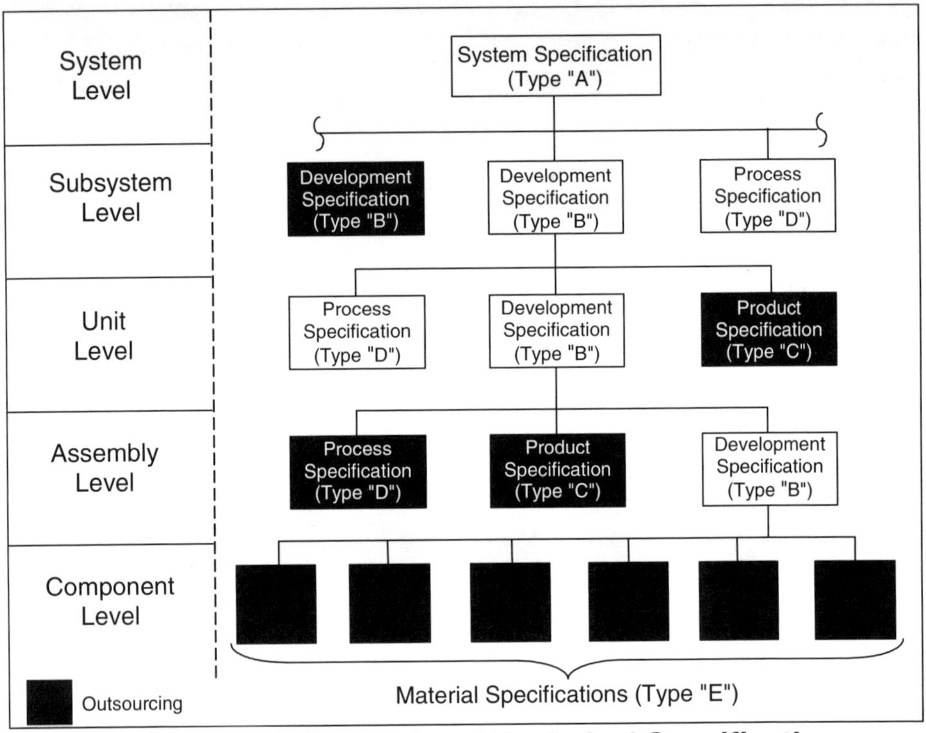

Figure 3.2.8-1. Hierarchy of Technical Specifications

As a final point (and referring to Figure 3.2.8-1), it should be noted that there may be quite a few system elements/components that are being provided by outside suppliers. Further, these suppliers may be dispersed and located throughout the world. Thus, it is critical that the requirements for the system (and its maintenance support infrastructure) be complete, and that good and effective "communications" exist among the various applicable specifications; i.e., the specifications must "talk to each other" as appropriate.

For More Information:

1. Blanchard, B.S. and W.J. Fabrycky, *Systems Engineering and Analysis,* 3rd Ed., Pearson Prentice Hall, Upper Saddle River, NJ, 1998.

2. Blanchard, B.S., *Logistics Engineering and Management*, 6th Ed., Pearson Prentice Hall, Upper Saddle River, NJ, 2004.

3. "Reliability Toolkit: Commercial Practices Edition," Reliability Analysis Center, 201 Mill Street, Rome, NY.

4. "Maintainability Toolkit: A Practical Guide for Designing and Developing Maintainable Products and Systems," Reliability Analysis Center, 201 Mill Street, Rome, NY.

5. "Quality Toolkit: A Practical Guide for Developing and Implementing a Quality Program for Products and Services," Reliability Analysis Center, 201 Mill Street, Rome, NY.

This page intentionally left blank

Section 4: Designing for Supportability

Insight

The successful incorporation of supportability features in the ultimate system configuration is, of course, highly dependent on (1) the establishment of a good set of *requirements* from the beginning; (2) the development of the appropriate design *criteria* as a result of these requirements; and (3) the incorporation of the necessary *characteristics* (or attributes) in the design, on a day-to-day basis, in response to the specified criteria. While the need for a well-defined set of requirements and the development of input design criteria have been covered in the earlier sections of this Toolkit, the actual incorporation of the appropriate characteristics in design occurs through the on-going day-to-day efforts of the responsible "design team," as the design process evolves from the conceptual design stage through preliminary system design and detail design and development. Such characteristics as accessibility, functional packaging and modularization, good diagnostics, standardization and the incorporation of reliable components, etc., must be "built-into" the design as one proceeds through this process.

Good system design is the result of a "team" approach, with the design team constituting an interdisciplinary mix of expertise from various technical fields. This is particularly true with regard to the *design for supportability.* In addition to the more traditional fields of electrical design, mechanical design, structural design, etc., selected expertise will be required in the areas of reliability, maintainability, human factors, safety, quality, and so on. While representation from all of these areas will not be required at the same time, the design team must be highly dynamic and flexible in make-up, incorporating the right level of effort and expertise at the time needed. In other words, the responsible design team must have the appropriate support from a wide variety of disciplines, often spread across and throughout one or more given organizational structures.

In order to gain a full understanding of the objectives and requirements in designing for system supportability, it seems appropriate (at this point) to provide some additional coverage of a few closely related disciplines, their objectives, the models/tools utilized in the accomplishment of such, and the many interfaces that exist as we evolve through the design process. Of specific interest are the requirements associated with the *design for reliability, design for maintainability, design for human factors, design for economic feasibility, design for quality,* and so on. For example, the utilization of such tools as Failure Mode, Effects, and Criticality Analysis (FMECA), Fault-Tree Analysis (FTA), and Reliability-Centered Maintenance (RCM) Analysis is required for the implementation of a reliability program; the Maintenance Task Analysis (MTA), Level-Of-Repair Analysis (LORA), and Life-Cycle Cost Analysis (LCCA) are required in fulfilling the requirements of a maintainability program; the Safety/

Insight (Cont'd)

Hazard Analysis is required for a safety program; and the Operator Task Analysis (OTA) and Operational Sequence Diagrams (OSDs) are required as part of a human factors program activity. At the same time, these program requirements (and tools) are also essential for accomplishing logistics and supportability program requirements. Further, reliability and maintainability predictions constitute a major input in the development of supportability analysis (SA) data. Thus, and referring to Figure 1.6.3-2, the interfaces are numerous. The objective in this section of the Toolkit is to describe some of these program requirements and the need for such in the *design for supportability.*

Topic 4.1: The System Design Process and Specific Design Objectives

The design process, as defined herein, includes the stages of *conceptual design* (Concept Refinement Phase), *preliminary system design* (Technology Development Phase), and *detail design and development* (System Development and Demonstration Phase). These stages of design basically fall within the life-cycle phases specified for major defense systems, as shown in Figure 1.7-1. Figure 4.1-1, which is an expansion of the top-down system engineering process illustrated in Figure 1.6.1-1, is included to establish a frame of reference for the various discipline-related activities discussed under Topic 4.2. Further, the design disciplines that are closely related to the *design for supportability*, and where the "interfaces" are numerous, are identified in Figure 4.1-2 and also covered in Topic 4.2.

Referring to Figure 4.1-1, *conceptual design* constitutes the first stage in the overall design process and is initiated in response to an identified consumer need. Given a need and a definition of the current problem (or deficiency), feasibility studies are conducted with the objective of identifying alternative technical approaches to solving the problem, and includes the conductance of trade-off studies to support early design decisions at the *system* level. System operational requirements and the maintenance concept are defined, a top-level functional analysis for the system is developed, applicable technical performance measures (TPMs) are identified and prioritized, and a system specification (Type "A") is prepared to describe the design requirements for the system. Specific performance measures and design-to criteria, applicable at the system level and for its major elements, are included. This stage, including requirements analysis and system definition, is represented by blocks 1 and 2 in Figure 4.1-1.

Logistics requirements for the system usually evolve from the initial problem definition and are defined through the criteria established from the feasibility analysis, from the operational requirements and the maintenance concept, and from the identification and prioritization of TPMs. The selection of technology applications in system design does have supportability implications; the operational requirements for the system lead to the identification of geographical locations where logistic support will be required and for how long; the maintenance concept defines levels, responsibilities, and the effectiveness of the support required; and the prioritized TPMs indicate critical areas that need to be addressed in design. It is at this point that the appropriate performance-based logistics (PBL) factors, discussed in earlier sections, are defined and included in the system specification. Logistics engineering representation, as part of the design team, is essential in the requirements analysis and definition process (i.e., from the beginning and throughout the Concept Refinement Phase).

Preliminary system design starts with a "functional baseline" system configuration, described in the system specification, and proceeds toward the translation of the established system-level requirements into detailed qualitative and quantitative design requirements for the various elements of the system. Preliminary system design, reflected by the activities in blocks 3 and 4 of Figure 4.1-1, includes the process of functional analysis, the allocation of system-level requirements downward to the applicable elements of the system, system synthesis and analysis, the accomplishment of trade-off studies and design optimization, and the definition of an "allocated baseline" configuration in the form of Development (Type "B"), Product (Type "C"), and Process (Type "D") Specifications, or combinations thereof (refer to Topic 3.2.8, Section 3).

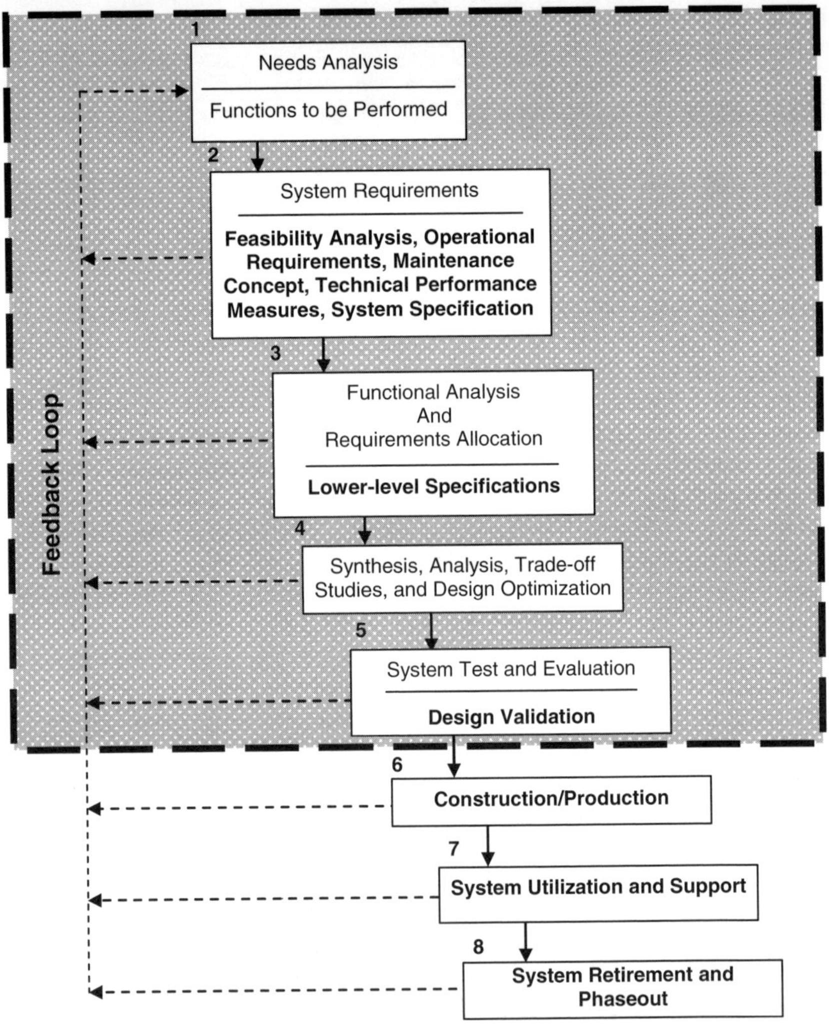

Figure 4.1-1. Steps in the System Life Cycle (Design Emphasis)

Referring to Figure 4.1-1 (block 3), the functional analysis leads to the identification of specific resource requirements to include different combinations of equipment, software, people, real estate and facilities, data/information, and so on. The process evolves from a functional baseline at the system level to the identification of specific physical elements that, when combined and properly integrated, will make up the system; i.e., evolving from the *WHATs* to the *HOWs*. Referring to Figure 3.2.6-4 (Section 3), each block in the functional description is evaluated, resources are identified and combined, and the results convey a breakdown of the system into its components such as shown in Figure 3.2.6-5.

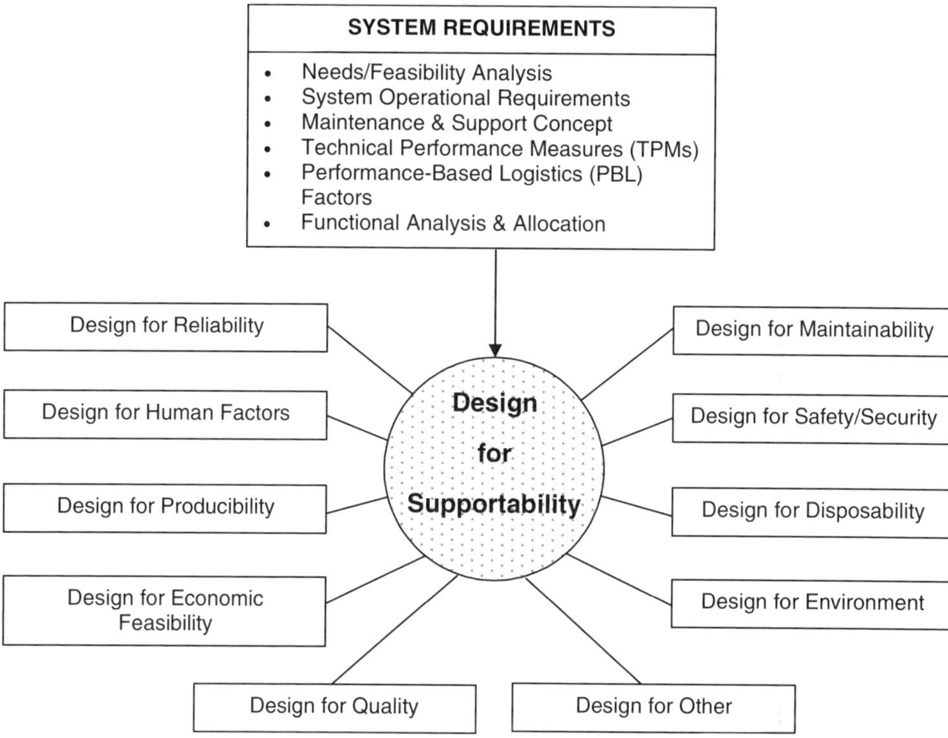

Figure 4.1-2. Design for Supportability (Major Design Interfaces)

As in conceptual design, logistics requirements must be considered as an integral part of the preliminary system design process. The functional analysis includes the development of maintenance and support functions (i.e., those logistics and maintenance support activities that are reflected in the forward and reverse flows in Figure 1.3-2); the subsequent allocation process must include a top-down distribution of applicable performance-based logistics (PBL) factors and the establishment of supportability criteria for design; the accomplishment of trade-off studies and design optimization includes logistics and the application of supportability criteria in the evaluation of alternatives; and the supportability analysis (SA) continues to be a major activity for designing the support infrastructure and ensuring that logistics (and the design for supportability) is being adequately addressed in the design process (and throughout the Technology Development Phase).

Detail design and development begins with the configuration developed through the preliminary system design stage and evolves through the development of a system prototype ready for production and/or construction. A system configuration has been defined, a functional analysis has been accomplished, and the requirements for detail design have been included in the appropriate specifications (Types *A, B, C,* and/or *D*). The next step is to convert the information provided into the proper mix of hardware, software, people, facilities, data, and specific items of support. The process from this point includes the following:

1. *Definition of system elements*: subsystems, units, assemblies, lower-level component parts, modules of software, facilities, and data. As logistics and the maintenance support infrastructure constitute a major element of the system, it is essential that this area be addressed as an integral part of the design process, which is basically accomplished through the day-to-day design participation activity and completion of the Supportability Analysis (SA).
2. *Preparation of design data*: specifications, drawings, databases, electronic data files, trade-off study reports, prediction and analysis reports, and so on, describing the system and all of its elements.
3. *Development of physical models of the system and/or major system components*: engineering laboratory models, service test models, mock-ups, and prototype models used for testing and system evaluation (primarily directed toward "validation" of system elements/components).
4. *Conductance of system integration and test*: the measurement and evaluation of system characteristics, verification that the various elements of the system operate effectively as an integrated entity, and verification that the overall requirements for the system have been met (i.e., that the design is in compliance with the required and specified TPM and PBL requirements). Any noted deficiencies may be corrected through redesign, system modification, and retest as necessary.

These activities, represented by blocks 4 and 5 in Figure 4.1-1, must, of course, be tailored to the system being developed. In instances where the design includes the utilization of new technologies and/or the incorporation of highly complex components, the level of design definition and subsequent test and evaluation requirements may be more extensive (i.e., a greater degree of design documentation, a greater depth of testing) than for cases where the design configuration is relatively state-of-the-art and data are available from earlier applications. Basically, this area of activity is applicable throughout the System Development and Demonstration Phase (see Figures 1.4-1 and 3.1-1).

Topic 4.2: Related Design Disciplines

Referring to Figure 4.1-2, the objective here is to briefly describe a few closely related design disciplines, their objectives, and associated activities that directly contribute to the goals pertaining to the *design for supportability.* In other words, the accomplishment of many of these activities is essential in order to meet the requirements described throughout this Toolkit. It is not intended (at this point) to recommend any particular type of organizational structure, but to cover some of the tools necessary in fulfilling supportability objectives. For example, the completion of an FMECA is required in the accomplishment of a Supportability Analysis (SA), whether it is performed by a reliability organization, a human factors organization, or by someone else. However, for the sake of presentation, the material in the sections to follow is organized and related to the disciplines that are characteristic for a typical large-scale project.

Topic 4.2.1: Design for Reliability

As described in Section 1 (Topic 1.6.4), *reliability* can be defined as the probability that a system or product will perform in a satisfactory manner for a given period of time, or in the accomplishment of a defined mission scenario, when used under specified operating conditions. Reliability is a "characteristic" of design and is heavily influenced by the selection and application of components, the degree of standardization incorporated into the design, and so on. The basic measures of reliability include *R*, *MTBF*, *MTTF,* and *Failure Rate* (λ), as described in Topic 2.1. These requirements directly impact the frequency and depth of logistics and maintenance support required, and the ultimate measures of reliability are a key input in the performance of the Supportability Analysis (SA).

To ensure that reliability is adequately addressed throughout the system design process, formal program requirements are often specified, particularly for large-scale systems, that will include activities such as those illustrated in Figure 4.2.1-1. Referring to the figure, reliability requirements (e.g., *R*, *MTBF*) must be specified, as part of the requirements for the overall system during conceptual design, and must be specified in the System Specification (Type *A*). These requirements, presented as applicable TPM/PBL factors, are then allocated, alternative design approaches are evaluated, system design definition evolves, and the resultant configuration is then validated through test and evaluation. In essence, reliability activities are inherent within each of the steps shown in Figure 4.1-1.

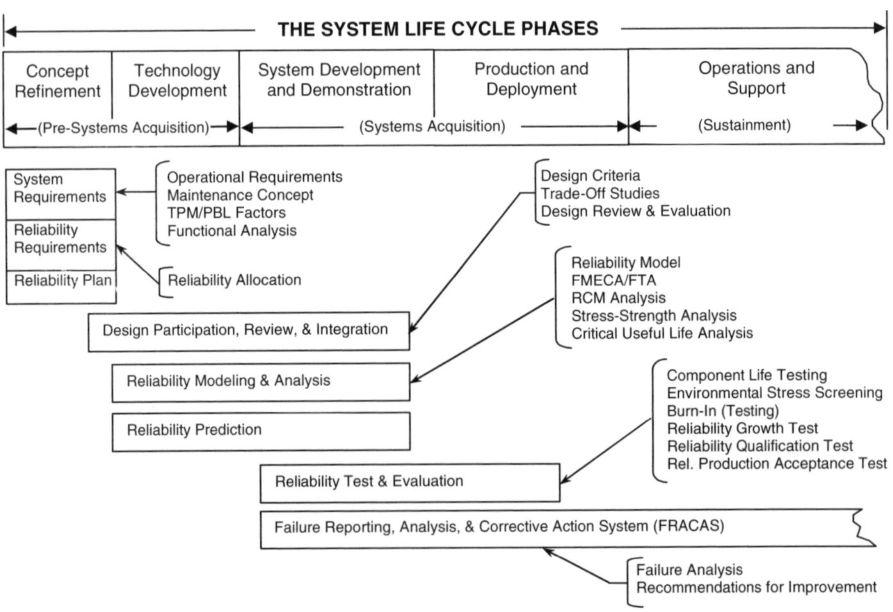

Figure 4.2.1-1. Reliability Program Requirements

With regard to logistics and the maintenance support infrastructure, critical design-related activities (and associated methods/tools) that are required in the fulfillment of supportability objectives, and as an input to the Supportability Analysis (SA), will include:

1. Specification of quantitative reliability requirements in the development of system operational requirements, the maintenance concept, in the definition and prioritization of TPM/PBL factors, and in the allocation of these requirements to the applicable lower-level system elements. The results constitute design-to goals for the logistics and maintenance support infrastructure.
2. Reliability predictions accomplished on a periodic basis, throughout the preliminary system design and detail design and development phases, as one evolves from one level of design definition to another. The results reflect the projected reliability measures for a given system configuration at a particular point in the design process.
3. The Failure Mode, Effects, and Criticality Analysis (FMECA); Fault-Tree Analysis (FTA); and the Reliability-Centered Maintenance (RCM) Analysis. The results will reflect the ways in which a system is likely to fail (if it does), the consequences of such failures, critical areas where design improvement is recommended, and the need for preventive maintenance and the resources required for such. These analyses, applied to varying degrees, constitute an inherent part of the Supportability Analysis (SA), and are illustrated through the examples presented in Section 5 of this Toolkit.
4. Reliability testing and evaluation at the system level. The results indicate actual experience in the laboratory, at a test facility, and/or in the field at a user's site, and constitute a relatively accurate measure of the "true" system reliability. The output should be applied in updating the Supportability Analysis (SA).

The program requirements for reliability and the interfaces throughout the design process are numerous, and the activities listed above represent only a few of the more critical areas where the accomplishment of reliability objectives and supportability objectives must be mutually supportive. [1]

[1] For additional coverage of typical reliability program requirements, refer to "Reliability Toolkit: Commercial Practices Edition," Reliability Analysis Center (RAC), 201 Mill Street, Rome, NY 13440-6916.

Topic 4.2.2: Design for Maintainability

Referring to Section 1 (Topic 1.6.5), *maintainability* can be described as a characteristic of design that pertains to the ease, accuracy, safety, and economy in the performance of system maintenance. The objective is to incorporate into the design configuration such characteristics as standardization and the use of standard/common components, functional packaging and interchangeability, accessibility, effective diagnostics and condition monitoring, labeling, and so on. The basic measures of maintainability relate to maintenance elapsed time, maintenance frequency, maintenance labor hours, and maintenance cost, and include such factors as $\overline{M}_{ct}$, $\overline{M}_{pt}$, MDT, MTBM, MLH/OH, and Cost/MA, noted in Table 1.6.5-1 and described in Topic 2.2. These requirements are closely related to the reliability measures described earlier, and have a significant impact on the frequency and depth of the logistics and maintenance support required for the system.

To ensure that maintainability characteristics are adequately addressed throughout the system design process, formal program requirements are often specified, particularly for large-scale systems, that will include activities such as those illustrated in Figure 4.2.2-1. Referring to the figure, maintainability requirements (e.g., $\overline{M}_{ct}$, MTBM, MLH/OH) must be specified, as part of the requirements for the overall system during conceptual design, and must be included in the System Specification (Type A). These requirements, presented as applicable TPM/PBL factors, are then allocated, alternative design approaches are evaluated, system design definition evolves, and the resultant configuration is then validated through formal test and evaluation. Basically, maintainability activities are inherent within the steps shown in Figure 4.1-1.

With regard to logistics and the maintenance support infrastructure, critical design-related activities (and associated methods/tools) that are required in the fulfillment of supportability objectives, and as an input to the Supportability Analysis (SA), will include:

1. Specification of quantitative maintainability requirements in the development of system operational requirements, the maintenance concept, in the definition and prioritization of TPM/PBL factors, and in the allocation of these requirements to the applicable lower-level system elements. The results constitute design-to goals for the logistics and maintenance support infrastructure.
2. Maintainability predictions accomplished on a periodic basis, throughout the preliminary system design and detail design and development phases, as one evolves from one level of design definition to another. The results, which depend on reliability predictions as an input, reflect the projected maintainability measures for a given system configuration at a particular point in the design process.

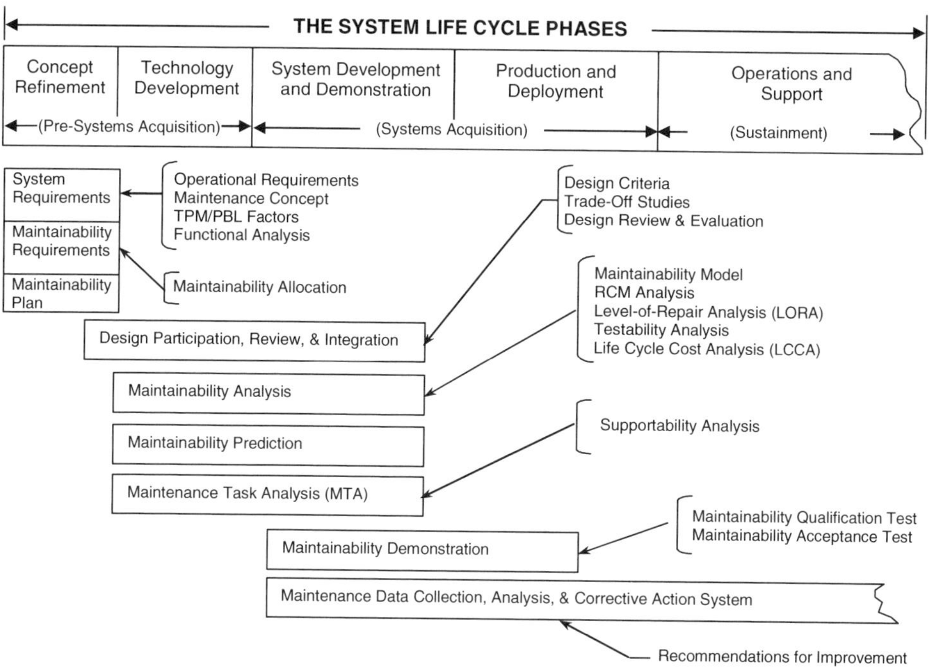

Figure 4.2.2-1. Maintainability Program Requirements

3. Level-Of-Repair Analysis (LORA), Reliability-Centered Maintenance (RCM) Analysis, Maintenance Task Analysis (MTA), and Life-Cycle Cost Analysis (LCCA). The LORA is accomplished to determine the depth to which maintenance will be performed and the level/location where it will be performed. RCM Analysis, which is based on the reliability FMECA, leads to the development of preventive (scheduled) maintenance requirements. The MTA, which is based on the maintenance functions identified as part of the functional analysis (refer to Topic 3.2.6), leads to the identification of all corrective and preventive maintenance tasks and the associated resources (e.g., personnel, spares, test equipment) that are required for the life-cycle support of the system. The LCCA, accomplished periodically, includes the prediction of all future costs over the entire system life cycle, including system maintenance costs which often constitute a large percentage of the total. These analysis efforts, which are heavily dependent on the development of good reliability data as inputs, are performed as an inherent part of the Supportability Analysis (SA), and are illustrated through the examples presented in Section 5 of this Toolkit.

4. Maintainability demonstration at the system level. The results will indicate actual experience at a formal test facility and/or in the field at a user's site, and constitute a relatively accurate measure of the "true" system maintainability and whether the initially specified TPM/PBL requirements

have been met. The output should be applied in updating the Supportability Analysis (SA).

The program requirements for maintainability and the interfaces throughout the design process are numerous, particularly with respect to reliability program requirements, and the activities listed above represent only a few of the more critical areas which are essential in fulfilling the supportability objectives described throughout this Toolkit.[2]

[2] For additional coverage of typical maintainability program requirements, refer to "Maintainability Toolkit: A Practical Guide for Designing and Developing Maintainable Products and Systems," Reliability Analysis Center (RAC), 201 Mill Street, Rome, NY 13440-6916.

Topic 4.2.3: Design for Human Factors

Referring to Section 1 (Topic 1.6.6), *human factors (ergonomics)* pertain to the human element of the system and the interfaces between the human being, the machine (equipment), facilities, associated software, data, and related items. The objective in design is to ensure the complete compatibility between the physical and functional features of the system and the human element in the operation, maintenance, and support of that system. More specifically, consideration in design must be given to the anthropometric, sensory, physiological, and psychological factors as they pertain to the human being. The intent is to minimize required personnel quantities, minimize high skill-level requirements, and minimize special training needs.

The emphasis here pertains not only to designing the prime elements of the system but the elements of the maintenance support infrastructure to be effectively and economically supportable. This, of course, requires the initial establishment of design-to criteria for the human during conceptual design, the evaluation of alternative design configurations and the selection of a preferred approach, and subsequent validation through system test and evaluation. As is the case for reliability and maintainability, formal program requirements are often established and which will include activities such as those illustrated in Figure 4.2.3-1.

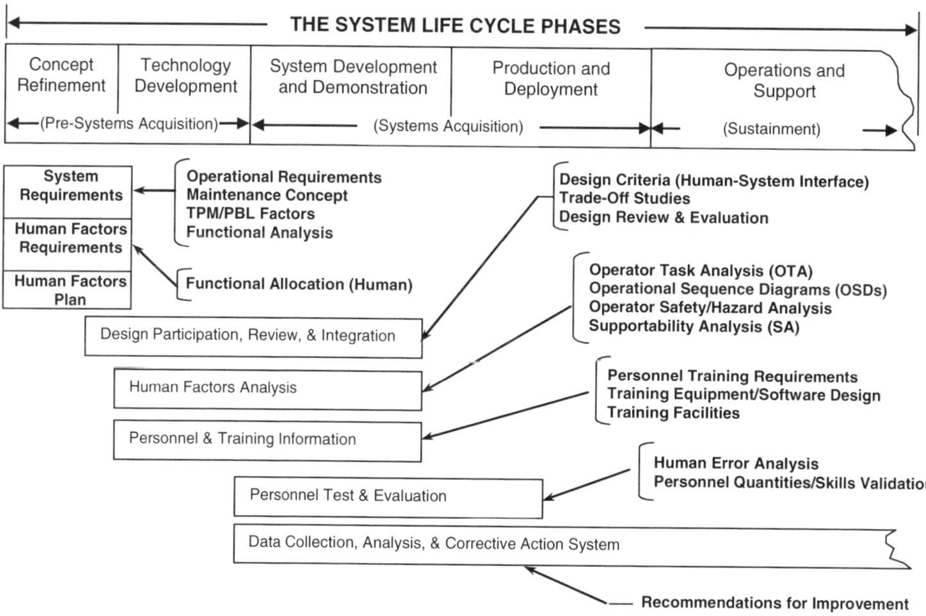

Figure 4.2.3-1. Human Factors Program Requirements

Referring to the figure, the activities that are of particular interest in fulfilling the supportability objectives described throughout this Toolkit include:

1. Specification of specific human-oriented requirements in the development of system operational requirements, the maintenance concept, in the definition and prioritization of TPM/PBL factors, and in the allocation of these requirements to the applicable lower-level system elements. The results constitute design-to goals for the logistics and maintenance support infrastructure.
2. Operator Task Analysis (OTA), Operational Sequence Diagrams (OSDs), and Operator Safety/Hazard Analysis. The OTA, which is based on the operator functions identified as part of the functional analysis (refer to Topic 3.2.6), leads to the identification of maintenance functions and the maintenance tasks in the MTA. OSDs are developed to help identify the design interfaces between the human and the equipment used in the operation of the system. While the application is primarily directed toward system "operation," OSDs can also be used to help define the design interface requirements between the maintenance person and the support equipment being utilized. The Operator Safety/Hazard Analysis is required to help identify potential areas in the design which could lead to hazardous conditions and possible injury to the system operator. While the emphasis is primarily directed with the system operator in mind, the same approach can be applied in identifying possible unsafe conditions in the performance of maintenance activities. These analysis efforts are not usually included as part of the Supportability Analysis (SA); however, they can be effectively applied in the enhancement of such.
3. Personnel test and evaluation at the system level. Operator and maintenance personnel requirements (personnel quantities and skill levels) have been determined through the OTA and MTA, the recommended training requirements have been implemented, and the results from a formal test program verify whether the initially specified personnel requirements have been fulfilled. While the emphasis here is on system "operation," there is a close relationship when evaluating the adequacy of personnel requirements in the performance of maintenance activities.

The activities shown in Figure 4.2.3-1 are not usually included in the accomplishment of supportability program requirements. However, the "human" is a major element of the system, and these requirements are considered as being critical and should be addressed in conjunction with reliability and maintainability program activities in fulfilling the supportability objectives described herein.[3]

---

[3] Three recommended references pertaining to human factors are (1) Chapanis, A., *Human Factors in Systems Engineering*, John Wiley & Sons, Hoboken, NJ, 1996; (2) Salvendy, G. (ed.), *Handbook of Human Factors and Ergonomics*, 2nd Ed., John Wiley & Sons, Hoboken, NJ, 1997; and (3) Blanchard, B.S. and W.J. Fabrycky, *Systems Engineering and Analysis*, 3rd Ed., Prentice Hall, Upper Saddle River, NY, Chapter 14, 1998.

Topic 4.2.4: Design for Safety and Security

Referring to Section 1 (Topic 1.6.7), the system must incorporate the necessary characteristics (attributes) to ensure it will be able to successfully accomplish its mission under safe and secure conditions. An objective in *system safety* is to incorporate features in the design that will enable operation and/or maintenance of the system without causing any damage to equipment, software, facilities, and/or injury to the human being.

An objective in *system security* is to incorporate the necessary features that will prevent sabotage and the planned introduction of faults/failures that will cause the system to be completely destroyed and/or the loss of life. This pertains to designing the system to counter the continuing threats of terrorism.

With these objectives in mind, the following activities are of particular interest:

1. Specification of specific design-to criteria in the development of system operational requirements and the maintenance concept during conceptual design (i.e., the Concept Refinement Phase). Such criteria must address such areas as the selection of materials (to avoid toxic effects), the placement and mounting of components (to avoid sharp edges and related unsafe conditions resulting in possible personnel injury), the use of dangerous fuels and other liquids (to avoid creating an explosive environment), the installation of condition monitoring and an alarm capability for the purposes of protection (to preclude unauthorized personnel from gaining access to the system), and so on. Such criteria, stated both in quantitative and qualitative terms, must be included in the System Specification (Type *A*).
2. The Failure Mode, Effects, and Criticality Analysis (FMECA) and the Safety/Hazard Analysis. The FMECA, which is generally accomplished as a reliability analysis task, identifies failure modes, determines cause-and-effect relationships, and facilitates other analyses and simulation. The Safety/Hazard Analysis, which is also important in the performance of human factors activities, is accomplished to identify possible hazardous conditions, the degree of severity, and areas where redesign is required. An example of each is included in Section 5 of this Toolkit.
3. System test and evaluation to verify that the design objectives have been met.

4.2.4: DESIGN FOR SAFETY AND SECURITY

These activities, while not always included in a supportability program, are obviously important to ensure not only the safety and security of the prime elements of the system, but the safety and security of the logistics and maintenance support infrastructure as well.[4]

[4] Four recommended references on system safety include: (1) MIL-STD-882C, Notice 1, "System Safety Program Requirements," Department of Defense, Washington, DC, January 1996; (2) MIL-HDBK-338B, Electronic Reliability Design Handbook," Department of Defense, Washington, DC, October 1998; (3) Bahr, N.J., *System Safety Engineering and Risk Assessment: A Practical Approach*, Taylor & Francis, New York, NY, 1997; and (4) Roland, H.E., and B. Moriarity, *System Safety Engineering and Management*, 2nd Ed., John Wiley & Sons, Hoboken, NJ, 1990.

Topic 4.2.5: Design for Producibility

Producibility is a measure of the relative ease and economy of producing a system or product (refer to Section 1, Topic 1.6.8). The characteristics of design must be such that an item can be produced (manufactured) easily and economically, using conventional and flexible manufacturing methods and processes without sacrificing function, performance, effectiveness, or quality. The objectives include simplicity and flexibility, and to minimize (if not eliminate) the use of critical and non-standard materials, the use of proprietary items, the use of special production tooling and facilities, the application of unrealistic tolerances in fabrication and assembly, the use of special test systems, and the need for highly skilled personnel in manufacturing. The degree of success in developing a "producible" entity can have a great impact on the degree of "maintainability" (and supportability) incorporated in the item; e.g., if the manufacturing assembly and test processes are easy and can be accomplished both rapidly and economically, then the maintenance and repair processes are likely to be performed easily and economically as well.

As is the case for reliability and maintainability, the design-to requirements for producibility must be determined during the conceptual design stage (i.e., Concept Refinement Phase), and expressed in qualitative and quantitative terms in the System Specification (Type *A*). Validation is ultimately accomplished through review and evaluation of the production/manufacturing processes during the Production and Deployment Phase.

Topic 4.2.6: Design for Disposability

Disposability pertains to the degree to which an item can be removed from the operational inventory, recycled for some other use, or disposed of economically and without causing any degradation to the environment; that is, without resulting the generation of any solid waste, toxic substances (air pollution), water pollution, noise pollution, and so on (refer to Section 1, Topic 1.6.9). Although there may be a variety of measures used in this area, one example relates to the specification of a given percentage of the components utilized in the system configuration that must be recyclable for other uses; for example, *The system shall be designed such that at least 95% of all of the components utilized in the design shall be directly recyclable for other applications.* As is the case for reliability and maintainability, the design requirements for disposability must be determined during conceptual design (i.e., Concept Refinement Phase).

Topic 4.2.7: Design for Economic Feasibility

Referring to Section 1 (Topic 1.6.12), *economic feasibility* as described herein pertains to the establishment of one or more cost goals in the form of "design-to" requirements. Additionally, such goals should address cost from a total system life-cycle perspective. For example, it can be specified that *the system shall be designed such that the unit life-cycle cost shall not exceed $1.5M* (see Table 3.2.4-1 where cost is included as one of the TPM requirements for the system). Such a goal should be established and specified early in conceptual design (Concept Refinement Phase), allocated to the various elements of the system (refer to Figure 3.2.7-1 for an example of cost allocation), and the subsequent design of the various elements of the system should be evaluated through prediction in order to determine compliance with the initially specified requirement(s). Thus, there is a need to accomplish a life-cycle cost analysis (LCCA) throughout the design and development process, not only in the evaluation of alternative design approaches but in the assessment of the *design for economic feasibility*. Of particular interest are the costs associated with the logistics and maintenance support infrastructure. The steps involved in accomplishing an LCCA are illustrated through the case study presented in Section 5 of this Toolkit.[5]

---

[5] An in-depth discussion of the life-cycle cost analysis (LCCA) process is included in Blanchard, B.S., *Logistics Engineering and Management*, 6th Ed., Pearson Prentice Hall, Upper Saddle River, NJ, Chapter 5, Section 5.2.1, and Appendix *D*, 2004.

Topic 4.2.8: Design for Quality

Referring to Section 1 (Topic 1.6.10), *quality* is defined in a rather broad context to include a combination of measures of "goodness," of the attributes in a given product, of the conformance to specifications in manufacturing, of value in terms of product cost, and of "satisfaction" as perceived by the customer. Again, the attainment of good quality for a system must evolve from the initial establishment of requirements and "expectations" during conceptual design (Concept Refinement Phase), and the subsequent monitoring and control activities applied throughout the overall design, development, and production/construction processes.[6]

[6] "Quality Toolkit: A Practical Guide for Developing and Implementing a Quality Program for Products and Services," Reliability Analysis Center (RAC), 201 Mill Street, Rome, NY 13440-6916. Also, refer to the Bibliography in Appendix *A*.

Topic 4.3: Design Integration, Design Review, and System Evaluation

As stated in Section 1 (Topic 1.2), supportability may be defined as *the capability of a total system design to support operations and readiness needs throughout the system's service life at an affordable cost. It provides a means of assessing the suitability of a total system design for a set of operational needs within the intended operations and support environment (including cost constraints). Supportability characteristics may be specified in terms of many different performance measures.*[7]

The nature of "supportability" is highly *interdisciplinary*, and the measures constitute a wide variety and mix of different performance parameters. A measure of supportability is dependent on reliability factors, maintainability factors, human factors, safety factors, quality factors, economic factors, and many other related factors as discussed throughout Section 2 of this Toolkit. Thus, it was felt necessary to address (in Topic 4.2) some of the objectives and design goals for these very critical and related disciplines, and their respective measures.

Given such, the implementation of supportability program requirements for a particular system design and development effort is highly dependent on the timely and effective integration of these disciplines from the beginning. Design integration activities commence with the establishment of a "design team" during the early stages of conceptual design (i.e., the Concept Refinement Phase) and extend through system development, production/construction, distribution, system utilization and sustaining maintenance support, and the ultimate retirement and disposal (or recycling) of materials. Initially, such activities are more associated with the design and development of a new system, whereas subsequently these activities may pertain to the evaluation, modification, and improvement of systems already in being. As requirements are established, the design team is formed.

In the early stages of conceptual design, the emphasis is on the system as an entity and the development of requirements for such. The design team may include only a small number of selected qualified individuals, with the objective of preparing a comprehensive System Specification (Type *A*). The applicable requirements described in Topic 4.2 must be properly integrated into this top-level specification. As system development progresses, the appropriate design specialists (with special expertise in reliability, maintainability, etc.) are added to the team as necessary. Additionally, supplier activities are introduced, and the team makeup may include representation from various parts of the world. There may be activities that must be accomplished in parallel and for which the principles of *concurrent engineering* should be implemented. The objective is to ensure that the right specialists are available at

[7] MIL-HDBK-502, "Department of Defense Handbook on Acquisition Logistics," Department of Defense, Washington, DC, May 1997.

the time required and that their individual contributions are properly integrated into the whole. The selection of domain specialists is highly dependent on the requirements developed through the functional analysis and allocation process. As the criteria for design will vary with the type of system and its mission, the emphasis in assigning the proper level of expertise to the team will differ from one project to the next.

Referring to Figure 4.3-1, the intent is to show major supportability activities in the system life cycle, and using the same approach (and layout) as illustrated in Figures 4.2.1-1, 4.2.2-1, and 4.2.3-1. The supportability activities in Figure 4.3-1 are "broader" in scope and should reflect the proper integration of reliability activities (shown in Figure 4.2.1-1), maintainability activities (shown in Figure 4.2.2-1), and so on. In other words, the activities pertaining to the "design for supportability" should include "design for reliability" activities as applicable. Additionally, the "supportability analysis" activities (in Figure 4.3-1) should include the utilization of the appropriate reliability tools/models (e.g., FMECA, FTA, RCM Analysis), maintainability tools/models (e.g., LORA, MTA, LCCA), etc. The important issue here is the implementation of a highly *integrated* approach from the beginning.

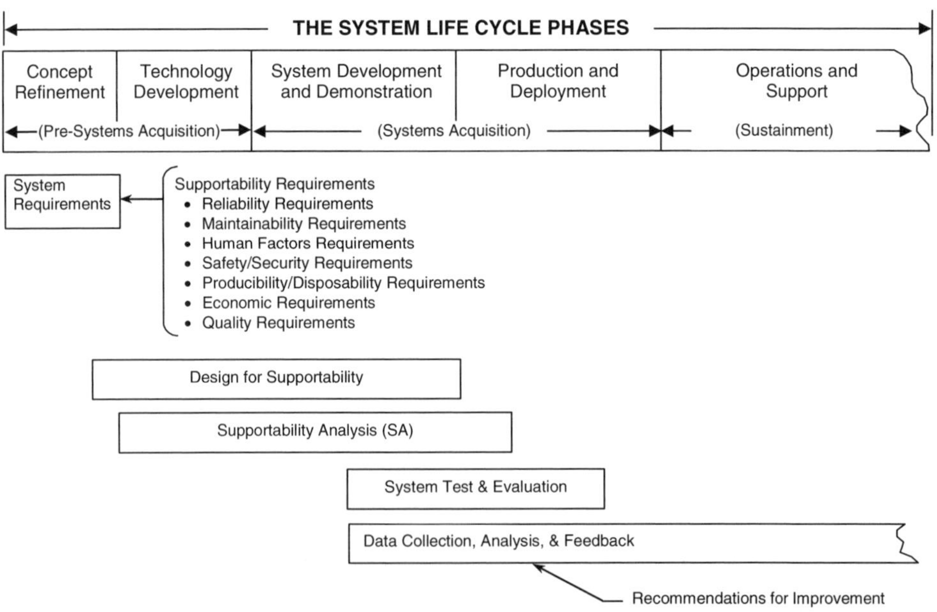

Figure 4.3-1. Supportability Requirements in the System Life Cycle

Referring to Figure 4.3-2, design is a progression from an abstract notion to something that has form and function, is fixed, and can be reproduced in designated quantities to satisfy a need. The design team produces a model that is used as a template for the replication of additional models. In the course of production, an error in any one model will result in a single rejection; however, an error in design, repeated in all subsequent models, may lead to a serious problem (e.g., complete recycling of all equipment for a major modification). Thus, the design team's responsibility is significantly large. Given such, it is appropriate (and highly recommended) to schedule a series of *design reviews* at critical times throughout the design process as the level of design definition evolves from one baseline to another.

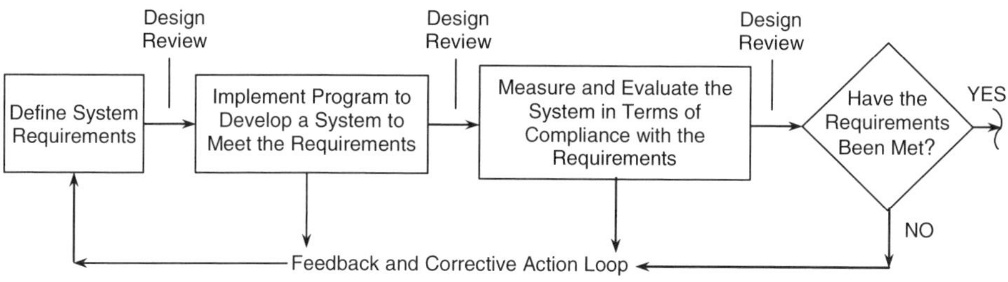

Figure 4.3-2. Program Design Reviews

Although the number and type of scheduled design reviews may vary from program to program, Figure 4.3-3 provides a sample design review and evaluation plan, which is presented for the purposes of further discussion. Referring to the figure, there is an on-going informal day-to-day review process where the design, at any given point in time, is reviewed in terms of compliance with the initially specified supportability and other requirements.

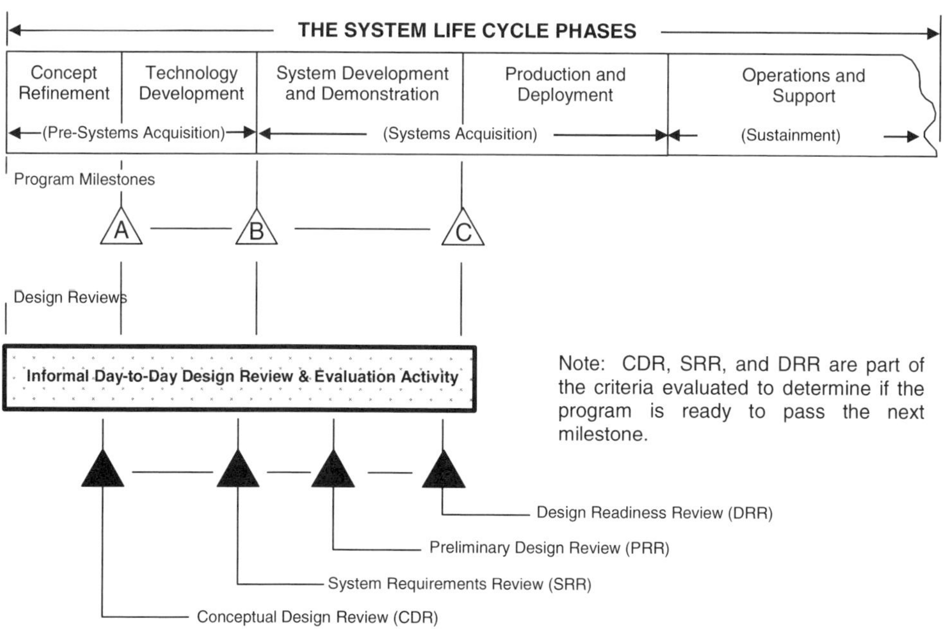

Figure 4.3-3. System Evaluation, Review, and Feedback Process

In addition, there is a series of formal design reviews scheduled at specific times prior to each major evolutionary step in the design process. In other words, a formal review is scheduled upon the completion of conceptual design and the Concept Refinement Phase and prior to entry into the Technology Development Phase, another review is conducted before entry into the System Development and Demonstration Phase, and so on. A brief description of the purpose and objectives of these formal reviews follows.

1. *Conceptual Design Review (CDR).* The CDR is generally scheduled early in a program after accomplishment of the needs analysis, feasibility studies, the definition of system operational requirements and the maintenance concept, and the identification and prioritization of TPM/PBL factors. Preliminary system design concepts and system-level design-to requirements have been specified and included in the System Specification (Type *A*). Logistics and related supportability requirements at this point generally include the development and specification of PBL design-to goals, associated supportability design criteria at the system level, and applicable constraints.
2. *System Requirements Review (SRR).* The SRR is generally scheduled toward the end of the Technology Development Phase and includes a review of system operational requirements, the maintenance concept, TPM/PBL requirements, a top-level functional analysis and allocation, and specific requirements for the various elements of the system. Early supportability analysis (SA) activities have been accomplished (to include R&M allocations/ predictions and an initial FMECA, LORA, OTA, MTA, and LCCA). The basic elements of the logistics and maintenance support infrastructure have been identified, and associated design requirements for these elements have been specified in the applicable lower-level specifications (Types *B*, *C*, and *D*). At this point, it is appropriate to review and evaluate the maintenance and support infrastructure to the levels (depth) illustrated in Figures 1.3-3, 1.3-4, 3.2.3-1, and 3.2.3-2.
3. *Preliminary Design Review (PDR).* There may be any number of PDRs conducted throughout the System Development and Demonstration Phase covering major items of equipment, software, facilities, data/documentation, elements of support, along with the drawings, databases, component part lists, etc., that describe the items being reviewed and evaluated. In addition, these reviews may cover engineering laboratory models, mock-ups, software programs, and prototype models as available. With regard to logistics, several iterations of supportability analysis (SA), along with the appropriate modeling activity (i.e., R&M predictions, FMECA, FTA, RCM, LORA, OTA, MTA, Safety Analysis, LCCA), have been accomplished. Such reviews may cover any number of critical items, and should be conducted prior to finalizing the detail design.
4. *Design Readiness Review (DRR).* The DRR is generally scheduled after detail design has been completed and prior to entering into the Production and Deployment Phase. A complete design data package has been

produced, design is essentially "frozen," early system test and evaluation activities have been completed, and the system configuration is ready to go into production. Further, a proposed production plan has been developed for review and evaluation. From a logistics perspective, the supportability analysis (SA), based on a final and fixed design configuration, has been completed, and all of the elements of logistics and the maintenance support infrastructure have been adequately defined and specified.

The purpose of the design review process is to ensure that the initially specified requirements are being realized, as one progresses through the system design and development process (refer to Figure 4.3-2). Additionally, if there are any problems (in terms of non-compliance with a given requirement), it is better to identify such as early in the life cycle as possible when the introduction of changes for corrective action are relatively easy and not too costly to implement. Thus, the conducting of periodic design reviews on a progressive basis is recommended.

When reviewing and evaluating the design, one might ask questions such as --- *What items require a review? When and why should such a review be conducted? How should the review be conducted? In conducting such a review what criteria should be applied in determining whether the item in question is good from a design perspective?* In response to the first three questions, an abbreviated plan should be prepared prior to the review stating purpose, the status of the design, drawings/documentation describing the item, the results of trade-off studies justifying the design approach, and so on. With regard to the last question, a recommended approach is to develop a comprehensive "design review checklist," tailored to and with specific questions based on the TPM/PBL requirements, input design criteria, desired characteristics (attributes) in design, and so on.[8]

Accomplishing formal design reviews serves a number of purposes and provides many advantages:

1. It provides a formalized check (audit) of the proposed system/product design with respect to the customer expectations and contractual and specification requirements. Major problem areas (if any) are discussed and corrective action is initiated as required.
2. It provides a common baseline for all project personnel. The responsible designer is provided the opportunity to explain and justify his/her design approach, and representatives from various supporting organizations (e.g., reliability, maintainability, logistics, etc.) are provided the opportunity to listen to the designer's problems. The review serves as an effective communications medium for all involved and creates a better understanding among design and support personnel.

[8] A rather extensive detailed design review checklist is presented in Blanchard, B.S., *Logistics Engineering and Management*, 6th Ed., Pearson Prentice Hall, Upper Saddle River, NJ, Figure 6.8 and Appendix *B*, 2004.

3. It provides a means for solving interface problems and promotes assurance that all system elements will be compatible. For example, major interface problems between engineering and manufacturing, relative to the lack of producibility, are often not detected until after design data are released and production is underway. Additionally, it is not uncommon (when downstream in the life cycle) to identify problems dealing with the lack of compatibility between a prime mission-related element of the system and some item of support. The objective of a formal design review is to identify and correct such problems early.
4. It provides a formalized record of what design decisions were made and the reasons for making them. Analyses, predictions, and trade-off study reports are noted and are available to support design decisions. Compromises (if any) to reliability, maintainability, human factors, and supportability are documented and included in the trade-off study reports.
5. It promotes a greater opportunity of mature design as well as the incorporated of the latest technologies, design techniques, etc., where appropriate. Group review may identify new ideas, possibly resulting in simplified processes and ultimate cost savings.
6. It promotes an approach in design where the "technical" and "management" risks are minimized. This is due primarily to the "visibility" that is created, relatively early in the system life cycle, as a result of the design reviews.

Formal "design reviews," when properly planned and conducted, can be very effective and can provide many benefits. It is essential that the design team include the appropriate representation from logistics and the associated design support disciplines, and that the system and all of its elements (to include the logistics and maintenance support infrastructure) be addressed through the design review process.[9]

In summary, the purpose of this section was to provide an overview of supportability requirements in the system design and development process, and to emphasize the importance of viewing the *design for supportability* from an "interdisciplinary" perspective. To understand and properly implement the requirements for supportability requires an understanding of the related requirements in reliability, maintainability, human factors, and so on. The key message throughout is that of *proper integration.*

[9] Additional coverage of design reviews can be found in: (1) Blanchard, B.S., *Logistics Engineering Management*, 6th Ed., Pearson Prentice Hall, Upper Saddle River, NJ, pages. 302-310, 2004; and (2) Blanchard, B.S., *System Engineering Management*, 3rd Ed., John Wiley & Sons, Hoboken, NJ, Chapter 5, 2004.

For More Information:

1. Blanchard, B.S., *Logistics Engineering and Management*, 6th Ed., Pearson Prentice Hall, Upper Saddle River, NJ, 2004.
2. Blanchard, B.S., *System Engineering Management*, 3rd Ed., John Wiley & Sons, Hoboken, NJ, 2004.
3. "Reliability Toolkit: Commercial Practices Edition," Reliability Analysis Center (RAC), 201 Mill Street, Rome, NY 13440-6916.
4. "Maintainability Toolkit: A Practical Guide for Designing and Developing Maintainable Products and Systems," Reliability Analysis Center (RAC), 201 Mill Street, Rome, NY 13440-6916.
5. Chapanis, A., *Human Factors in Systems Engineering*, John Wiley & Sons Hoboken, NJ, 1996.
6. Salvendy, G. (ed), *Handbook of Human Factors and Ergonomics*, 2nd Ed., John Wiley & Sons, Hoboken, NJ, 1997.
7. MIL-STD-882C, "System Safety Program Requirements," Department of Defense, Washington, DC, January 1996.
8. Roland, H.E., and B. Moriarty, *System Safety Engineering and Management*, 2nd Ed., John Wiley & Sons, Hoboken, NJ, 1990.
9. "Quality Toolkit: A Practical Guide for Developing and Implementing a Quality Program for Products and Services," Reliability Analysis Center (RAC), 201 Mill Street, Rome, NY 13440-6916.
10. MIL-HDBK-502, "Department of Defense Handbook on Acquisition Logistics," Department of Defense, Washington, DC, May 1997.

This page intentionally left blank

Section 5: Supportability Analysis (SA)

Insight

Throughout system design and development, there is an ongoing and iterative process of system synthesis, analysis, and design optimization, which leads to definition of the ultimate design configuration (as described in Section 1, Topic 1.6.3). As part of the evolutionary design process, many different design alternatives are identified and evaluated using various analytical methods/models to facilitate the trade-off and evaluation process, resulting in the selection of a preferred approach (refer to Figure 1.6.2-1). Inherent within this iterative and continuous analytical process is that part of the effort directed to the design and evaluation of the logistics and maintenance support infrastructure, which is identified herein as the *supportability analysis (SA).*[a]

The objective of this section is to briefly describe the SA process, include a representative mix of "case-study" examples of analytical methods/tools that are inherent and utilized within the process, and to discuss some examples where the SA can be effectively applied in the overall system development and evaluation effort.

a. The principles and concepts associated with the supportability analysis (SA), presented in this section, are not new and have been previously covered under such titles as the *logistics support analysis (LSA), maintenance engineering analysis (MEA), maintenance level analysis (MLA), maintenance task analysis (MTA), maintenance engineering analysis record (MEAR)*, and so on. Although the nomenclature may change from time to time, the objectives and the processes utilized in each instance are basically the same.

This page intentionally left blank

Topic 5.1: The Analysis Process

The supportability analysis (SA) process commences with the problem identification and the needs analysis, and the subsequent definition of analysis goals, ground rules, constraints, and so on. As the name infers, a significant proportion of the effort comprises *analysis.* A wide variety of problems must be addressed and require the application of the proper mix of analytical methods. For example, the analyst may wish to use computerized simulation methods in the development of a maintenance concept and the identification of major support functions. A Monte Carlo analysis may be employed to aid in the determination of maintenance frequency factors, or a queuing analysis may be used in developing test and support equipment and maintenance shop requirements. Reliability and maintainability analyses and predictions may be accomplished through the use of statistical processes. Poisson distribution factors and inventory control theory may be required in determining spares requirements. Linear and dynamic programming may be used in the determination of transportation and material handling requirements. Network theory may be applied in the development of a computerized information system capability. Accounting methods are utilized in the accomplishment of life-cycle cost analyses. Basically, there are different techniques that are available and can be appropriately applied throughout the supportability analysis (SA) process.[1]

In essence, the SA constitutes a continuous effort involving the use of different analytical methods in solving a wide variety of problems; those dealing with the initial definition of requirements, the evaluation of various system design alternatives, the determination of logistics resource requirements for a given configuration, and the ultimate assessment of a system and its maintenance support capability while in use. During the early phases of conceptual design (Concept Refinement Phase), the level of analysis is limited in depth (using rough estimates as input data) because the various elements of the system have not been adequately defined. As design progresses, the analysis assumes a more refined approach because firm design data and/or engineering models are available. Detailed predictions, maintenance task analyses, and test results are often available and provide an input to the SA effort.[2] The critical issue is to properly "tailor" the level of analysis to the problem(s) at hand. Quite often there is a tendency to do either too little or too much. If the level of analysis is inadequate, then the risks associated with follow-on decisions that are based on the analysis results may be high. Conversely, going to a depth beyond that required may result in the generation of a lot of meaningless data, provided too late in many

[1] Many of these factors were discussed in Section 2 of this Toolkit. Additionally, it is highly recommended that the analyst become familiar with some of the operations research (OR) and related techniques that can be applied as a design analysis aid. Several recommended references are included in Section I, Appendix A.

[2] Many of the input requirements for the SA were described in Section 4 (e.g., reliability and maintainability predictions, and so on).

instances, and that can be very costly. The key issue is first to gain a good understanding of the problem (task) to be solved and then be able to define analysis goals in a specific and complete manner.

When the SA is related to the overall system development process shown in Figure 1.6.3-1, each of the blocks on the left in the flowchart is affected; however, the major areas of emphasis are represented by blocks 4, 6, 8, 10, and 15. Referring to Figure 5.1-1, in the early stages of system development the emphasis is on *design influence*, whereas the objective later is on *resource identification.* Further, as the system is utilized in an operational environment, the emphasis is on *supportability assessment.*

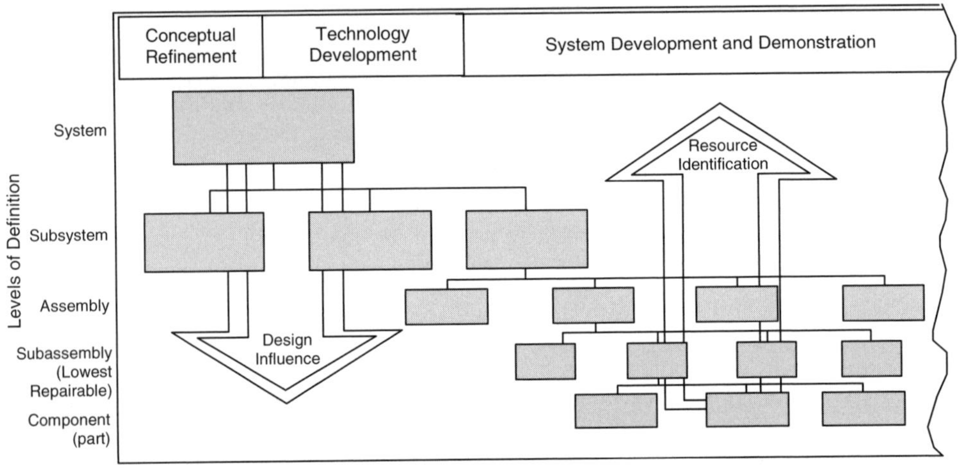

Figure 5.1-1. Major Supportability Analysis Thrusts

The overall process in accomplishing a supportability analysis (SA) effort basically follows the steps shown in Figure 5.1-2. Referring to block 3 in the figure, a critical step in the SA process is the establishment of the proper measures, or metrics, for evaluation. These measures will, of course, vary in terms of magnitude and degrees of importance depending on the system and its intended mission. What may be an excellent high-level measure for the system in one case may not be as critical in another situation. The determination of the proper measures is facilitated through application of such as a Quality Function Deployment (QFD) analysis and stems from the identification and prioritization of Total Productive Maintenance/Performance Based Logistics (TPM/PBL) factors in conceptual design, included in the System Specification (Type "A").

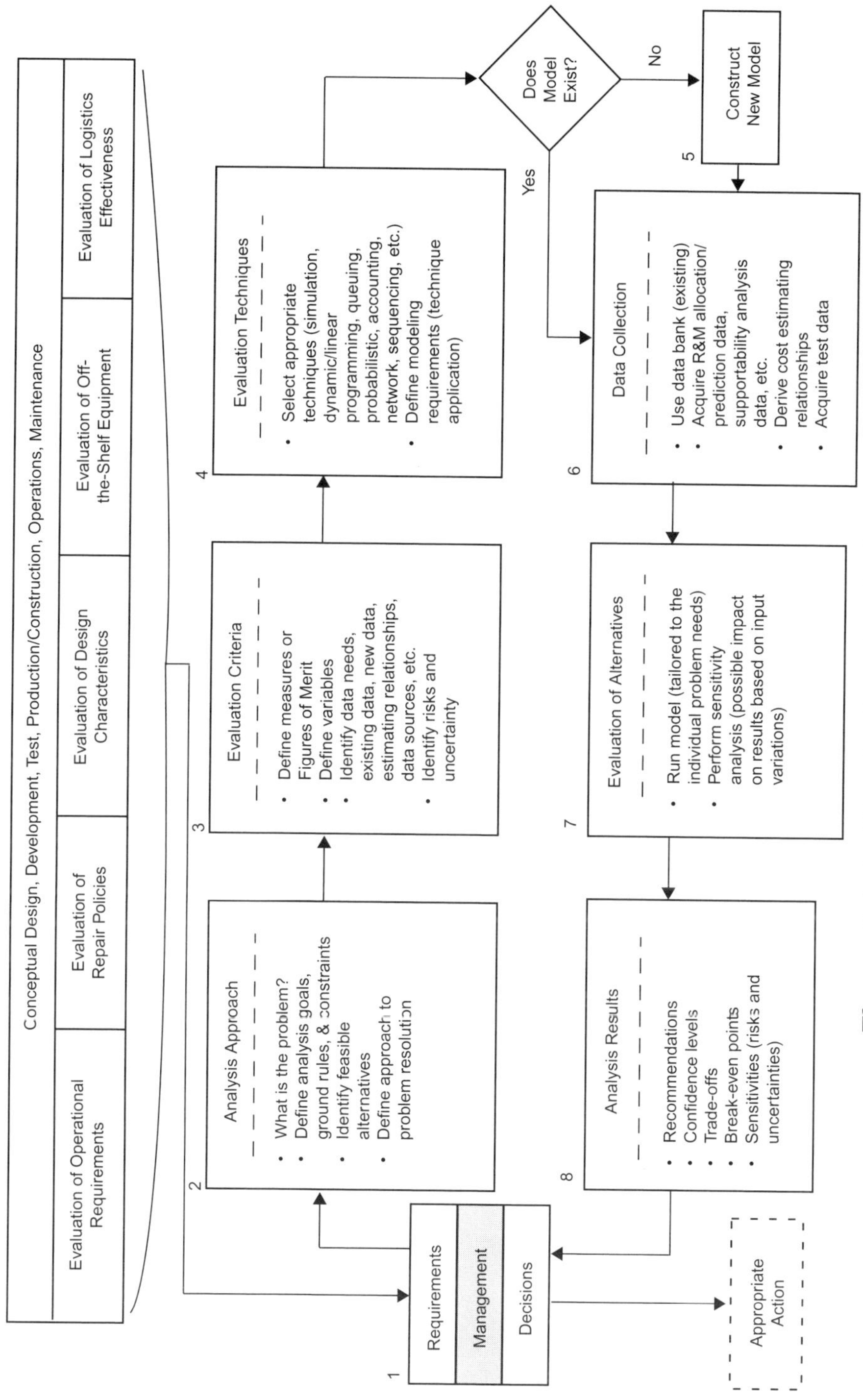

Figure 5.1-2. The System Analysis Approach

Given the basic quantitative requirements for the system (and the allocation of such to the various system elements), it is then necessary to determine the type and amount of data that will be required as an input to the analysis process. At the same time, one needs to assess the availability of the proper techniques/tools that can be applied in the processing of the data to provide the correct analysis results. It is at this point where familiarity with various analytical processes is important (refer to block 4, Figure 5.1-2). The question is -- *What analytical processes are available that can be applied to produce valid analysis results, and how can they be effectively applied?*

Having identified the appropriate evaluation techniques (and given the analysis objectives), the next step is to combine these into some form of analytical "model." A model, presented in the context of Figure 5.1-2, may include either a single computer-based entity or an integrated set of tools as shown in Figure 5.1-3. The objective is to select a model that incorporates the necessary characteristics, and technique relationships, that is sensitive to the problem at hand and will allow the analyst to evaluate the system as a whole or any of its elements on an individual basis. More specifically, the selection of a good model may offer the following benefits:

1. In terms of system application, a number of considerations exist – operational considerations, design considerations, production/construction considerations, testing considerations, and logistics and maintenance support considerations. There are many interrelated elements that must be integrated as a system and not treated on an individual basis. The model makes it possible to deal with the problem as an entity, and allows for consideration of all major variables of the problem on a simultaneous basis.
2. The model enables a comparison of *many* possible solutions and aids in selecting the best among them rapidly and efficiently. At the same time, the model often helps to explain situations that have been left unexplained in the past by indicating *cause-and-effect relationships.*
3. The model readily indicates the type of data (and in what format) that should be collected (as an input) to deal with the problem in a quantitative manner and provide effective analysis results.
4. The model facilitates the prediction of future events such as effectiveness factors, reliability and maintainability factors, logistics requirements, life-cycle cost profiles, and so on. In addition, the model aids in identifying areas of risk and uncertainly.

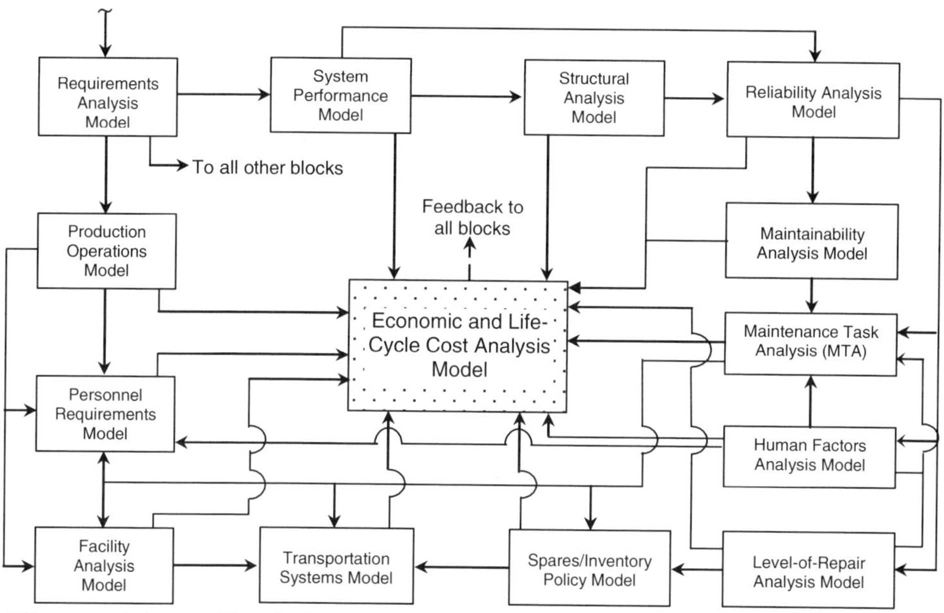

Figure 5.1-3. The Integration of Models – An Overall Approach

When one is assessing the problem in terms of selecting a computerized analytical model for evaluation purposes, it is desirable to first investigate the tools that are currently available. If a model already exists, is proven, and is commercially available, then it may be feasible to adopt that model for your specific application. However, extreme care must be exercised to relate the right technique(s) with the problem being addressed and to be able to apply it to the depth necessary to provide the desired sensitivity required in arriving at a final solution. Through the years, there have been many analytical "models" available in the commercial market place, promoted through good advertising, that will "solve all of your problems." The question is -- *Does the model incorporate the right characteristics and project the proper parameter relationships when applied to solving the problem at hand?* Improper application may not provide the results desired and may be costly.

On the other hand, it may be necessary to construct a new model. In accomplishing such, one should generate a comprehensive list of system/equipment parameters that will describe the situation to be simulated. Next, the analyst should develop a matrix showing parameter relationships, with each parameter being analyzed with respect to every other parameter to determine the magnitude of the relationship. Model input/output factors and parameter feedback relationships need to be established. The model is constructed by combining the various factors and then tested for validity. Testing may be difficult to accomplish since the problems addressed primarily deal with actions in the future that are impossible to verify; however, it may be possible to select a known system or equipment, with similar mission requirements and that has been in the operational inventory for a number of years, and exercise the model using established parameters. In any event, the analyst should be able to respond to the following questions:

1. Can the model describe known facts, situations, and relationships sufficiently well?
2. When major input parameters are varied, do the results remain consistent and are they realistic? Have the appropriate sensitivities been verified?
3. Relative to system application, is the model sensitive to changes in operational requirements, the maintenance concept, design, production/ construction, and logistics support?
4. Can the proper cause-and-effect relationships be established (verified)?

In accomplishing a supportability analysis (SA), there may be a number of individual computer-based tools integrated as shown in Figure 5.1-3. However, the specific arrangement may vary depending on the problem at hand. In some instances, it may be appropriate to start off by accomplishing a life-cycle cost analysis (LCCA) to identify the "high-cost contributors" and then accomplish a Failure Mode, Effects and Criticality Analysis (FMECA) to help determine the "causes" for such. In another case, it may be appropriate to accomplish a maintenance task analysis (MTA) and then conduct a level-of-repair analysis (LORA) to verify the depth to which MTA coverage is necessary. The example presented in Figure 5.1-4 illustrates the relationships of models in the development of preventive maintenance requirements. In any case, the analyst needs to have a good understanding of these various tools and how they can best be applied in responding to a given requirement.

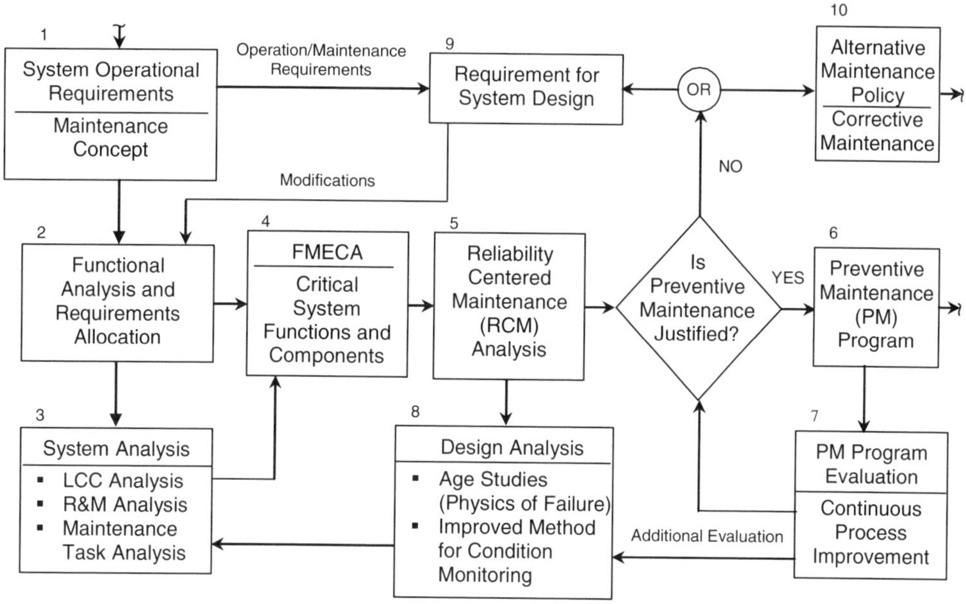

Figure 5.1-4. The Integration of Models for Preventive Maintenance Requirements

Referring back to Figure 5.1-2 (the system analysis approach), after the selection of the proper tools for application in the SA, the process continues with the activities identified in blocks 6, 7, and 8. Overall, this is an ongoing and iterative analytical process leading to the documentation of trade-off study results and design-input recommendations, and ultimately to the preparation of *supportability analysis summaries (SASs)* and the development of a *logistics management information (LMI)* database.

Topic 5.2: Analysis Methods and Models

Within the context of the system analysis approach (Figure 5.1-2), and in particular the "model" as shown, there may be a number of different lower-level models that can be combined to provide the desired output. Referring to Figure 1.6.3-2 (as an example), there are 14 different supplemental analyses identified that, when applied in varying combinations, facilitate the accomplishment of the supportability analysis (SA). The illustration in the figure is just an example demonstrating the concept where the successful accomplishment of the SA requires the integration of a wide variety of tools used in different applications. This concept is further amplified through the illustrations in Figures 5.1-3 and 5.1-4.

For the purposes of further illustration, eight abbreviated "case-study" examples have been selected and presented in a summary form. For each of the selected examples, identified in Table 5.2-1, there is a problem statement, a description of the analysis approach or the process, and the subsequent analysis results. While not considered as being all-inclusive by any means, a review of the material presented in Topics 5.2.1 through 5.2.8 should provide an excellent example of a few of the tools/models that are considered as being necessary and critical in accomplishing a supportability analysis.[3]

Table 5.2-1. A Selection of Supportability Analysis Tools

Analysis Tool	Description of Application
1. Life-Cycle Cost Analysis (LCCA) (Topic 5.2.1)	Determination of the system/product/process/service life-cycle cost (design and development, production and/or construction, system utilization, maintenance and support, and system retirement); high-cost contributors; cause-and-effect relationships; potential areas of risk; identification of areas for improvement (i.e., cost reduction).
2. Failure Mode, Effects, and Criticality Analysis (FMECA) (Topic 5.2.2)	Identification of potential product/process/service failures, the expected or anticipated modes of failure and causes, failure effects and mechanisms, anticipated frequency of occurrence, criticality, and steps required for compensation (i.e., the requirement for redesign or the accomplishment of preventive maintenance).

[3] Many of these examples were taken from Blanchard, B., *Logistics Engineering And Analysis*, 6th Ed., Pearson Prentice Hall, Upper Saddle River, NJ, 2004, Chapter 5.

Table 5.2-1. A Selection of Supportability Analysis Tools (Cont'd)

Analysis Tool	Description of Application
3. Fault Tree Analysis (FTA) (Topic 5.2.3)	A deductive approach involving a graphical enumeration and analysis of different ways in which a particular system failure can occur, and the probability of occurrence. A separate fault tree may be developed for every critical failure mode or undesired top event.
4. Safety/Hazard Analysis (Topic 5.2.4)	Identification of potential hazards and safety problems as they pertain to equipment, software, facilities, and personnel; description of hazard; case-and-effect of hazard; hazard classification; anticipated frequency of occurrence; and recommended action or preventive measures.
5. Maintenance Task Analysis (MTA) (Topic 5.2.5)	Evaluation of those maintenance functions that are to be allocated to the human; identification of maintenance functions/tasks in terms of task times and sequences, anticipated frequency, number of personnel and their skill levels, and supporting resources (i.e., spares, repair parts, inventories, tools and test equipment, facilities, transportation and handling requirements, technical data, training, and computers and associated software); identification of high-resource consumption areas.
6. Reliability Centered Maintenance (RCM) (Topic 5.2.6)	Evaluation of the system, in terms of the life cycle, to determine the best overall program for preventive (scheduled and on-condition) maintenance (PM). Emphasis is on establishing a cost-effective PM program based on reliability data derived from the FMECA.
7. Level-of-Repair Analysis (LORA) (Topic 5.2.7)	Evaluation of maintenance policies in terms of level of repair (i.e., should an item be repaired in the event of failure or discarded; and, given the repair option, where the repair should be accomplished). Decision factors include economic, technical, social, environmental, and political. The emphasis here is on life-cycle cost.
8. Evaluation of Design Alternatives (Topic 5.2.8)	Evaluation of alternative design configurations using multiple criteria. Weighting factors are established to specify relative importance.

Topic 5.2.1: Life-Cycle Cost Analysis (LCCA)

In addressing the *economic* side of the balance in Figure 1.6.11-1, one must deal with both revenues and cost. The emphasis here is on the element of *cost*, as conveyed in Figure 1.6.12-1, although the loss of revenues can constitute a "cost." In any case, one must address cost from a *life-cycle* perspective if one is to properly assess the risks associated with the day-to-day engineering and management decision-making processes. Although individual decisions may be based on some smaller aspect of cost (e.g., product procurement cost, component purchasing price), the analyst is remiss unless he/she views the consequences of those decisions in terms of the *total* cost.

As described in Section 1, Topic 1.6.12, life-cycle costing addresses the *total cost* of the system and its supporting activities throughout its planned life cycle. LCC includes all future costs associated with research, design and development, construction and/or production, system utilization, maintenance and support, and system retirement and material recycling and/or disposal activities. Historically, experience has indicated that a large proportion of the projected LCC for a given system has been associated with the logistics and maintenance support activities accomplished during the period of system utilization. Further, much of this downstream cost has been the consequence of design and management decisions made during the early stages of conceptual and preliminary design. Thus, the supportability analysis (SA) objectives described throughout this Toolkit are critical in system design and development (i.e., during the Concept Refinement, Technology Development, and System Development and Demonstration Phases), and the use of LCCA methods is essential if one is to assess whether the system can be operated and supported in a cost-effective manner.

In accomplishing a typical LCCA for any system (or an element thereof), the analyst may wish to follow the process illustrated in Figure 5.2.1-1. Referring to the figure, this process is somewhat "generic" and conveys an overall approach. More specifically, the basic steps presented in Table 5.2.1-1 provide detailed guidance in the accomplishment of an LCCA. However, the process and these steps must be appropriately "tailored" to the problem at hand. In the abbreviated example that follows, these steps were followed although the information is presented in the form of a summary.

<u>Example: Vehicle Communications System</u>

Definition of the Problem. A ground vehicle currently in the development stage requires the incorporation of a new communications system. There is a need to design and develop a new system configuration, to be installed in the vehicle and operational two years from this point, and the per-unit LCC of the system shall not exceed $20,000. Two different supplier proposals have been received, and the objective is to evaluate each of the proposed design configurations and to select a preferred approach.

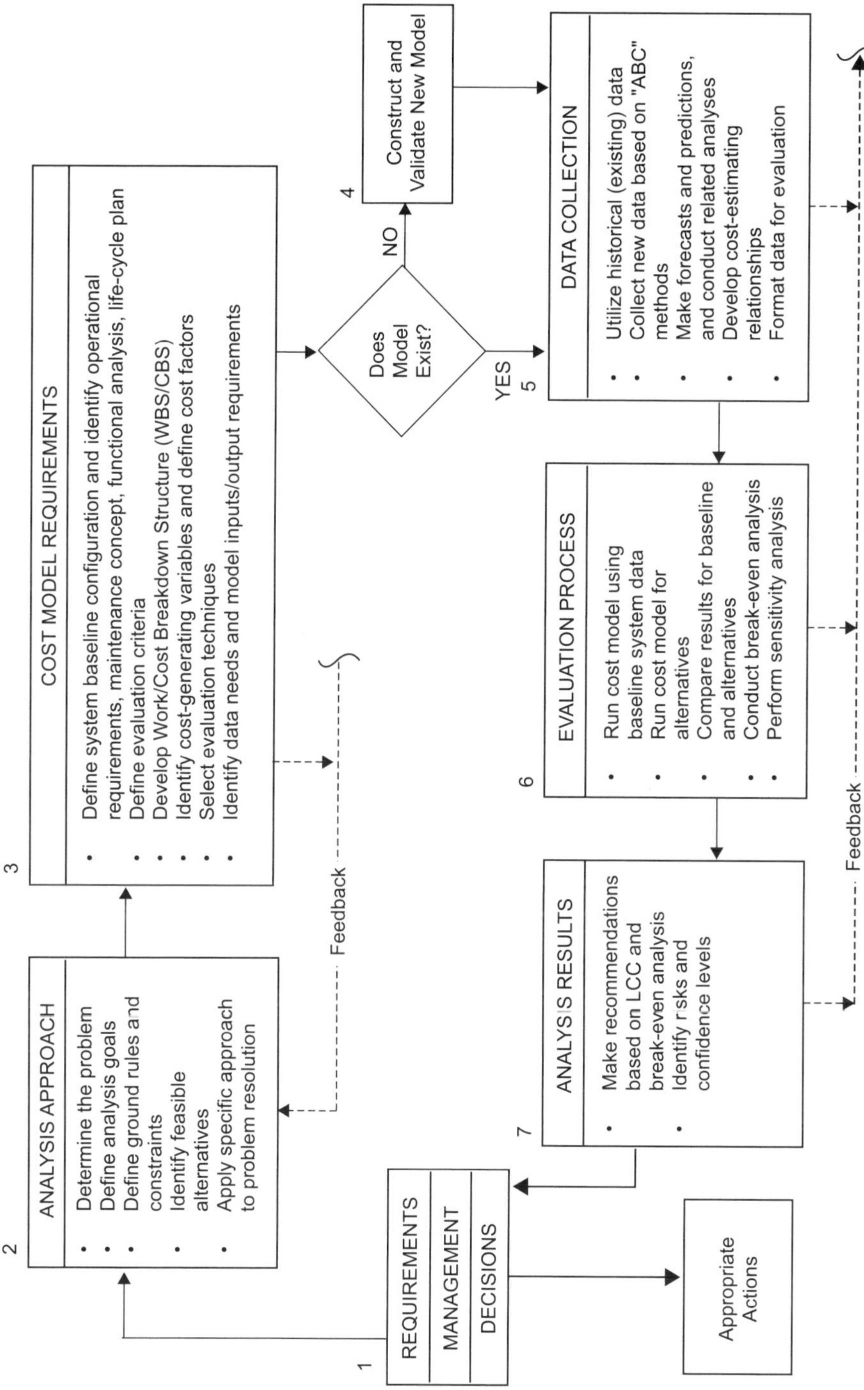

Figure 5.2.1-1. Life-Cycle Cost Analysis Process

Table 5.2.1-1. The Basic Steps in a Life-Cycle Cost Analysis (LCCA)

1. **Define system requirements** – Define system operational requirements and the maintenance concept, identify the applicable technical performance measures (TPMs), and describe the system in functional terms (functional analysis at the system level).
2. **Describe the system life cycle phases and identify the activities in each phase** – Establish a "baseline" for developing a cost breakdown structure and estimating cost for each year of the projected life cycle.
3. **Develop a cost breakdown structure (CBS)** – Provide a top-down/bottom-up structure to include all categories of cost, for the purpose of initially allocating costs (top-down) and subsequently collecting and summarizing costs (bottom-up). All life-cycle activities must be covered within.
4. **Identify data input requirements** – Identify the input data requirements and the possible sources from which the data can be acquired. The type and amount of data will, of course, depend on the nature of the problem being addressed, the phase of the system life cycle, and the depth of the analysis.
5. **Establish the costs for each category in the CBS** – Develop the appropriate cost-estimating relationships (CERs) and estimate the costs for each category in the CBS on a year-by-year basis (using activity-based costing [ABC] methods).
6. **Select a cost model for the purposes of analysis and evaluation** – Select or develop a computer-based model to facilitate the LCCA process. The model must be sensitive to the factors affecting the specific system cost.
7. **Develop a cost profile and summary** – Construct a cost profile (i.e., cost stream) showing the flow of costs over the entire life cycle and the percent contribution of each category to the total cost.
8. **Identify the high-cost contributors and establish cause-and-effect relationships** – Highlight those functions and elements of the system, or segments of a process, that should be investigated for possible improvement.
9. **Conduct a sensitivity analysis** – Evaluate the model, input-output data relationships, and the results of the baseline analysis to ensure that: (1) the overall LCCA approach is valid, and (2) the model itself is well-constructed and sensitive to the factors affecting the problem at hand. Given (1) and (2), the sensitivity analysis can aid in identifying major risk areas as part of a risk analysis.
10. **Construct a Pareto diagram and identify priorities for problem resolution** – Conduct a Pareto analysis, construct a Pareto diagram, and identify priorities for problem resolution (i.e., those problems that require the most management attention).
11. **Identify feasible design alternatives for evaluation** – Having developed an approach for the LCC evaluation of a given baseline design, extend the analysis to evaluating multiple design alternatives.
12. **Evaluate feasible alternatives and select a preferred approach** – Develop a cost profile for each of the alternatives being evaluated, compare the alternatives considering the time value of money, conduct a break-even analysis, and select a preferred design approach.

Analysis Approach. The first step is to expand the problem definition by describing the system operational requirements and the maintenance concept. Figure 5.2.1-2 illustrates these basic concepts. The communications system is installed in a light vehicle, with 65 vehicles being distributed in three different geographical locations. The vehicles will be used an average of 4 hours per day. The system shall enable the day-to-day communication with other vehicles up to a range of 200 miles, with overhead aircraft at an altitude of up to 10,000 feet, and with a centralized communications facility located within 250 miles. The system, as installed in the vehicle, shall have a reliability MTBF of at least 450 hours, a $\bar{M}_{ct}$ not to exceed 30 minutes, and a MLH/OH requirement of 0.2 or less. The system shall be completely operational in two years, and it shall be retired from the inventory in 12 years.

The next step is to develop a cost breakdown structure (CBS), tailored to this particular problem, as shown in Figure 5.2.1-3. The life-cycle plan, shown at the bottom of Figure 5.2.1-2, leads to the identification of the pertinent activities that must be covered by the CBS. Costs are estimated for each activity in the life cycle, on a year-to-year basis, and are reflected in the form of a cost profile, as illustrated in Figure 5.2.1-4 (and as described in Section 2, Topic 2.8.3, and Figure 2.8.3-3). Individual profiles are developed for each of the two alternative design configurations being evaluated. The costs for each configuration, Configuration *A* and Configuration *B*, are summarized in Table 5.2.1-2, indicating the cost for each significant category in the CBS. Note that the costs are presented in terms of the *present value (PV)* so that the alternatives can be compared on an equivalent basis.[4]

[4] As discussed in Section 2, Topic 2.8.3, in performing an economic analysis, one must compare alternatives and convert them to some form of "equivalence," and the applicable profiles must be related in terms of the point in time when the decision for selection is to be made. If the decision is to be made *today*, then the profiles being evaluated must be converted to the *present value (PV)*. Refer to Equation (2.48) for determining the PV cost, given some future inflated cost, and Tables 2.8.3-1 and 2.8.3-2 for an example of tabulated PV costs.

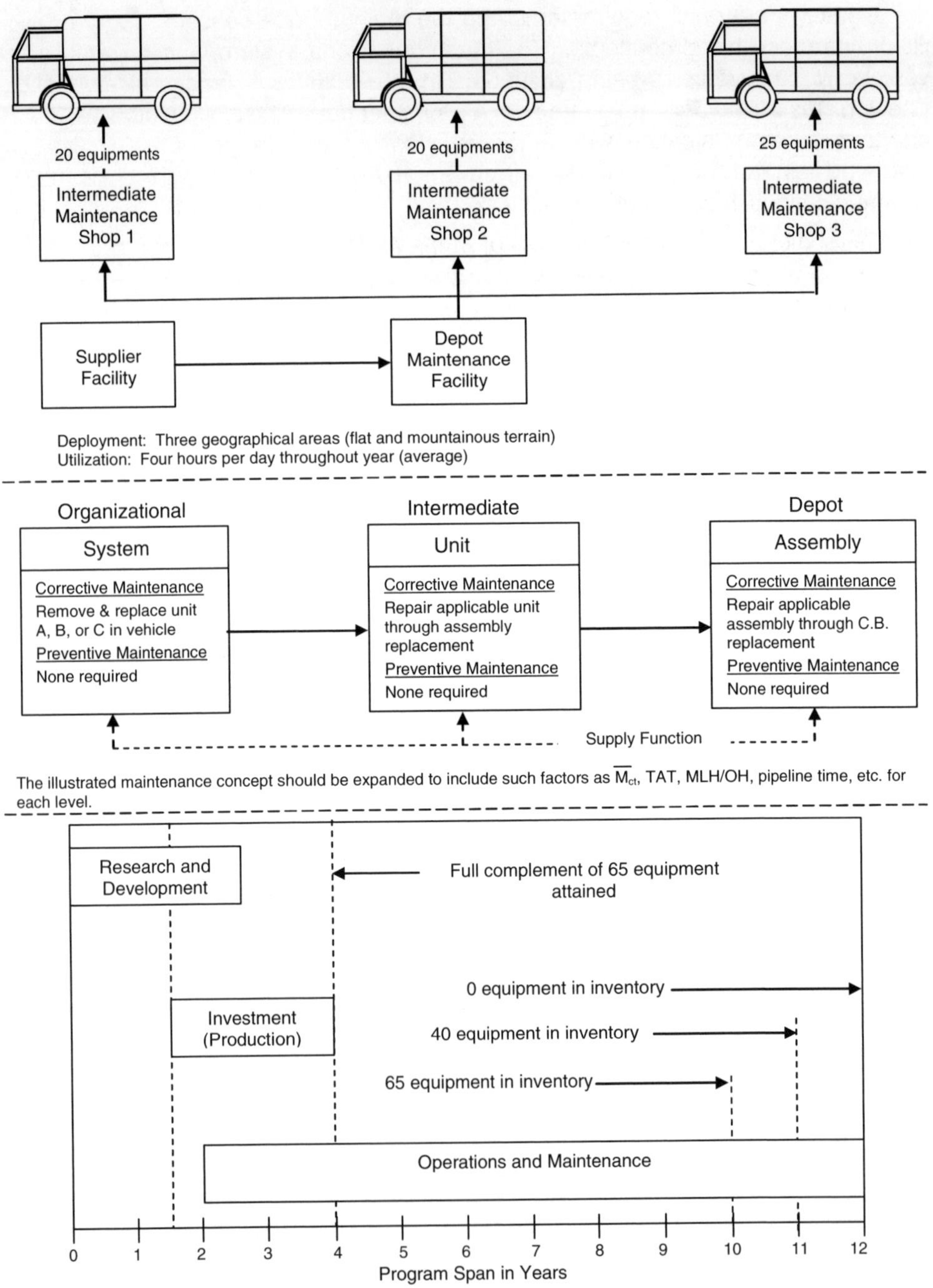

Figure 5.2.1-2. Basic System Concepts (Operations and Support)

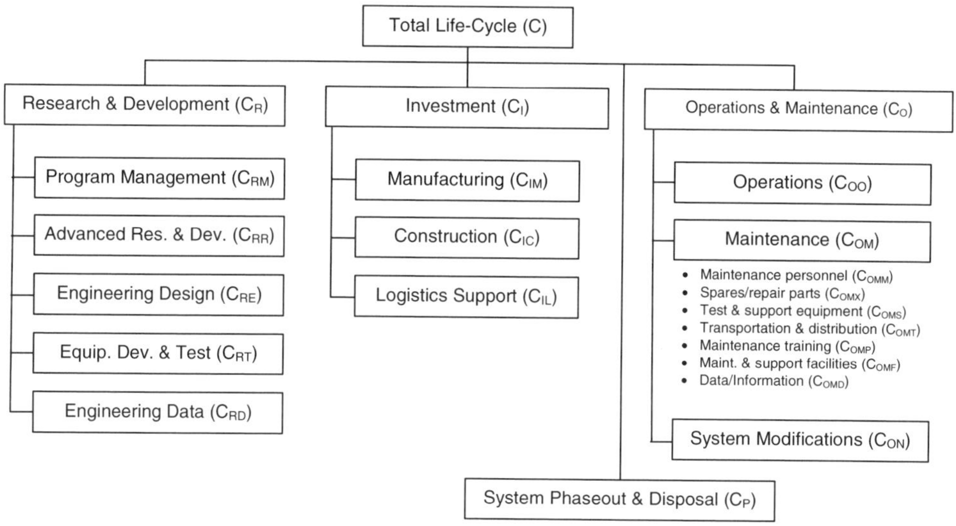

Figure 5.2.1-3. Cost Breakdown Structure (CBS)

Cost Category	Program Year												Total
	1	2	3	4	5	6	7	8	9	10	11	12	
R&D	32,119	38,100	-	-	-	-	-	-	-	-	-	-	70,219
Invest.	-	94,110	156,852	156,852	-	-	-	-	-	-	-	-	407,814
O&M	-	-	12,180	32,480	60,492	57,472	53,480	50,584	50,470	50,494	37,480	17,185	422,217
Total	32,119	132,210	169,032	189,332	60,492	57,472	53,480	50,484	50,470	50,494	37,480	17,185	900,250

Notes: R&D – Research & Development
Invest. – Investment
O&M – Operations and Maintenance
All costs in dollars

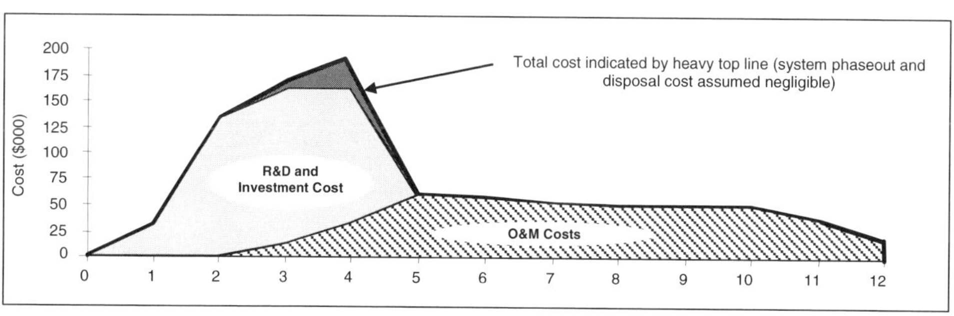

Figure 5.2.1-4. Cost Profile for Configuration "A" (Example)

Table 5.2.1-2. Life Cycle Cost Summary

Life-Cycle Phase	Cost Category	Configuration A		Configuration B	
		P.V. Cost ($)	% of Total	P.V. Cost ($)	% of Total
Acquisition	1. Research and Development (C_R) (Total)	70,219	7.8	53,246	4.2
	(a) Program management (C_{RM})	9,374	1.1	9,252	0.8
	(b) Advanced R&D (C_{RR})	4,152	0.5	4,150	0.4
	(c) Engineering design (C_{RE})	41,400	4.5	24,581	1.9
	(d) Equipment development and test (C_{RT})	12,176	1.4	12,153	0.9
	(e) Engineering data (C_{RD})	3,117	0.3	3,110	0.2
	2. Investment (C_I) (Total)	407,814	45.3	330,885	26.1
	(a) Manufacturing (C_{IM})	333,994	37.1	262,504	20.8
	(b) Construction (C_{IC})	45,553	5.1	43,227	3.4
	(c) Initial logistic support (C_{IL})	28,267	3.1	25,154	1.9
Operation and Maintenance	3. Operations and Maintenance (C_O) (Total)	422,217	46.9	883,629	69.7
	(a) Operations (C_{OO}) (Total)	37,811	4.2	39,301	3.1
	(b) Maintenance (C_{OM}) (Total)	384,406	42.7	844,328	66.6
	• Maintenance personnel (C_{OMM})	210,659	23.4	407,219	32.2
	• Spare/repair parts (C_{OMX})	103,520	11.5	228,926	18.1
	• Test and support equipment (C_{OMS})	47,713	5.3	131,747	10.4
	• Transportation & distribution (C_{OMT})	14,404	1.6	51,838	4.1
	• Maintenance training (C_{OMP})	1,808	0.2	2,125	-
	• Maintenance & support facilities (C_{OMF})	900	0.1	1,021	-
	• Data/Information (C_{OMI})	5,402	0.6	21,452	1.7
	(c) System/equipment modifications (C_{ON})	-	-	-	-
Phaseout and Disposal	4. System Phaseout and Disposal (C_P)**	-	-	-	-
	Grand Total*	$900,250	100	$1,267,760	100

*A 10% discount factor was used to convert to present value (PV).

**In this example, it was assumed that the phaseout and disposal costs were negligible. Of course, these costs are not negligible for all systems and must be evaluated on a system by system basis.

Note in Table 5.2.1-2 that the acquisition cost (R&D and Investment) is higher for Configuration *A* than for Configuration *B* ($478,033 versus $384,131). This difference is due primarily to a better design using more reliable components and added redundancy in critical areas of operation. Although the initial cost is higher, the overall LCC is lower due to a reduction in maintenance actions, resulting in lower O&M costs. These characteristics in equipment design have a tremendous impact on LCC. Nevertheless, and in spite of the higher initial cost, Configuration *A* is preferred based on total LCC. Prior to a final decision, however, the analyst should accomplish a break-even analysis to determine the point at which Configuration *A* becomes more cost-effective than Configuration *B*. Figure 5.2.1-5 illustrates a break-even point that is approximately 6 years and 5 months, or a little more than 2 years after a full complement of systems has been acquired in the operational inventory. This point is assumed to be early enough in the life cycle to support the decision of selection Configuration *A*. On the other hand, if the break-even point were much further out in time, the decision might be questioned. Additional considerations such as possible obsolescence and the continuing validity of the initially specified operational requirements need to be addressed.

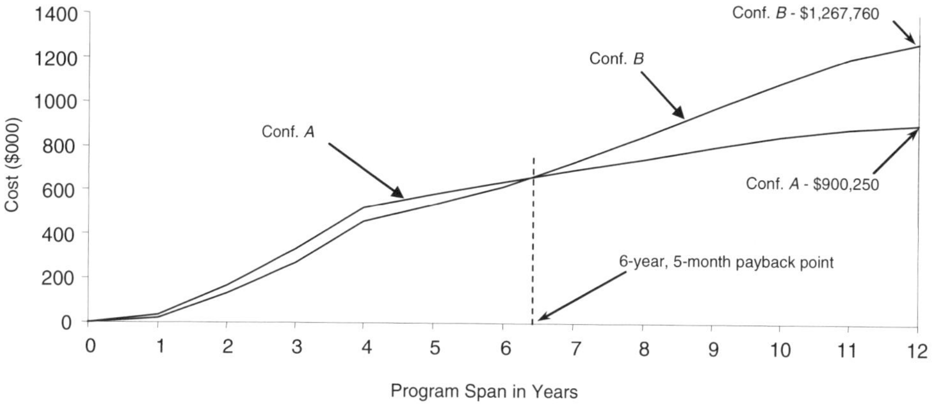

Figure 5.2.1-5. Breakeven Analysis (Two-System Configurations)

The next step (i.e., Table 5.2.1-1, step 9) is to conduct a sensitivity analysis. Referring to the life-cycle cost summary in Table 5.2.1-2, the analyst can readily pick out the *high contributors* (e.g., those categories that contribute more that 10% of the total cost). These reflect areas where a more refined analysis is required and additional valid input data needs to be provided. For instance, maintenance personnel and support cost (Comm) and spare/repair parts cost (Comx) contribute 23.4% and 11.5%, respectively, of the total cost for Configuration *A*. This leads the analyst to re-evaluate the design in terms of the impact on personnel and spares; the prediction methods used in determining maintenance frequencies and inventory requirements; the analytical model to ensure that the proper parameter relationships are established; and the cost factors such as personnel labor hours, spares material costs, inventory holding cost; and so on. If the analyst wishes to determine the sensitivity of these areas to input variations, he/she may wish to perform a sensitivity analysis. In this instance, it is appropriate to vary the "input" MTBF factor as a function of maintenance personnel and support cost (Comm) and spare/repair parts cost (Comx). Referring to Figure 5.2.1-6, if a small variation at the "input" stage results in a large difference at the "output," then MTBF becomes a highly sensitive parameter. On the other hand, if a large MTBF variation at the "input" results in a small change at the "output," then the MTBF factor is not as critical in terms of sensitivity.

The analyst or decision-maker should review the break-even analysis in Figure 5.2.1-5 and determine how far out in time he/she is willing to go and remain with Configuration *A*. For example, if the selected maximum break-even point is seven years, the difference in alternatives is equivalent to approximately \$65,000 (the present value difference between the two configurations at the seven-year point). The result indicates the range of input variations allowed. For instance, if the design of Configuration *A* changes or if the reliability prediction was in error (resulting in an MTBF of the required 450 hours versus a MTBF of 675 hours that was predicted by the supplier and was the basis for determining the reported LCC values), the maintenance personnel and support cost (C_{OMM}) will increase to approximately \$324,000, an increase of about \$113,340 above the baseline value. Thus, although the system reliability is within the specified requirements, the cost increase due to the input MTBF variation causes a decision shift in favor of Configuration *B*. In any event, the analyst must assess the sensitivity of significant input parameters and determine their impact on the ultimate decision.

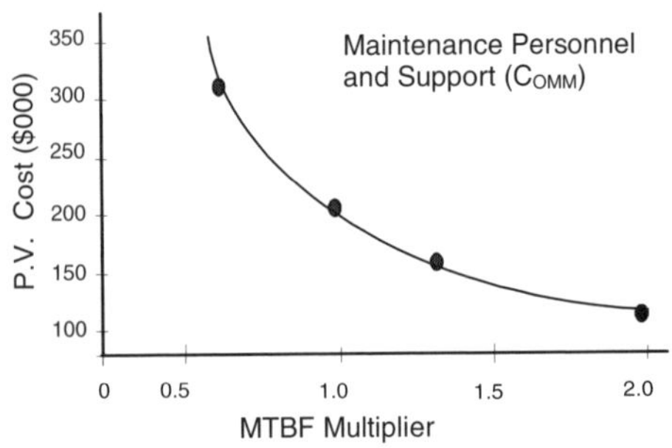

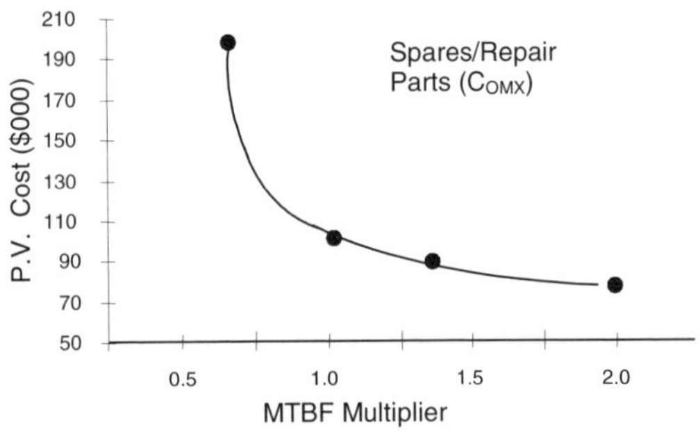

MTBF Multiplier	P.V. Cost (\$) (C_{OMM})
0.67	323,140
1.00*	210,659
1.33	162,325
2.00	112,565

*Baseline Configuration A

MTBF Multiplier	P.V. Cost (\$) (C_{OMX})
0.67	199,576
1.00**	103,520
1.33	92,235
2.00	80,130

**Baseline Configuration B

Figure 5.2.1-6. Sensitivity Analysis (Partial Example)

Analysis Results. The decision in this case is to select Configuration *A* based on the delta cost, the results of the break-even analysis, and the relative relationships of each in terms of reliability and unit cost, as illustrated in Figure 5.2.1-7. Given that both configurations will meet the specified reliability requirement, the one with the lowest predicted LCC is selected.

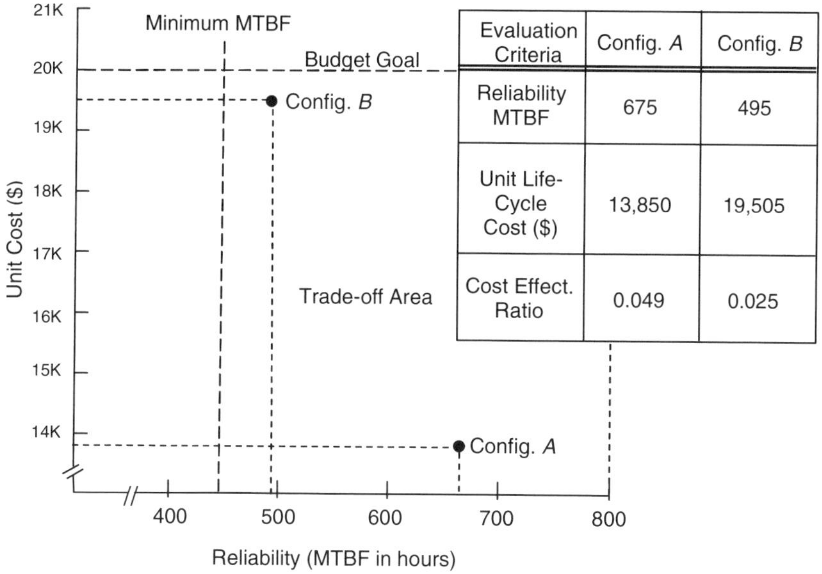

Evaluation Criteria	Config. *A*	Config. *B*
Reliability MTBF	675	495
Unit Life-Cycle Cost ($)	13,850	19,505
Cost Effect. Ratio	0.049	0.025

Figure 5.2.1-7. System Reliability Versus Cost

Topic 5.2.2: Failure Mode, Effects, and Criticality Analysis (FMECA)

Given a system description, both in *functional* and *physical* attributes, the designer needs to evaluate the system with respect to possible failures, the anticipated modes and expected frequency of failure, their causes, their consequences and impact(s) on the overall system, and areas where preventive measures can be incorporated to preclude such failures in the future. The objective is to identify the weak links in the system, the criticality and impact of failures on the system in accomplishing its mission, potential safety problems, and areas of high risk. The FMECA is an effective design tool which can be used initially during the conceptual and preliminary design stages, and can be subsequently applied during detail design and development. Although the analysis is best utilized to influence "before the fact" enhancements to system design, it can also be used as an "after the fact" tool to evaluate and improve existing systems on a continuing basis. The FMECA can be applied to either a functional or a physical entity. To illustrate, Figure 5.2.2-1 portrays an application of the FMECA to a packaging handling system.

The sequential actions involved in the approach used in conducting a Failure Mode, Effects, and Criticality Analysis are shown in Figure 5.2.2-2.[5]

[5] MIL-STD-1629 provided for a similar FMECA process used within the Department of Defense (DoD) for military systems. For a description of the DoD process, refer to Langford, J.W., *Logistics Principles and Applications*, McGraw-Hill, New York, NY, 1995, pages 487-508.

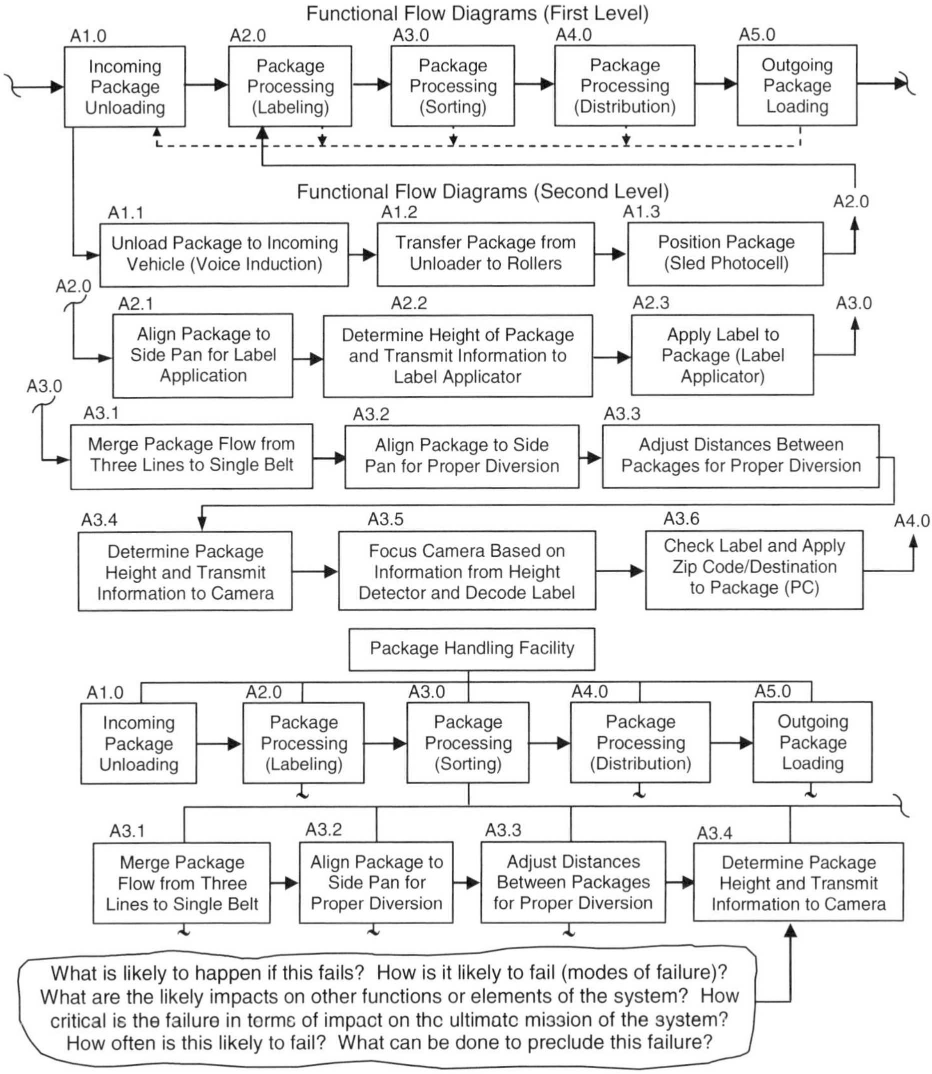

Figure 5.2.2-1. Application of the FMECA to a Packaging Handling System

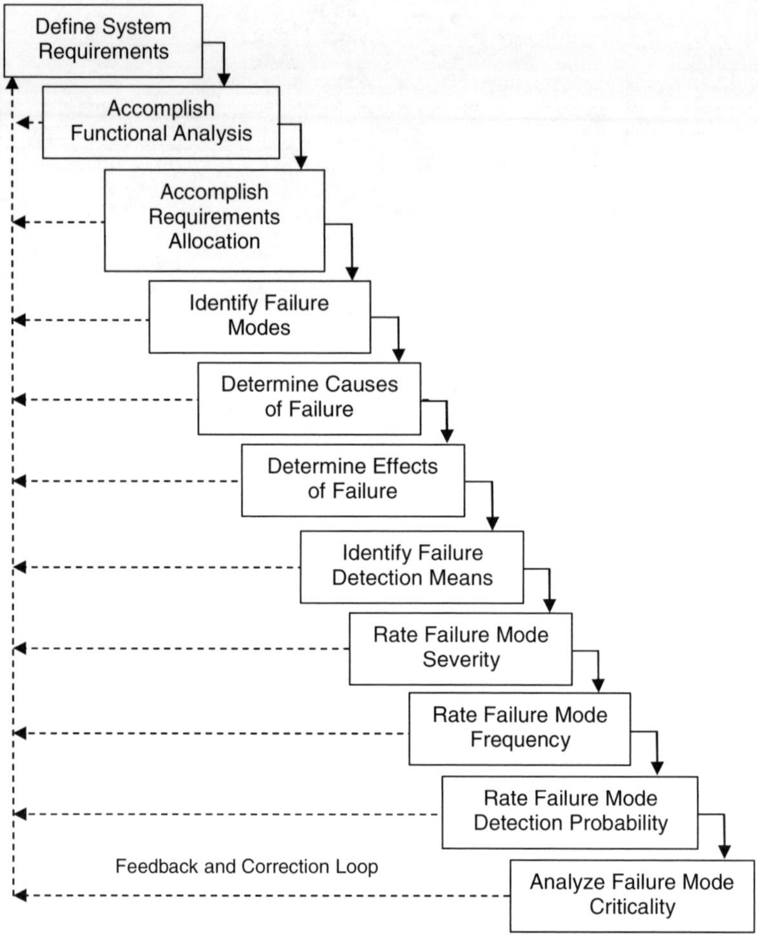

Figure 5.2.2-2. Approach to Conducting a FMECA

The following is a brief, step-by-step description of the sequence of actions in the FMECA process, illustrated with an automobile example:[6]

1. *Definition of System Requirements.* Describe the system (product or process) in terms of the expected outcomes and the relevant technical performance measures (TPMs). Figure 5.2.2-3 shows an example of how the Failure Mode, Effects, and Criticality Analysis can be applied to the manufacturing process for a gasket, as well as addressing the gasket within the automobile.

[6] This process is derived from a number of references used by the automobile industry. For further information, refer to Blanchard, B.S. and Fabrycky, W.J., *Systems Engineering and Analysis*, 3rd Ed., Upper Saddle River, NJ, 1998, p 377.

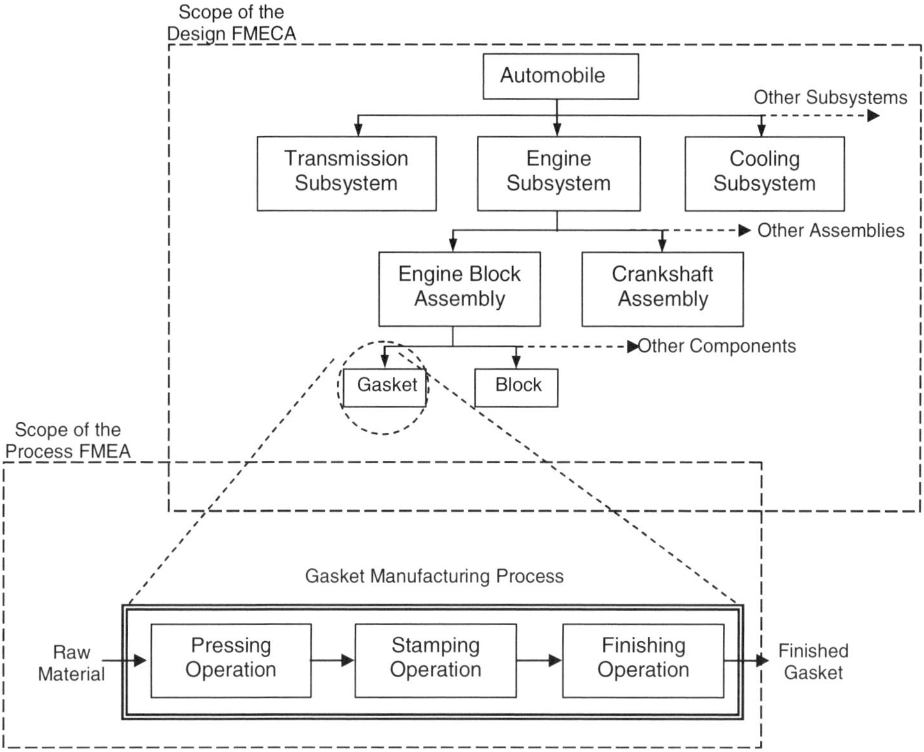

Figure 5.2.2-3. Example of the FMECA Process

One can first conduct an FMECA on the "stamping operation", determine the modes and effects of failure on the other operations within the process, and assess the impacts on the gasket. On basis of this, a product-oriented FMECA can be accomplished assessing the effects of the gasket failure on the engine block and on the automobile as an entity.

2. *Accomplishment of Functional Analysis.* This entails defining the system in functional terms as described in Topic 3.2.6. A system may be broken down into functional entities early in the life cycle (refer to Figure 3.2.6-2) and subsequently organized into a physical packaging scheme as illustrated for hypothetical System XYZ by Figure 5.2.2-4.
3. *Accomplishment of Requirements Allocation.* This is a top-down breakout of system level requirements as described in Topic 3.2.7.
4. *Identification of Failure Modes.* A "failure mode" is the manner in which a system fails to accomplish its function. For example, a switch may fail in an "open" position; a pipe may "rupture;" a given material may "shear" because of stress; a document may fail to be delivered on time; etc.

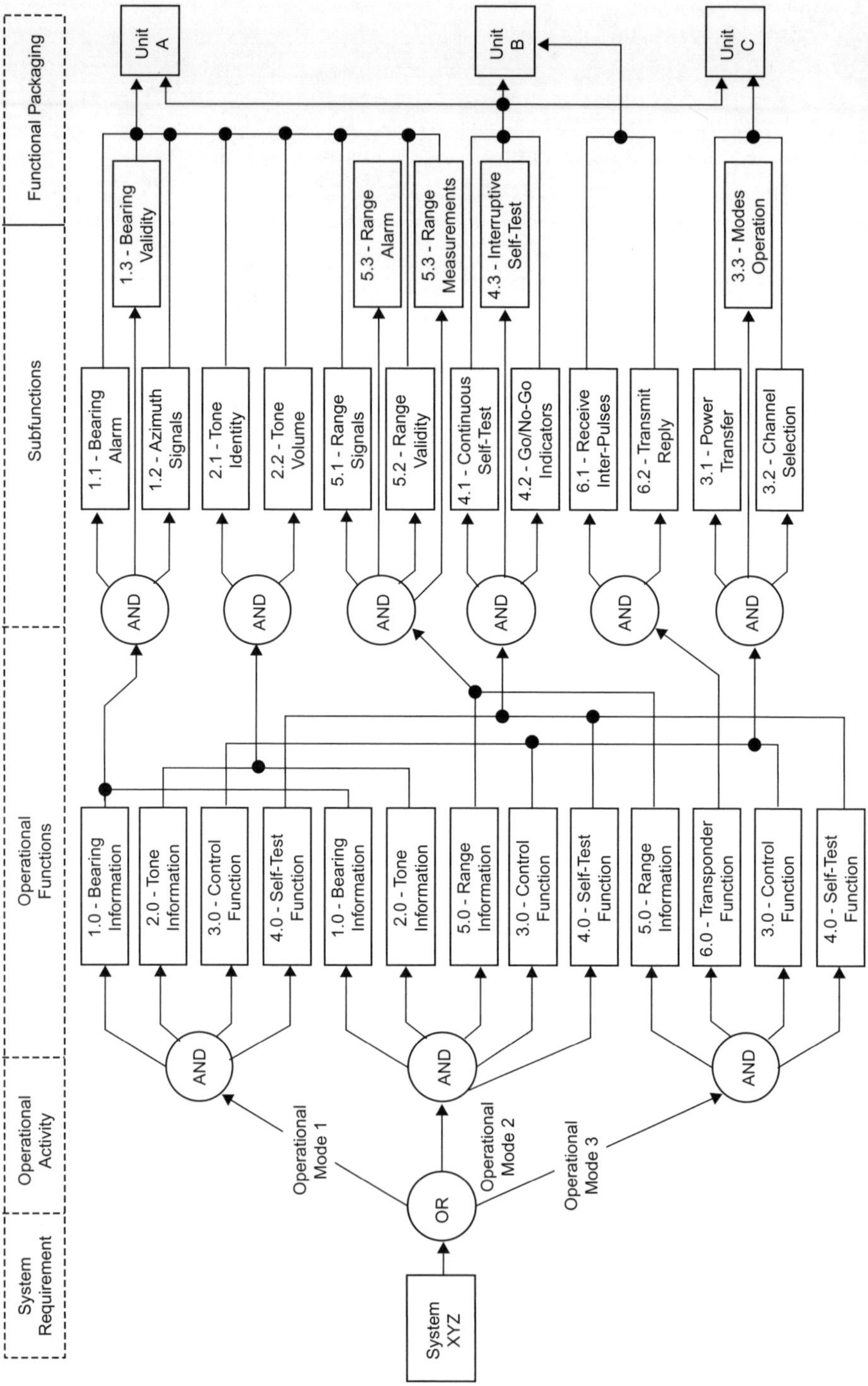

Figure 5.2.2-4. Functional Packaging of a System into Physical Elements for System XYZ

5. *Determination of the Cause of Failure.* This step involves analyzing the process or product to determine the actual cause(s) responsible for the occurrence of failure. Typical causes might include abnormal equipment stresses during operation, aging and wearout, a software coding error, poor workmanship, defective materials, damage due to improper transportation and handling, operator-induced faults, or maintenance-induced faults. Although experience with similar systems or the availability of good data from the field is preferred, using an Ishikawa "cause-and-effect" diagram can prove to be highly effective in delineating potential failure causes. Figure 5.2.2-5 illustrates such a "cause-and-effect" diagram.[7]

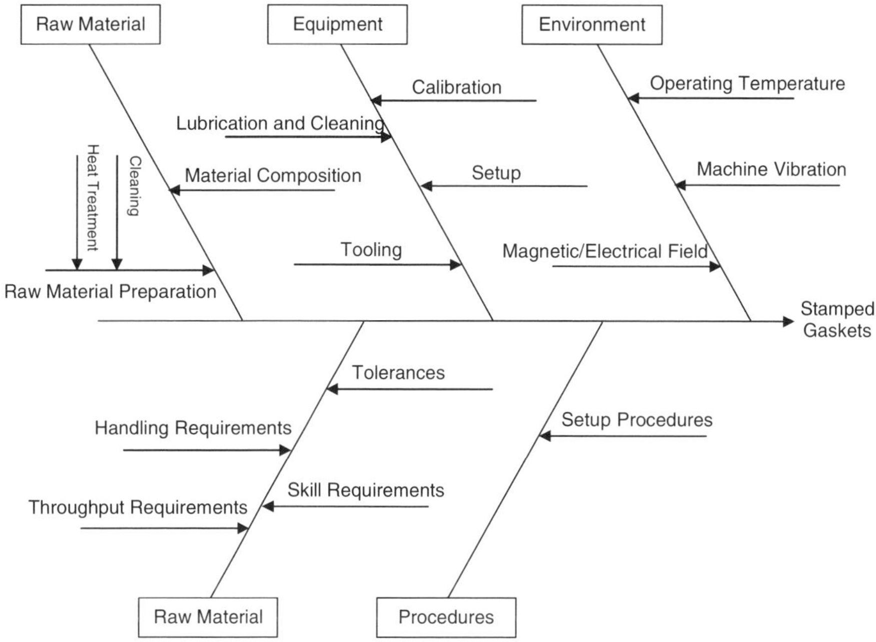

Figure 5.2.2-5. Ishikawa Cause-and-Effect "Fishbone" Diagram

6. *Determination of the Effects of Failure.* Failures impact, often in multiple ways, the performance and effectiveness of not only the associated functional elements but the overall system as well. It is important to consider the effects of failure on other elements at the same level in the system hierarchal structure, at the next higher level, and on the overall system. Referring to Figure 5.2.2-3, if the stamping operation fails, it is then necessary to consider the effects on the finishing operation, on the gasket itself, on the automobile, and ultimately on the customer?

[7] K. Ishikawa, *Guide to Quality Control*, 2nd revision, English edition, Tokyo, Japan, Asian Productivity Organization, 1982; and K. Ishikawa, *Introduction to Quality Control*, London, England, Chapman and Hall, 1991.

7. *Identification of Failure Detection Means.* For a process-oriented FMECA, this refers to the current process controls which may detect the occurrence of failures or defects. When the FMECA has a design focus, this concerns the existence of any design features, aids, gauges, readout devices, condition monitoring provisions, or evaluation procedures that will result in the detection of potential failures.
8. *Determination of Failure Mode Severity.* This refers to the *seriousness* of the effect or impact of a particular failure. If a failure occurs, the scope of the consequences could range from only a slight degradation in system performance to the death of the operator and/or destruction of the system. For purpose of illustration, the degree of severity may be expressed quantitatively (using a checklist approach) on a scale of 1 to 10 with *minor* effects assigned a value of 1, *low* effects being 2 to 3, *moderate* effects being 4 to 6, *high* effects being 7 to 8, and *very high* effects being graded as 9 to 10. The level of severity can be related to issues pertaining to safety or the degree of customer dissatisfaction.
9. *Rating of Failure Mode Frequency.* Given that a function or physical component within the system may fail in a variety of ways, this step addresses the frequency of occurrence of each individual failure mode. The sum of all modal failure frequencies for a system element must equal the system failure rate. For the purposes of quantification, a scale of 1 to 10 may be used with *remote* (failure is unlikely) being 1, *low* (relatively few failures) being 2 to 3, *moderate* (occasional failure) being 4 to 6, *high* (repeated failures) being 7 to 8, and *very high* (failure is almost inevitable) being rated from 9 to10. These ratings can be based on the expected number of failures per unit of time, or an equivalent parameter.
10. *Rating of Failure Mode Detection Probability.* This relates to the probability that process controls, design features/aids, verification procedures, etc., will detect potential failures in time to prevent a major system catastrophy. For a process application, this refers to the probability that a set of process controls currently in place will be in a position to detect and verify a failure before its effects are transferred to a subsequent process, or to the ultimate consumer/customer. For the purposes of quantification, a scale of 1 to 10 may be used with *very high* being 1 to 2, *high* being 3 to 4, *moderate* being 5 to 6, *low* being 7 to 8, *very low* being 9, and *absolute certainty of nondetection* rated as 10.
11. *Analysis of Failure Mode Criticality.* The objective for this process is to consolidate the outputs of the preceding steps to delineate the more critical aspects of the system design. *Criticality,* in this context, is a function of *severity, frequency,* and *probability of detection,* and may be expressed in terms of a *risk priority number* (RPN). The RPN can be determined as follows:

$$\text{RPN} = (\text{severity rating})(\text{frequency rating})(\text{probability of detection rating}) \quad (5.1)$$

The RPN indicates failure mode criticality. On examination, one can observe that a failure mode with a high frequency of occurrence, with significant impact on system performance, and that is difficult to detect is likely to have a very high RPN. A typical RPN hierarchy is illustrated by Table 5.2.2-1.

12. *Initiation of Recommendations for Product/Process Improvement.* This step pertains to the iterative process of identifying areas with high RPNs, evaluating the causes, and the subsequent initiation of recommendations for product/process improvement (refer to the feedback loop in Figure 5.2.2-2). A Pareto analysis, as illustrated by Figure 5.2.2-6, can be performed to highlight the high-priority items that need to be addressed.

Table 5.2.2-1. Typical Risk Priority Number Hierarchy

Causes	Risk Priority Number
Chip Breaker Angle Ground Incorrectly	273
Hold-down Incorrectly Set	210
Undersize Sensor Fails	200
Undersize Sensor Dirty	200
Undersize Sensor Incorrectly Positioned	200
Loose Load Block	161
Sharp Die Edge	120
Improper Projection Angle/Resharpening of Punch	108
Oversize Sensor Fails	105
Oversize Sensor Dirty	105
Oversize Sensor Incorrectly Positioned	105
Improper Sharpening of Insert	93
Misaligned Pusher	80
Worn Tooling	72
Adapter Reground to Wrong Dimension	60
Insert Loose	60
Slivers in Adapter	60
Insert Off Location	60
Worn/Loose Insert	60
Burrs from Punch Process Caught	40
Ram Strokes too Long	36
Ram Strokes too Short	21
Broken/Loose Punch	12
Setscrew Fault	12
Insufficient Stroke by Ram/Punch	12
Broken Pressure Spring	10
TOTAL	2,475

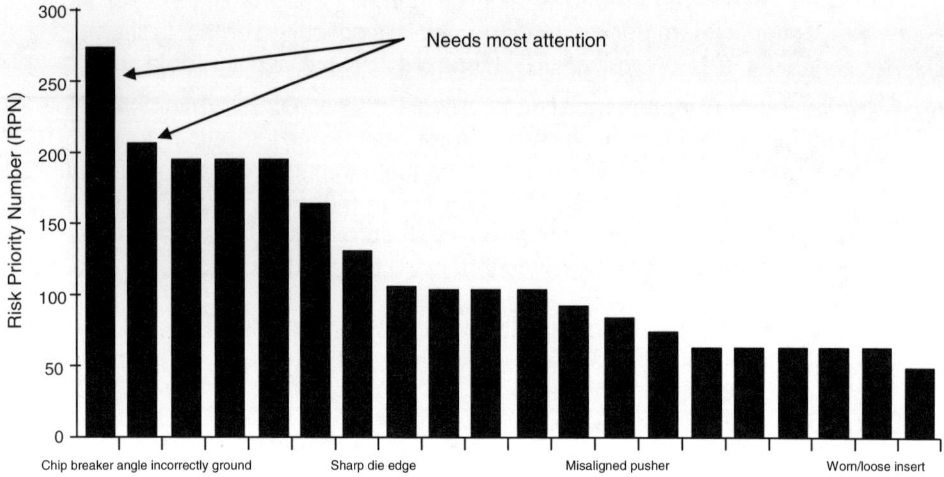

Figure 5.2.2-6. Pareto Analysis-Potential Causes of Failure (Partial)

Figure 5.2.2-7 shows a partial example of a popular format that is utilized for recording the results of the FMECA. The information was derived from the functional flow analysis and expanded to include the results from the steps described by Figure 5.2.2-2.

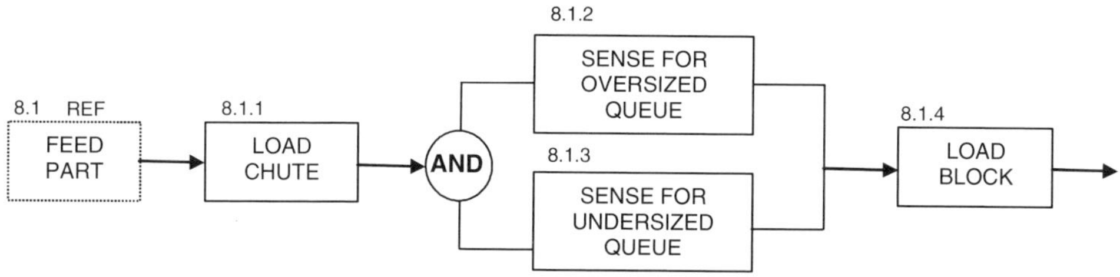

Reference Number	Process Description	Potential Failure Mode	Potential Cause of Failure	Potential Effect(s) of Failure - Producer	Potential Effect(s) of Failure - Customer	Current Controls	Occurrence	RPN	Severity - Producer	Severity - Consumer	Detection	Recommended Action(s) and Status	Responsible Activity
8.1.2	Sense for oversized queue	Change in freespread	Sensor fails	Lip process jams	Bearing loose	Machine stops	1	1	-	1	1		
				Height variation	when installed	1 Pc/5 min.	1	7	-	3	21		
					in engine	2 Pcs/0.5 hr.	1	-	7	5	35		
			Sensor dirty	Lip process jams	Bearing loose	Machine stops	1	1	-	1	1		
				Height variation	when installed in engine	1 Pc/5 min.	1	7	-	3	21		
			Improper setup	Lip process jams	Bearing loose	Machine stops	1	1	-	1	1		
				Height variation	when installed	1 Pc/5 min.	1	7	-	3	21		
					in engine	2 Pcs/0.5 hr.	1	-	7	5	35		
8.1.3	Sense for undersized queue	Lip mislocated	Sensor fails	Lip process jams	Filler ride	100% visual	2	1	-	1	2		
						1 Pc/5 min.	2	-	3	7	42		
			Sensor dirty	Lip process jams	Filler ride	100% visual	2	1	-	1	2		
						1 Pc/5 min.	2	-	3	7	42		
			Improper setup	Lip process jams	Filler ride	100% visual	2	1	-	1	2		
						1 Pc/5 min.	2	-	3	7	42		
8.1.4													

Figure 5.2.2-7. FMECA Process Results (Partial)

Topic 5.2.3: Fault-Tree Analysis (FTA)

Fault-Tree Analysis (FTA) is a deductive approach involving a graphical enumeration and analysis of different ways by which a particular system failure can occur, as well as the probability of occurrence. A separate fault tree may be developed for every critical mode or undesired event. Events must be defined in rather specific terms, inasmuch as a generic or nonspecific definition is likely to result in a broad-based fault tree too wide in scope, with a lack of focus. Attention is directed to the top level event (using a technique such as Ishikawa's *cause-and-effect* diagram as portrayed by Figure 5.2.2-5) and its associated first-tier causes. Each of these causes is next investigated for *its* causes, and so on. This top-down hierarchy, and associated probabilities, is described as a fault tree.[8]

The FTA, which is narrower in focus than the FMECA, is often used in the early phases of conceptual and preliminary development when feasible design data are limited and when the analyst needs to gain insight relative to the feasibility of a selected technological application or to investigate potential safety issues. For large and complex systems, which are highly software intensive and where there are many interfaces, the use of the FTA is usually preferable in lieu of the FMECA. Conversely, accomplishing the FMECA provides more information for design evaluation, the establishment of "cause-and-effect" relationships across the spectrum of the system, and for pinpointing high-cost and high-risk contributors. The following is an illustration of the FTA process.

1. *Definition of the Problem.* During the early stages of the system design process and in absence of the information required to complete an FMECA, a fault-tree analysis is conducted to gain insight into critical aspects of system design. The top-down hierarchy is portrayed in Figure 5.2.3-1. Figure 5.2.3-2 depicts some of the symbology used in the development of such a structure.
2. *Analysis Approach.* One of the outputs of the FTA is the probability of occurrence of the top-level event or failure. If the probability factor is unacceptable, the causal hierarchy developed provides engineers with insight into aspects of the system to which redesign efforts may be directed or compensatory provisions provided. The logic used in developing and analyzing a fault tree has its foundations in Boolean algebra. Axioms from Boolean algebra are used to collapse the initial version of the fault tree to an equivalent reduced tree with objectives of deriving *minimum cut sets.* Minimum cut sets are unique combinations of basic failure events that can cause the undesired top-level event to occur. These minimum cut sets are

[8] For further discussion on FTA, refer to Blanchard, B., *Logistics Engineering And Analysis*, 6th Ed., Pearson Prentice Hall, Upper Saddle River, NJ, 2004, pages 216-222.

necessary to evaluate a fault tree from a qualitative and quantitative perspective. The basic steps in conducting an FTA are as follows:

a. *Identify the top-level event.* It is essential that the analyst be specific in defining this event. For example, it could be delineated as "the system catches fire" rather than "the system fails". Further, the top-level event should be clearly observable, unambiguously definable, and measurable. A generic and nonspecific definition is likely to result in a broad-based fault tree too wide in scope and lacking in focus.

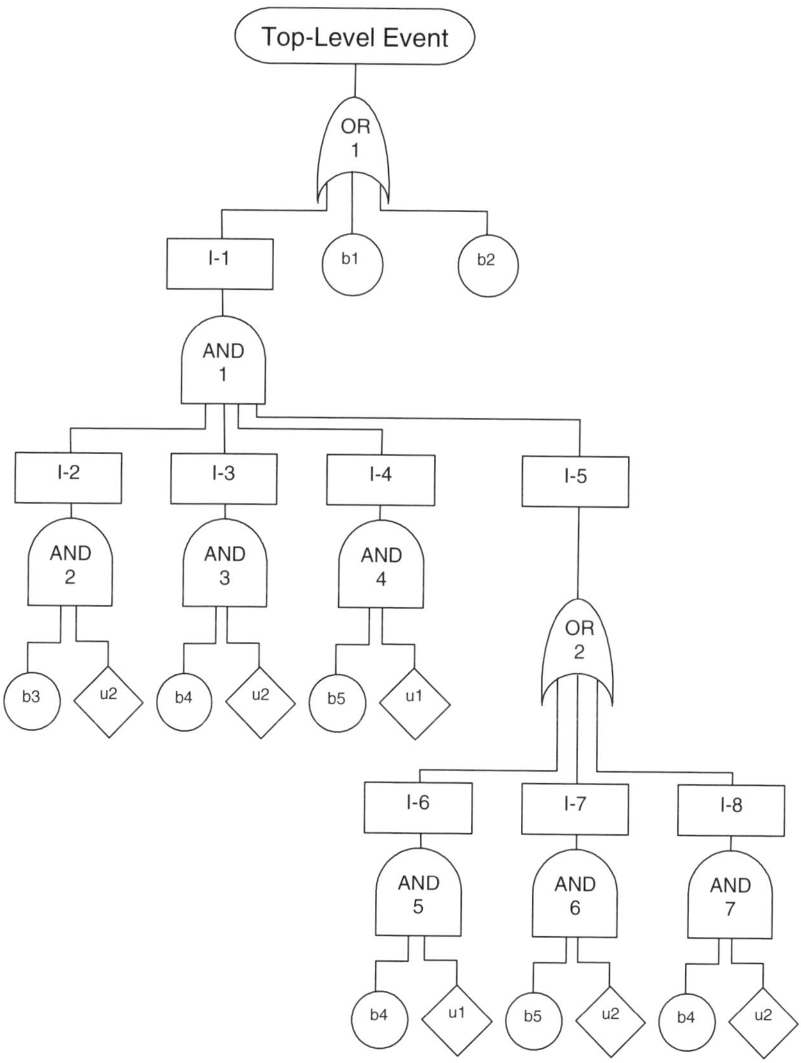

Figure 5.2.3-1. An Illustrative Fault Tree

b. *Develop the fault tree.* Once the top-level event has been satisfactorily defined, the next step is to construct the initial causal hierarchy in the form of a fault tree. As previously addressed, a technique such as the Ishikawa cause-and-effect diagram can prove to be beneficial (refer to Figure 5.2.2-5). While developing the fault tree, the analyst must consider and incorporate all hidden failures. For the sake of consistency and communication, a standard symbology to develop the fault tree is recommended. Figure 5.2.3-2 depicts and defines the symbology to comprehensively represent the causal hierarchy and interconnects associated event(s) with a particular top-level event. In Figure 5.2.3-1, the symbols OR1 and OR2 represent the two OR gates, AND1 through AND7 represent seven AND logic gates, I-1 through I-8 represent eight intermediate fault events, b1 through b5 represent five basic events, and u1 and u2 represent two undeveloped failure events. While constructing a fault tree, it is important to break each branch down to a consistent and reasonable level of detail.

c. *Analyze the fault tree.* The third step in conducting the FTA is to analyze the initial fault tree developed. A comprehensive analysis of a fault tree involves both a quantitative and a qualitative perspective. The important steps in completing the analysis of a fault tree are:

- *Delineate the minimum cut sets:* As part of the analysis process, the minimum cut sets in the initial fault tree are first delineated. These are necessary to evaluate a fault tree from a qualitative or quantitative perspective. The purpose of this step is to reduce the initial tree to a simpler equivalent reduced fault tree. The minimum cut sets can be derived using two different approaches. The first approach involves a graphical analysis of the initial tree, an enumeration of all the cut sets, and the subsequent delineation of the minimal cut sets. The second approach, conversely, involves translating the graphical fault tree into an equivalent Boolean expression. This Boolean expression is then reduced to a simpler equivalent expression by eliminating all the redundancies , and so on. As an example, the fault tree depicted in Figure 5.2.3-1 can be translated into a simpler and equivalent fault tree, through Boolean reduction, as depicted in Figure 5.2.3-3.
- *Determine the reliability of the top-level event:* This is accomplished by first determining the probabilities of all relevant events; then, subsequently consolidating these probabilities in accordance with the underlying logic of the tree. The reliability of the top-level event is computed by taking the product of the reliabilities of the individual minimum cut sets.
- *Review analysis output:* If the derived top-level probability is unacceptable, necessary redesign or compensation efforts will

need to be initiated. The development of the fault tree and subsequent delineation of minimum cut sets provide engineers and analysts with the foundation needed for making sound decisions.

Fault Tree Symbol	Discussion
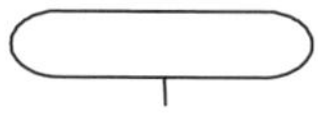	The ellipse represents the *top-level event*. Obviously, the ellipse always appears art the very top level of the fault tree.
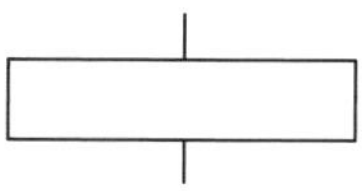	The rectangle represents an *intermediate fault event*. A rectangle can appear anywhere in a tree except at the lowest level in the hierarchy.
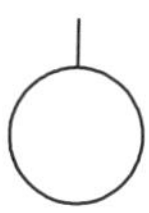	A circle represents the lowest level failure event, also called a basic event. Basic events are likely to appear at the lowest level in a fault tree.
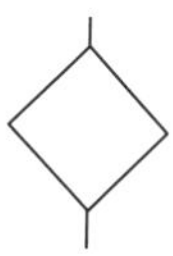	The diamond represents an *undeveloped event*. Undeveloped events could be further broken down but are not for the sake of simplicity. Very often, complex undeveloped events are analyzed through a separate fault tree. Undeveloped events appear at the lowest level in a fault tree.
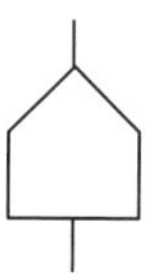	This symbol, sometimes called the house, represents an *input event*. An input event refers to a signal or input that could cause a system failure.
	This symbol represents the AND logic gate. In this case, the output is realized only after all of the associated inputs have been received.
	This symbol represents the OR logic gate. In this case, any one or more of the inputs need to be received for the output to be realized.
	This symbol represents the ORDERED AND logic gate. In this case, the output is realized only after all of the associated inputs have been received in a particular predetermined order.
	This symbol represents the EXCLUSIVE OR logic gate. In this case, one and only one of the associated inputs needs to be received for the output to be realized.

Figure 5.2.3-2. Fault-Tree Constructive Symbology

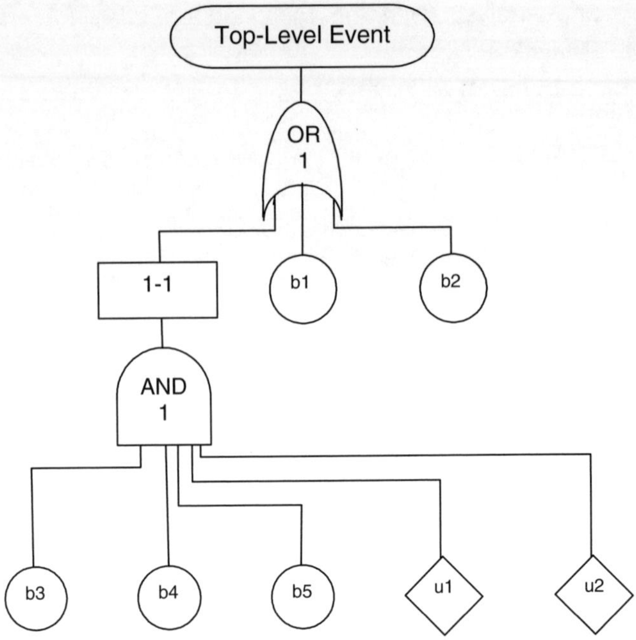

Figure 5.2.3-3. A Reduced Equivalent Fault Tree

Topic 5.2.4: Safety/Hazard Analysis (SHA)

Safety Engineering is a companion discipline to and closely interrelated with Human Factors Engineering. It is also aligned with the FMECA as described in Topic 5.2.2. Safety pertains to personnel, equipment, and other elements of the system. The measures pertinent to safety/hazard analysis are set forth in Topic 2.4. Based on the metrics specifically applicable to the system, the following describes the key terms and approach to analysis of potential risks and hazards that constitute the critical elements of the Safety Risk and Hazard Analysis.[9]

1. *Safety Terms and Definitions*. Measuring of system safety encompasses primarily the following terminology.
    a. *Hazard Probability*: The aggregate probability of occurrence of individual events that create a specific hazard.
    b. *Hazard Severity*: An assessment of the worst credible mishap that could be caused by a specific hazard.
    c. *Risk*: An expression of the possibility of a mishap in terms of hazard severity and hazard probability.
    d. *Safety*: The absence of those conditions that cause death, injury, occupational illness, or damage to or loss of equipment or property.
2. *Safety Hazard Severity* and *Probability of Occurrence*. MIL-STD-882D provides measurement guidelines for accomplishing a *Safety Hazard and Risk Analysis*. The basic quantitative approach involves (1) determining the severity of each of the hazards, and (2) determining the probability of each of the hazards occurring.
    a. *Hazard Severity*. Hazard severity is categorized as *catastrophic*, *critical*, *marginal*, or *negligible*. These categories are displayed by Table 5.2.4-1.
    b. *Hazard Probability*. Hazard probability is alphabetically defined by levels, i.e., *frequent*, *probable*, *occasional*, *remote*, and *improbable*. The levels of probability of occurrence of a hazard are displayed by Table 5.2.4-2.

    The *Hazard Severity* and *Hazard Probability* values for all potential hazards are combined to provide a Hazard and Risk Analysis to support management oversight for a safety program. Through statistical methods, it is possible to estimate the probability of occurrence of the anticipated hazard frequency in terms of calendar

[9] For detailed discussion see MIL-STD-882D, *Department of Defense Standard Practice for System Safety*, Department of Defense, Washington, DC, 19 January 2000.

time, system operating cycles, equipment operating hours, or the equivalent.

Table 5.2.4-1. Hazard Severity Categories

Description	Category	Environmental, Safety, and Health Result Criteria
Catastrophic	I	Could result in death, permanent disability, loss exceeding $1M, or irreversible severe environmental damage that violates law or regulation.
Critical	II	Could result in permanent partial disability, injuries or occupational illness that may result in hospitalization of at least three personnel, loss exceeding $200K but less than $1M, or reversible environmental damage causing a violation of law or regulation.
Marginal	III	Could result in injury or occupational illness resulting in one or more lost work day(s), loss exceeding $10K but less than $200K, or mitigatible environmental damage without violation of law or regulation where restoration activities can be accomplished.
Negligible	IV	Could result in injury or illness not resulting in a lost work day, loss exceeding $2K but less than $10K, or minimal environmental damage not violating law or regulation.

Table 5.2.4-2. Hazard Probability Levels

Description*	Level	Specific Individual Item	Fleet or Inventory**
Frequent	A	Likely to occur often in the life of an item, with a probability of occurrence greater than 10^{-1} in that life.	Continuously experienced.
Probable	B	Will occur several times in the life of an item, with a probability of occurrence less than 10^{-1}, but greater than 10^{-2} in that life.	Will occur frequently.
Occasional	C	Likely to occur some time in the life of an item, with a probability of occurrence less than 10^{-2}, but greater than 10^{-3} in that life.	Will occur several times.
Remote	D	Unlikely but possible to occur in the life of an item, with a probability of occurrence less than 10^{-3}, but greater than 10^{-6} in that life.	Unlikely, but can be reasonably expected to occur.
Improbable	E	So unlikely, it can be assumed occurrence may not be experienced, with a probability of occurrence of less than 10^{-6} in that life.	Unlikely to occur, but possible.

* Definitions of descriptive words may have to be modified based on quantity of items involved.
**The expected size of the fleet or inventory should be defined prior to accomplishing an assessment of the system.

c. *Formulation of the Safety Hazard and Risk Analysis.* The initial action is the preparation a hazard analysis report data sheet. This data will include the following basic information:[10]

- *Description of hazard.* System operational requirements and the maintenance concept are reviewed to identify possible hazardous conditions. Past experiences with similar systems or products used in comparable environments serve as a logical starting point. Hazardous conditions may include acceleration and motion, electrical shock, chemical reactions, explosion and fire, heat and temperature, radiation, pressure, moisture, vibration and noise, and toxicity.
- *Cause of hazard.* Possible causes should be described for each identified hazard. This should address what events are likely to occur in creating the hazard.
- *Identification* of *hazard effect.* Describe the effects of each identified hazard on both personnel and equipment. Personnel effects may include injuries, such as cuts, bruises, broken bones, punctures, heat exhaustion, asphyxiation, trauma, and respiratory or circulatory damage.

  Table 5.2.4-3 provides a typical structure of the data sheet for the hazard analysis and risk analysis. The critical data factor is the *risk assessment.*

Table 5.2.4-3. Hazard Analysis Report Data Elements/Format

Item	Specification
Item identification code	Unique number used for tracking purposes
Subsystem	Subsystem most directly affected by the identified hazard
Risk assessment*	Alpha-numeric rating of the identified hazard
Hazard description	Succinct description of the hazard
Cause of hazard	Possible cause(s) of the hazard, including events and verification that are likely to occur leading to the hazard
Hazard effect	More detailed description of the hazard with potential effects of inadvertent action
Recommended controls and verification	Present and/or future actions planned to control, identify, and verify the closure or containment of the hazard
Comments	Description of known and/or probable interfaces between the potential hazard and interactive hardware and software
Status	Indication of requires solution to problem is *closed* or *open.*

*Refer to Table 5.2.4-6 (Hazard Risk Assessment Scoring Matrix)

[10] Blanchard, B., *Logistics Engineering And Analysis*, 6th Ed., Pearson Prentice Hall, Upper Saddle River, NJ, 2004, pages 291-292.

3. *Preparation of the Safety Hazard and Risk Analysis Report.* The hazard data are organized and portrayed based on the severity categories (Table 5.2.4-1) and occurrence probabilities (Table 5.2.4-2). This will be documented in the form of a management-level summary, which includes (1) hazard risk assessment severity definitions/weight factors, (2) hazard risk frequency definitions/weight factors, (3) hazard risk assessment scoring matrix, and (4) hazard risk summary matrix.
    a. Table 5.2.4-4 displays the hazard severity categories, along with the weighted values assigned to each category.

Table 5.2.4-4. Hazard Risk Assessment Severity Definitions/ Weight Factors

Description	Category	Mishap Definition	Weighting Factor
Catastrophic	I	Death to personnel or operators or loss of system	4
Critical	II	Severe injury, severe occupational illness to personnel or major damage to system	3
Marginal	III	Minor injury, minor occupational illness or minor system damage	2
Negligible	IV	Less than minor injury, less than minor occupational illness to personnel or less than minor system damage	1

    b. Table 5.2.4-5 shows the hazard risk frequency definition, along with the weighted values assigned to each designated frequency.

Table 5.2.4-5. Hazard Risk Frequency Definitions/Weight Factors

Description	Level	Individual Systems	Total Systems Population	Weighting Factor
Frequent	A	Likely to occur frequently.	Continuously experienced.	5
Probable	B	Will occur several times in the life of a system.	Will occur frequently.	4
Occasional	C	Likely to occur some time in the life of a system.	Will occur several times.	3
Remote	D	Unlikely but possible to occur in the life of a system.	Unlikely, but can be reasonably expected to occur.	2
Improbable	E	So unlikely, it can be assumed occurrence may not be expected.	Unlikely to occur, but possible.	1

c. Table 5.2.4-6 portrays the hazard risk assessment scoring matrix, which combines the hazard severity and frequency ratings from Table 5.2.4-4 and Table 5.2.4-5.

Table 5.2.4-6. Hazard Risk Assessment Scoring Matrix

Frequency		Hazard Severity Categories			
		I (Catastrophic)	**II (Critical)**	**III (Marginal)**	**IV (Negligible)**
A	Frequent	I-A	II-A	III-A	IV-A
B	Probable	I-B	II-B	III-B	IV-B
C	Occasional	I-C	II-C	III-C	IV-C
D	Remote	I-D	II-D	III-D	IV-D
E	Improbable	I-E	II-E	III-E	IV-E

d. The data collected from the reported hazard data (Table 5.2.4-3) are combined with the weight values in Table 5.2.4-4 and Table 5.2.4-5 to assign risk values based on the scoring factors in Table 5.2.4-6. For example, a hazard designated with severity categorized as *catastrophic (I)* and a frequency classified as *probable (B)* would be assigned a risk assessment of *I-B* with a matrix score of 4x4, or 16. All identified risks are scored on the basis of the same weight-scoring methodology. The matrix scores are normally organized and grouped in a hierarchy of four levels to display a *hazard risk index* such as:

15-20 (I-A, I-B, II-A)
8-12 (I-C, I-D, II-B, II-C, III-A, III-B)
4- 8 (I-E, II-D, III-C, III-D, IV-A, IV-B)
1- 3 (II-B, III-E, IV-C, IV-D, IV-E)

Table 5.2.4-7 illustrates the resultant hazard risk assessment summary matrix, along with safety management criteria for actions based on the corresponding risk levels in the hazard risk index.

Table 5.2.4-7. Hazard Risk Assessment Summary Matrix

Frequency	Hazard Severity Categories			
	I (Catastrophic)	II (Critical)	III (Marginal)	IV (Negligible)
A (Frequent)	I-A (4)(5)=20	II-A (3)(5)=15	III-A (2)(5)=10	IV-A (1)(5)=5
B (Probable)	I-B (4)(4)=16	II-B (3)(4)=12	III-B (2)(4)=8	IV-B (1)(4)=4
C (Occasional)	I-C (4)(3)=12	II-C (3)(3)=9	III-C (2)(3)=6	IV-C (1)(3)=3
D (Remote)	I-D (4)(2)=8	II-D (3)(2)=6	III-D (2)(2)=4	IV-D (1)(2)=2
E (Improbable)	I-E (4)(1)=4	II-E (3)(1)=3	III-E (2)(1)=2	IV-E (1)(1)=4

Matrix Score	Hazard Risk Index	Management Review Criteria
15-20	I-A I-B II-A	Risks that are unacceptable-must be reviewed by top management.
8-12	I-C I-D II-B II-C III-A III-B	Risks that are undesirable – must be reviewed by the principal for safety.
4-6	I-E II-D III-C III-D IV-A IV-B	Risks that are acceptable – must be reviewed by the system safety working group.
1-3	II-E III-E IV-C IV-D IV-E	Risks that are acceptable – must be reviewed by the system safety engineer.

Topic 5.2.5: Maintenance Task Analysis (MTA)

The maintenance task analysis (MTA) stems from the system maintenance concept (Section 3.2.3) and the subsequent identification of maintenance and support *functions* developed as part of the functional analysis (Section 3.2.6). Maintenance functions are broken down into sub-functions, job operations, duties, and detailed *tasks.* Maintenance tasks are then evaluated in terms of task sequences, task times, and the resources required in the accomplishment of these task; i.e., personnel quantities and skill levels, spares and replacement parts, test and support equipment, facility requirements, maintenance software, technical data, and information. While the specific data format for accomplishing a maintenance task analysis (MTA) may vary significantly, an example of one approach is illustrated in the paragraphs that follow.[11]

Definition of the Problem. A given company designed and developed System *XYZ*, and has been producing the system and its components for the past few years. Throughout the production process, there has been a persistent problem associated with the final manufacturing system inspection and test function within the factory. This, in turn, has caused a lot of downtime, and the resulting costs have been high. Thus, company management decided to evaluate the overall production capability and, more specifically, identify the problems associated with the manufacturing and test function.

Analysis Approach. A review of maintenance data covering the past performance of the manufacturing test capability indicated a frequent loss of power during the final test of a major sub-system. From this data, a typical symptom of failure was identified, and a sample logic troubleshooting flow diagram was developed, as shown in Figure 5.2.5-1.

The applicable go/no-go tasks, identified in Figure 5.2.5-1, are converted to the task analysis format in Figures 5.2.5-2 and 5.2.5-3. The tasks are analyzed on the basis of determining the resource requirements for each; i.e., task durations, series-parallel relationships, personnel quantities and skill levels, spare/repair part requirements, test and support equipment requirements, special facilities, technical data requirements, and so on. The intent of Figures 5.2.5-2 and 5.2.5-3 is to layout the applicable maintenance tasks required, determine the anticipated frequency of occurrence, and identify the logistics support resources that are likely to be necessary in the performance of the required maintenance.

[11]A detailed description of the steps followed in accomplishing a maintenance task analysis (MTA) using the format presented in this illustration is included in Appendix *E,* Blanchard, B.S., *Logistics Engineering and Management*, 6th Ed., Pearson Prentice Hall, Upper Saddle River, NJ, 2004, pages 496-519.

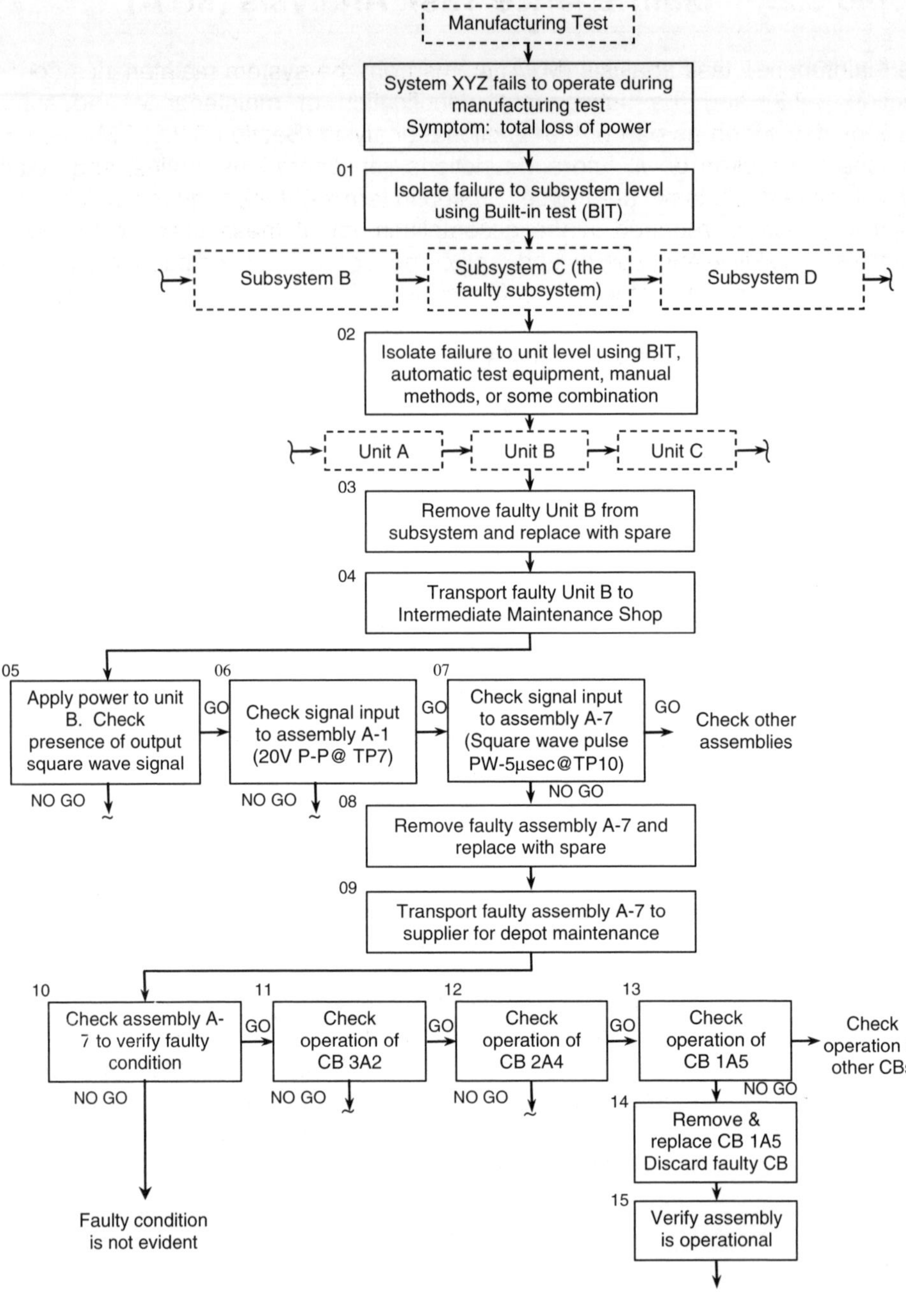

Figure 5.2.5-1. Partial Logic Troubleshooting Flow Diagram (System XYZ)

1. System: XYZ	2. Item Name/Part No: Manufacturing Test A4321		3. Next Higher Assembly: Assembly and Test		4. Description of Requirement: During manufacturing and test of product 12345 (serial no 654) System XYZ failed to operate. The symptom of failure was "Loss of total power output" Requirement: Troubleshoot and repair System.
5. Req No. 01	6. Requirement: Diag./Repair	7. Reg. Freq. 0.00450	8. Maint. Level Org/Inter	9. Ma. Cont. No.: A12B100	

10. Task No.	11. Task Description	12. Elapsed Time - Minutes (0 2 4 6 8 10 12 14 16 18 20 22 24 26 28 30 32 34 36 38 40)	13. Total Elapsed Time	14. Task Freq	Personnel man-minutes 15. B	16. I	17. S	18. Total
01	Isolate Failure to Subsystem Level		5	0.00450	5	-	-	5
	(Subsystem C is faulty)							
02	Isolate failure to Unit level	(1)	25			25		25
	(Unit B is faulty)	(2)					25	25
03	Remove Unit B from System and replace with a spare Unit B	(2nd Cycle)	15		15			15
04	Transport faulty unit to int. shop		30		30			30
05	Apply power to faulty unit. Check for output square wave signal	(3th Cycle)	20			20		20
06	Check signal input to Assembly A-1 (20V P P @ TP7)		15			15		15
07	Check signal input to Assembly A-7 (Square wave, PW-5µsec @ TP2)	(4th Cycle)	20			20		20
08	Remove faulty A-7 and replace		10		10			10
09	Transport faulty Assembly A-7 to supplier for Depot Maintenance	14 Calendar Days in Transit						25
10	Check A-7 and verify faulty condition	(5th Cycle)	25				25	25
11	Check operation of CB-3A2		15				15	15
12	Check operation of CB-2A4		10				10	10
13	Check operation of CB-IA5	(6th Cycle)	20				20	20
14	Remove and replace faulty CB-1A5. Discard faulty circuit board	(7th Cycle)	40			40		40
15	Verify that assembly is operational and return to inventory		15				15	15
		Total:	265	0.00450	60	120	110	290

Figure 5.2.5-2. Partial Maintenance Task Analysis (System XYZ – Part 1)

1. Item Name/Part No. Manufacturing Test/A4321		2. Req. No. 01		3. Requirement: Diagnostic Troubleshooting and Repair		4. Req Freq: 0.00450	5. Maint. Level: Organization, Intermediate, Depot	6. Ma. Cont. No.: A12B100
	Replacement Parts			Test and Support/Handling Equipment			16. Description of Facility Requirements	17. Special Technical Data Instructions
7. Task No.	8. Qty per Assy	9. Part Nomenclature / 11. Part Number	10. Rep Freq.	12. Qty.	13. Item Part Nomenclature / 15. Item Part Number	14. Use Time (Min)		
01	-	-	-	1	Built-in Test Equip A123456	5	-	Organizational Maintenance
02	-	-	-	1	Special System Tester 0-2310B	25	-	Organizational Maintenance
03	1	Unit B / B180265X	0.01866	1	Standard Tool Kit STK-100B	15	-	Organizational Maintenance
04	-	-	-	1	Standard Cart (M-10)	30	-	Intermediate Maintenance
05	-	-	-	1	Special System Tester 1-8891011A	20	-	Intermediate Maintenance
06	-	-	-	1	Special System Tester 1-8891011A	15	-	Intermediate Maintenance
07	-	-	-	1	Special System Tester 1-8891011A	20	-	Intermediate Maintenance
08	1	Assembly A-7 / MO-2378A	0.00995	1	Special Extractor Tool EX20003-4	10	-	Refer to Special Removal Instructions
09	-	-	-	1	Container Special Handling T-300A	14 days	-	Normal Trans. Environment
10	-	-	-	1	Special System Tester 1-8891011B	25	Clean Room Environment	Supplier (Depot) Maintenance
11	-	-	-	1	C.B. Test Set D-2252-A	15	Clean Room Environment	Supplier (Depot) Maintenance
12	-	-	-	1	C.B. Test Set D-2252-A	10	Clean Room Environment	Supplier (Depot) Maintenance
13	-	-	-	1	C.B. Test Set D-2252-A	20	Clean Room Environment	Supplier (Depot) Maintenance
14	1	CB-1A5 / GDA-2210565C	0.00450	1	Special Extractor Tool/EX45111-63 Standard Took Kit STK-200	40	Clean Room Environment	Supplier (Depot) Maintenance
15	-	-	-	1	Special System Tester 1-8891011A	15	Clean Room Environment	Return Operating Assy to Inventory

Figure 5.2.5-3. Partial Maintenance Task Analysis (System XYZ – Part 2)

Given the preliminary results of the analysis in terms of the layout of expected maintenance tasks, the next step is to evaluate the information presented in the figures and to suggest possible areas for improvement. This type of analysis can be accomplished for each and every anticipated maintenance function, and/or for just those that are considered as being the most critical relative to both system mission-related requirements and the consumption of resources which, in turn, can be quite costly.

Analysis Results. By reviewing the information presented in Figures 5.2.5-2 and 5.2.5-3, one can see that there are a number of areas that should be re-evaluated from the design perspective. For instance:

1. With the extensive resources required for the repair of Assembly *A-7* (e.g., the variety of special test and support equipment, the special clean room facility for maintenance, the extensive amount of time required for the removal and replacement of CB-1A5, etc.), it may be more feasible to identify Assembly *A-7* as being "non-repairable" when a failure occurs.
2. From Tasks 01 and 02 we see that a built-in test capability exists at the organizational level for fault isolation to the sub-system; however, fault isolation to the unit requires a Special System Tester (0-2310B), and it takes 25 minutes of testing plus a highly-skilled individual to accomplish the function. One should investigate the feasibility of extending the built-in test capability down to the unit level and eliminating the need for the special tester and the high-skilled individual.

Review of the figures indicates that there are other areas, currently incorporated within the existing design configuration, which should be challenged! For example, can re-design be accomplished to eliminate the need for the clean room, the need for special test equipment, the need for high-skilled personnel, the need for the special handling container, and can result in less downtime and overall cost? This type of evaluation can be accomplished as an iterative process and the subsequent incorporation of design changes for improvement.

Topic 5.2.6: Reliability-Centered Maintenance (RCM)

Reliability-centered maintenance (RCM) is a systematic approach to develop a focused, effective, and efficient preventive maintenance (PM) program and control plan for a system. The RCM technique was initiated in the 1960s and subsequently developed through the 1970s, primarily through the concerted efforts of the commercial airline industry. The approach is facilitated by the application of a structured decision tree that leads the analyst through a tailored logic flow to delineate the most significant preventive maintenance tasks, their nature, and their frequency. Figure 5.2.6-1 presents a simplified RCM logic, in which safety, performance, and cost are prime considerations.[12]

Analysis Approach. The major steps in accomplishing a RCM analysis are:

1. The first step is to identify the critical system functions or components. Criticality, in this analysis, is a function of failure frequency, the failure effect severity, and the probability of detection of the relevant failure modes. This step is facilitated through outputs from the system functional analysis and the FMECA. Basically, the RCM analysis is heavily dependent on the FMECA as an input.
2. Apply the RCM decision logic illustrated in Figure 5.2.6-1. The objective here is to better understand the nature of failures associated with the critical system functions and components. In each case, and whenever feasible, this knowledge is translated into a set of preventive maintenance tasks, or a recommendation for re-design. At the same time, there is concern as to whether the failure is likely to compromise personal safety and/or system functionality.
3. As one proceeds down through the logic diagram, there are a series of questions which are addressed, with the objective of delineating a feasible set of compensatory provisions in design, or in the identification of preventive maintenance tasks. This leads to the development and implementation of an overall Preventive Maintenance (PM) Implementation Plan for the system.[13]

---

[12] A maintenance steering group (MSG) was formed in the 1960s to undertake the development of the technique. The result was "747 Maintenance Steering Group Handbook: Maintenance Evaluation and Program Development (MSG-1)," published in 1968. This effort, focused on a particular aircraft, was generalized and published in 1970 as "Airline/Manufacturer Maintenance Program Planning Document – MSG-2." The MSG-2 approach was further developed and published in 1978 as "Reliability Centered Maintenance," Report Number A066-579, prepared by United Airlines, and in 1980 as "Airline/Manufacturer Maintenance Program Planning – MSG-3."

[13] For a more in-depth description of the specific objectives, steps in the process, and questions that are addressed, refer to Blanchard, B.S., *Logistics Engineering and Management,* 6th Ed., Pearson Prentice Hall, Upper Saddle River, NJ, 2004, pages 227-232.

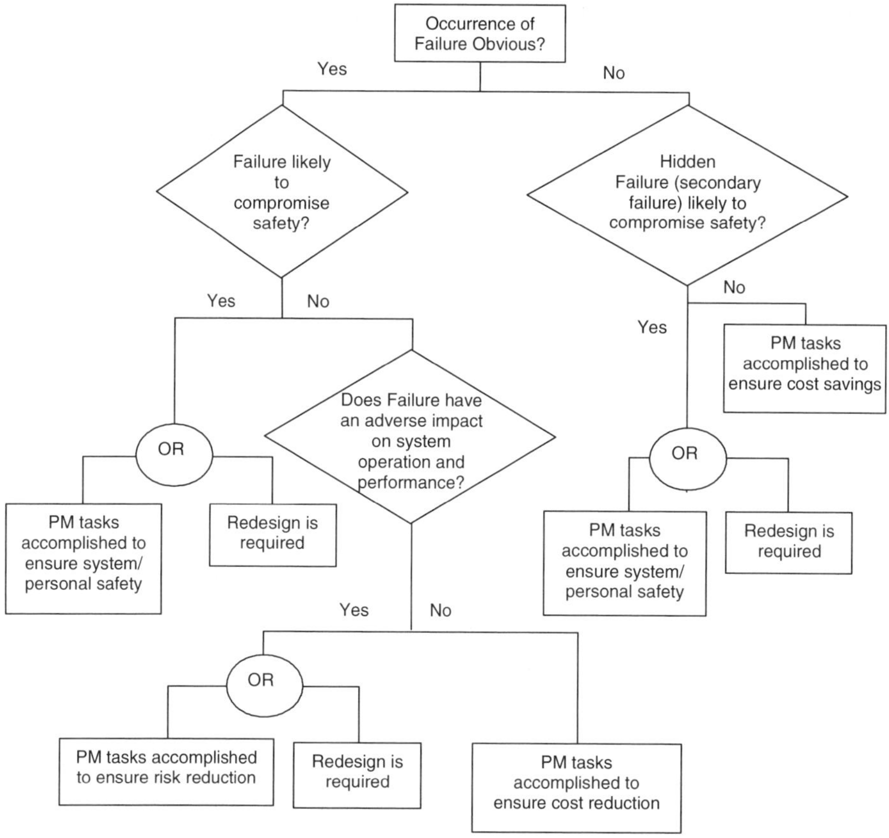

Figure 5.2.6-1. Simplified RCM Decision Logic

Analysis Results. Quite often in the design process there is a tendency to either: (1) ignore the issue of preventive maintenance all together; or (2) to "over-specify" the requirements for preventive maintenance, with the thinking that the more PM that is accomplished the better the system reliability. Experience has indicated that although the accomplishment of some selected PM is essential, the over-specification of PM activities can actually cause a degradation of system reliability and can be quite costly. The objective is to specify the correct amount of PM, to the depth required, and at the proper frequency; that is, not too much or too little.[14]

[14] Two recommended references to RCM include: (1) Moubray, J., *Reliability-Centered Maintenance*, 2nd Ed., Industrial Press, Boca Raton, FL, 1997; and (2) Smith, A.M., *Reliability-Centered Maintenance*, McGraw-Hill, New York, NY, 1992.

Topic 5.2.7: Level-of-Repair Analysis (LORA)

As the system maintenance concept is first developed early in design during the Concept Refinement Phase, gross estimates are made pertaining to whether or not an element of a given system is to be *repaired* or *discarded* in the event of failure. If the item is to be repaired, at what level should it be performed: i.e., at the intermediate level of maintenance or at the manufacturer, supplier, or at some depot? If discarded, the question pertains to whether it (or various parts of the item) can be recycled or disposed of through some form of decomposition? Such early decisions, applicable to high-level system elements, are based on a combination of economic, technical, safety, security, political, environmental, and other criteria. The case-study example discussed herein is based primarily on the use of economic (life-cycle cost) factors. Figure 5.2.7-1 illustrates the process to be implemented, first at a very gross level during the early aspects of conceptual and preliminary system design and later at a much more detailed level.

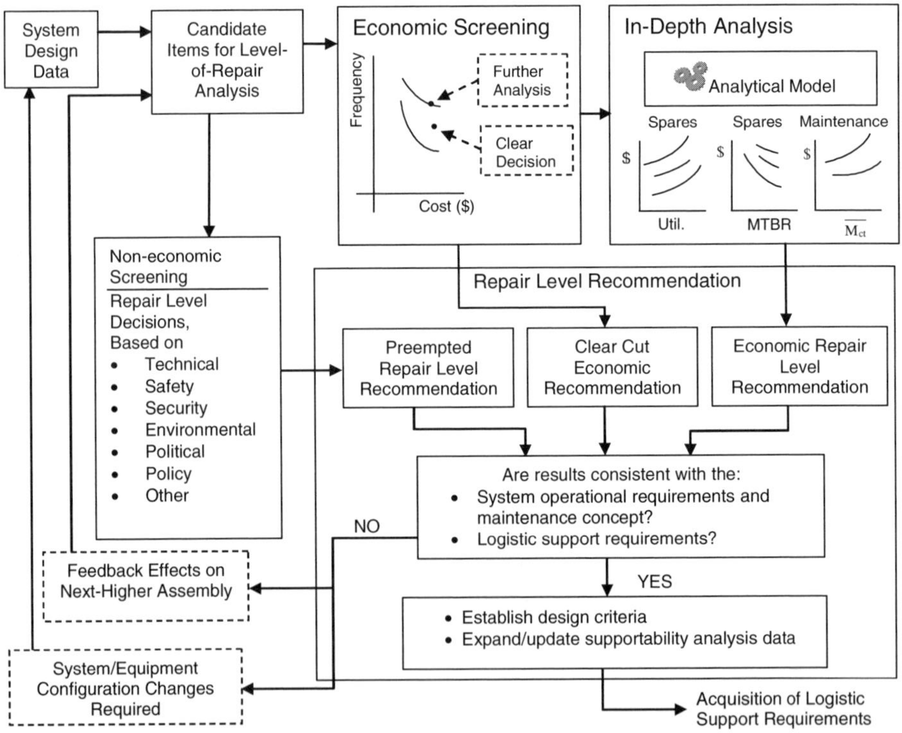

Figure 5.2.7-1. Level-Of-Repair Analysis (LORA) Process

Definition of the Problem. A computer sub-system is to be distributed in a quantity of 65 throughout three major geographical areas. Average daily usage is assumed to be 4 hours for a 360-day year. The computer sub-system is currently in early development, should be in production in 18 months, and will be operational in 2 years. All 65 computer sub-systems are expected to be operational and in use in 4 years, and will be utilized throughout the eighth year of the program before sub-system phase-out commences. The planned life cycle is 10 years.

Based on early design data, the computer sub-systems will be packaged in five major units, with a built-in test capability that will allow for fault isolation to the unit level in the event of failure. Faulty units will be removed and replaced at the organizational level (customer's facility) and sent to the intermediate maintenance shop for repair. Unit repair will be accomplished through assembly replacement, and the applicable assemblies will either be repaired or discarded. A total of 15 assemblies are being considered, and the requirement is to justify the repair versus discard decision on the basis of life-cycle cost criteria.

Analysis Approach. Referring to Table 5.2.7-1, the first of the 15 assemblies, Assembly *A-1*, was selected for the purposes of analysis, with the three alternatives being *repair at intermediate*, *repair at supplier*, and *discard at failure.* The cost categories, considered as being significant in this case, include *acquisition cost*, *unscheduled maintenance cost*, *supply support cost*, *test and support equipment cost*, and so on, as shown in the left column of the table. The applicable costs were determined for each alternative and, based on the evaluation of Assemble *A-1*, it appears as though the decision favors *repair at intermediate.*

As Assembly *A-1* is only one of the 15 assemblies in the unit being evaluated, it was decided to repeat the same process for Assemblies *A-2* through *A-15.* Referring to Table 5.2.7-2, the results of the individual evaluations for each of the 15 assemblies are shown. Note that there are variations as to the decision when viewing the results of the analysis on an individual-by-individual basis; i.e., *repair at intermediate* for Assembly *A-1*, *repair at supplier* for Assembly *A-2*, *discard at failure* for Assembly *A-3*, and so on.

Table 5.2.7-1. Repair Versus Discard Evaluation (Assembly A-1 Example)

Evaluation Criteria	Repair at IM[1] Cost ($)	Repair at SM[2] Cost ($)	DAF[3] Cost ($)	Description and Justification
1. Estimated acquisition costs for Assembly A-1 (to include R&D and production costs)	550/assy or 35,750	550/assy or 35,750	475/assy or 30,875	Acquisition cost includes all applicable costs in categories C_R, C_{PI}, C_{PM}, and C_{PQ} allocated to each Assembly A-1, based on a requirement of 65 systems. Assembly design and production are simplified in the discard area.
2. Unscheduled maintenance cost (C_{OLA})	6,480	8,100	NA	Based on the 8 year useful life, 65 systems, a utilization of 4 hours per day, a failure rate (λ) of 0.00045 for Assembly A-1 and a $\bar{M}_{ct}$ of 2 hours, the expected number of maintenance actions is 270. When a repair is accomplished, one technician is required on a full time basis. The labor rates are $12/hour for IM and $15/hour for SM.
3. Supply support – spare assemblies (C_{PLS} and C_{OLS})	3,300	4,950	128,250	For intermediate maintenance 6 spare assemblies are required to compensate for transportation time, the maintenance queue, TAT, etc. For supplier/depot maintenance, 9 spare assemblies are required. 100% spares are required in the discard case.
4. Supply support – spare modules or parts for assembly repair (C_{PLS} and C_{OLS})	6,750	6,750	NA	Assume $25 for materials per repair action.
5. Supply support – inventory management (C_{PLS} and C_{OLS})	2,010	2,340	25,650	Assume 20% of the inventory value (spare assemblies, modules and parts).

1. Intermediate Maintenance   2. Supplier Maintenance   3. Discard at Failure

IM – Intermediate Maintenance
SM – Supplier Maintenance
DAF – Discard at Failure

Table 5.2.7-1. Repair Versus Discard Evaluation (Assembly A-1 Example) (Cont'd)

Evaluation Criteria	Repair at IM[1] Cost ($)	Repair at SM[2] Cost ($)	DAF[3] Cost ($)	Description and Justification
6. Test and support equipment (C_{PLT} and C_{OLE})	5,001	1,667	NA	Special test equipment in the repair case. The acquisition and support cost is $25,000 per installation. The allocation for Assembly A-1 per installation is $1,667. No special test equipment is required in the discard case.
7. Transportation and handling (C_{OLH})	Not applicable	4,725	NA	Transportation costs at the intermediate level are negligible. For supplier maintenance, assume 540 one-way trips at $175/100 pounds. One assembly weighs 5 pounds.
8. Maintenance training (C_{OLT})	260	90	NA	Delta training cost to cover maintenance of the assembly based on the following: intermediate – 26 students, 2 hours each, $200/student week. Supplier: 9 students, 2 hours each, $200/student week.
9. Maintenance facilities (C_{OLM})	594	810	NA	From experience, a cost estimating relationship of $0.55 per direct maintenance labor hour is assumed for the intermediate level and $0.75 is assumed for the supplier level.
10. Technical data (C_{OLD})	1,250	1,250	NA	Assumes 5 pages for diagrams and text covering assembly repair at $250 per page.
11. Disposal (C_{DTS})	270	270	2,700	Assume $10 per assembly and $1/module or part as the cost of disposal.
Total estimated cost	61,665	66,702	187,475	

1. Intermediate Maintenance 2. Supplier Maintenance 3. Discard at Failure

IM – Intermediate Maintenance
SM – Supplier Maintenance
DAF – Discard at Failure

Table 5.2.7-2. Repair Versus Discard Cost Summary (15 Assemblies)

Assembly Number	Maintenance Status				Decision
	Repair at Intermediate Cost ($)	Repair at Supplier Cost ($)	Discard at Failure Cost ($)	Optimum Policy Cost ($)	
A-1	61,665	66,702	187,475	61,665	Repair – intermediate
A-2	58,149	51,341	122,611	51,341	Repair – supplier
A-3	85,115	81,544	73,932	73,932	Discard
A-4	85,778	78,972	65,072	65,072	Discard
A-5	66,679	61,724	95,108	61,724	Repair – supplier
A-6	65,101	72,988	89,216	65,101	Repair – intermediate
A-7	72,223	75,591	92,114	72,223	Repair – intermediate
A-8	89,348	78,204	76,222	76,222	Discard
A-9	78,762	71,444	89,875	71,444	Repair – supplier
A-10	63,915	67,805	97,212	63,915	Repair – intermediate
A-11	67,001	66,158	64,229	64,229	Discard
A-12	69,212	71,575	82,109	69,212	Repair – intermediate
A-13	77,101	65,555	83,219	65,555	Repair – supplier
A-14	59,299	62,515	62,005	59,299	Repair – intermediate
A-15	71,919	65,244	63,050	63,050	Discard
Policy Cost	1,071,267	1,037,362	1,343,449	983,984	

Analysis Results. Based on the results of this particular analysis, the decision favors a "mixed" policy, identified in the fifth column in Table 5.2.7-2; i.e., *optimum policy.* This decision was based on economic criteria only. However, as conveyed in Figure 5.2.7-1, there may be other non-economic factors that will influence the final decision.

Topic 5.2.8: Evaluation of Design Alternatives

There may be numerous instances when the evaluation of alternatives involves the selection of many different factors, against which each of the applicable alternatives is rated. Table 5.2.8-1 shows the results of an evaluation of three different system configurations, considering such factors as *performance*, *operability*, *effectiveness*, *support*, *design characteristics*, *design data*, *flexibility*, *life-cycle cost*, and so on.

Table 5.2.8-1. Evaluation Summary (Three Design Configurations)

Item	Evaluation Parameter	Weighting Factor	Config. A		Config. B		Config. C	
			Base Rate	Score	Base Rate	Score	Base Rate	Score
1	Performance-Compatibility, maintenance capability, maintenance load, method/mode of testing, versatility of testing	14	6	84	9	126	3	42
2	Operability – Simplicity of operation	4	10	40	7	28	4	16
3	Effectiveness – A_O, MTBM, $\bar{M}_{ct}$, $\bar{M}_{pt}$, MLH/OH	12	5	60	8	96	7	84
4	Design Characteristics – Reliability, maintainability, human factors, safety, transportability, producibility, interchangeability	9	8	72	6	54	3	27
5	Design Data – Design Drawings, Specifications, Logistics, Provisioning Data, Technical Manuals, Reports	2	6	12	8	16	5	10
6	Test Aids – General test equipment, calibration standards, maintenance tapes	3	5	15	8	24	3	9
7	Facilities and Utilities – Space, weight, volume, environment, power, heat, water, air conditioning	5	7	35	8	40	4	20
8	Spare/Repair Parts – Part type and quantity, standard parts, procurement time	6	9	54	7	42	5	30
9	Flexibility/Growth Potential – Test accuracies, performance range, space, reconfiguration, design change acceptability	3	4	12	8	24	6	18
10	Schedule – R&D, production	17	7	119	8	136	9	153
11	Cost – Life Cycle (R&D, investment, O&M)	25	10	250	9	225	5	125
Subtotal		-	-	753	-	811	-	534
Derating Factor (Development Risk)		-	-	113 15%	-	81 10%	-	107 20%
Grand total		100	-	640	-	730	-	427

Definition of the Problem. Company *ABC* is responsible for the design and development of a major system that, in turn, comprises a number of large sub-systems. Sub-system *XYZ* is to be procured from an outside supplier, and three different configurations are being evaluated for selection. Each configuration represents an existing design, with some re-design and additional development necessary to be compatible with the requirements for the new system.

Analysis Approach. The first step is to develop a list of evaluation parameters, as shown in the second column in Table 5.2.8-1. In this instance, there is no single parameter (or figure of merit) that is appropriate by itself, but there are 11 factors that must be considered on an integrated basis. Given the evaluation parameters, the next step is to determine the level of importance of each. Quantitative weighting factor from zero to 100 are assigned to each parameter in accordance with the degree of importance. The QFD, Delphi method, or some equivalent technique, may be used to help establish the appropriate weighting factors.[15]

For each of the 11 evaluation parameters, one should develop a special checklist, including the criteria against which to evaluate each of the three proposed configurations. For instance, the parameter *performance* may be described in terms of "highly desirable," "desirable," or "less desirable." Although each configuration must comply with a minimum set of requirements, one alternative may be more desirable than the next in terms of the proposed performance characteristics.

Each of the three proposed configurations of Sub-system *XYZ* is evaluated independently using the special checklist criteria. Base-rating values from 0 to 10 are applied according to the degree of compatibility with the desired goals. If a "highly desirable" evaluation is realized, a rating of 10 is assigned. The base-rating values are multiplied by the weighting factors to obtain a score, and a total score is then obtained for each option.

Analysis Results. Referring to Table 5.2.8-1, Configuration *B*, represents the preferred approach based on the highest score of 730 points. This configuration is recommended on the basis of its inherent built-in features (attributes) relating to performance, operability, effectiveness, design characteristics, design data, and so on.

[15] The selection of evaluation criteria should reflect the results from the identification and prioritization of the TPM/PBL factors described in Topic 3.2.4.

Topic 5.3: Logistics Management Information (LMI)

To accomplish the supportability analysis (SA) objectives described in Topic 1.6.3, a robust and comprehensive integrated database must be established from the beginning. First, it is essential that the appropriate system design data be generated, processed, and made available in a timely manner as the system development process evolves from the Concept Refinement Phase and through the Technology Development and System Development and Demonstration Phases. The designer and the logistics analyst must have access to a good database in order to be an effective member of the design team, and be able to contribute to the design process, in a timely manner, and on a day-to-day basis. Such data must not only include that data which are generated within the organizational group at the same geographical location, but also key supplier data as well. In particular, the emphasis here pertains to the first three objectives described in Topic 1.6.3; i.e., Items 1, 2, and 3.

Second, such data must be complete enough in describing the system design configuration at the time when an evaluation is required to determine the logistics resources that will be necessary to support the system throughout its planned life cycle. While the design configuration will likely change, as one evolves through the overall system development process, there must be an established "baseline" for the purposes of evaluation, assessment, and to enable the determination of the requirements for maintenance personnel, spares/repair parts, test and support equipment, facilities, maintenance software, technical documentation, information, etc.; i.e., the elements of logistics identified in Figure 1.3-5. The emphasis here is in response to the last three SA objectives described in Topic 1.6.3; i.e., Items 4, 5, and 6.

While there is no set format specified for logistics data for all activities across-the-board, there are some guidelines relative to the development of Logistics Management Information (LMI) for major defense systems.[16] Referring to Appendix *A* of MIL-PRF-49506, LMI data should include the following "SA Summaries:"

1. *Maintenance planning.* Includes maintenance planning information that may be used to develop initial fielding plans for the various items within the logistics and maintenance support infrastructure, and to verify those maintenance actions that have been specified through the development of the initial maintenance concept.
2. *Repair analysis.* Includes the results from the level-of-repair analysis (LORA), and the identification of the various elements and components of the

[16] MIL-PRF-49506, "Performance Specification, Logistics Management Information (LMI)," Department of Defense, Washington, DC, 11 November 1996. Also, refer to MIL-HDBK-502, "Department of Defense Handbook, Acquisition Logistics," Department of Defense, Washington, DC, 30 May 1997.

system that are recommended for repair, and where the repair is to be accomplished.

3. *Support and test equipment.* Includes a listing and a brief description of the recommended test and support equipment identified through the SA, where such items will be required, and for how long.
4. *Manpower, personnel, and training.* Includes a listing of recommended personnel quantities, skill-levels, and training requirements for the operation and support of the system throughout its planned life cycle.
5. *Facilities.* Includes a listing of and the basic requirements for facilities/utilities at all levels of maintenance (i.e., organizational, intermediate, depot/supplier, training facilities, testing facilities, etc.).
6. *Packaging, handling, storage, and transportation.* Includes a description of all of the requirements for packing, shipping, transportation, and handling in support of all maintenance activities throughout the system life cycle.
7. *Post production support.* Includes the identification of the elements of logistics support (e.g., equipment, software, spares/repair parts and their sources of supply) needed for the sustaining maintenance of the system throughout its life cycle and after the initial prime production capability has terminated its activity.

While these seven areas of activity, as listed in MIL-PRF-49506, represent the major items of logistics support, and particularly those relevant to the defense sector, the objective in developing a good LMI database is to cover all of the needed information for the overall maintenance and support of various categories of systems throughout their respective life cycles.

For More Information:

1. Blanchard, B.S., *Logistics Engineering and Management,* 6th Ed., Pearson Prentice Hall, Upper Saddle River, NJ, 2004.
2. Blanchard, B.S. and W.J. Fabrycky, *Systems Engineering and Analysis*, 3rd Ed., Pearson Prentice Hall, Upper Saddle River, NJ, 1998.
3. Langford, J.W., *Logistics Principles and Applications*, McGraw-Hill, New York, NY, 1995.
4. MIL-STD-882D, "Department of Defense Standard Practice for System Safety," Department of Defense, Washington, DC, 19 January 2000.
5. MIL-PRF-49506, "Performance Specification, Logistics Management Information (LMI)," Department of Defense, Washington, DC, 11 November 1996.
6. MIL-HDBK-502, "Department of Defense Handbook, Acquisition Logistics," Department of Defense, Washington, DC, 30 May 1997.

Section 6: System Test and Evaluation (Validation)

Insight

Test and Evaluation refers to the examination and judgment of a system (or elements of a system) to ascertain its worth, quality of performance, degree of effectiveness, condition, and related essential characteristics. Evaluation is a critical purpose of the testing process which begins during conceptual design and extends through the period of customer use of the system (or product), including the evaluation of the logistics and maintenance support infrastructure.

As the system evolves through design and development, ongoing measurement and evaluation (or validation) efforts are necessary. In actuality, a complete evaluation of the system in terms of meeting the initially specified user requirements cannot be realistically accomplished until the system is produced, delivered, and functioning in an operational environment. It is underscored that, if problems occur, performing the corrective modifications too far downstream in the life cycle may turn out to be very costly and perhaps technically impractical. In essence, the earlier the problems are detected and corrected, the better off one is in terms of both incorporating the required changes and decreasing the costs of the required modifications. When addressing the subject of evaluation, it must be recognized that the objective is to acquire a high degree of confidence, as early as possible in the life cycle, that the system will perform as intended. In this regard, a certain amount of formal testing is necessary to properly verify that the system performance requirements have been met.

This page intentionally left blank

Topic 6.1: System Test Requirements

The process of determining the needs for test and evaluation begins with the initial specifications of system requirements in conceptual design. As specific *technical performance measures (TPMs)* are defined, it is necessary to determine the methods by which compliance with these parameters will be verified. This entails determining how the pertinent system TPMs will be measured as well as the resources required. The approach may be the use of simulation and related analytical methods, employing an engineering model for test and evaluation purposes, testing a production model, evaluating an operational configuration in the user's environment, or using a combination of the aforementioned techniques. It is necessary to review the requirements for the system, determine the methods that can be used in the evaluation as well as the anticipated effectiveness of those methods, and develop a comprehensive plan for an overall integrated test and evaluation. Figure 6.1-1 provides a picture of the conventional categories of testing as they are applicable in system test and evaluation. The types of testing will be detailed in Topic 6.1.2. A proper test and evaluation program entails a process of preparation and a sequential series of individual categories of tests governed by the phases in the system life cycle.

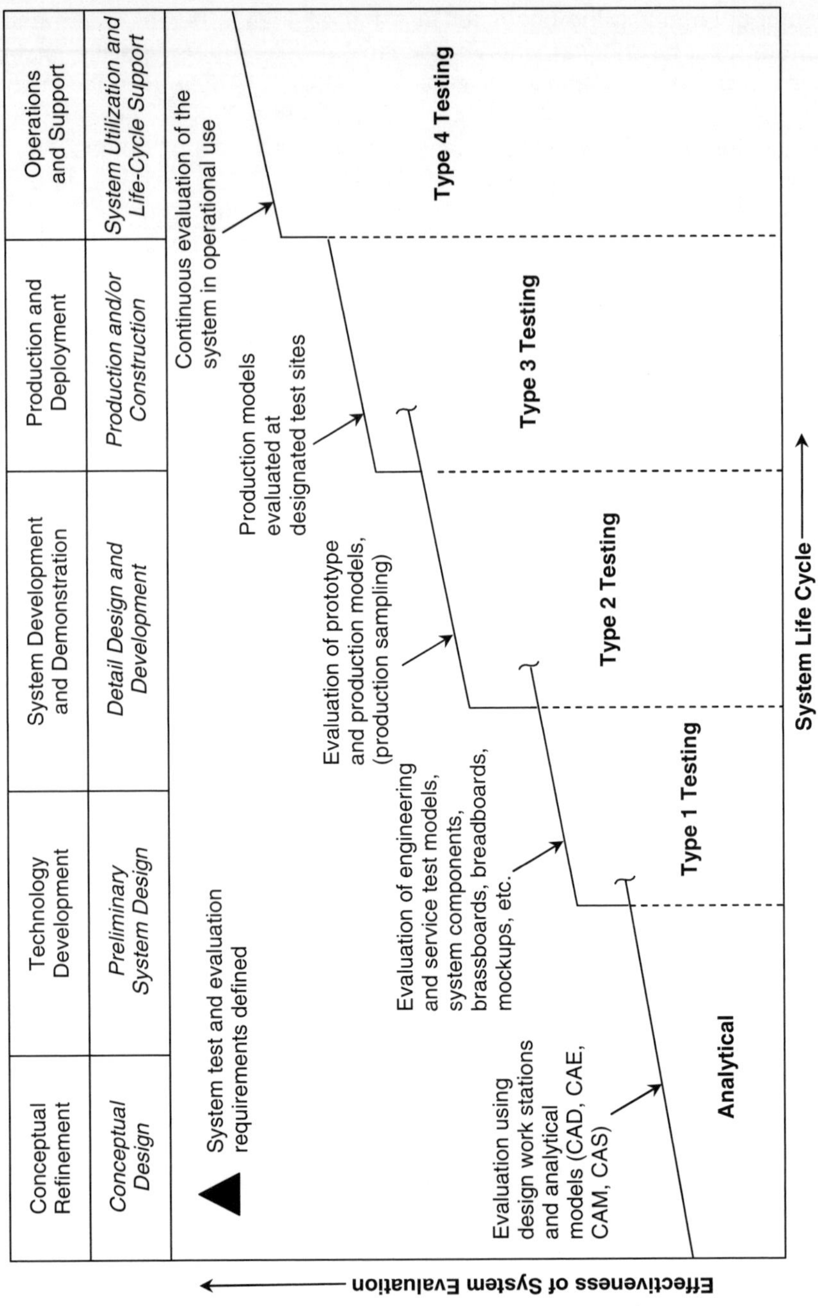

Figure 6.1-1. Stages of System Test and Evaluation during the System Life Cycle

Topic 6.1.1: Preparation for System Test and Evaluation

Prior to the start of formal testing an appropriate time period is designated for preparation. During this period, the proper conditions must be established to ensure valid results. These conditions may vary depending upon the category of testing. During the early phases of design and development, as analytical evaluations and Type 1 testing are performed, the extent of test preparation is minimal. Conversely, performance of Type 2 and Type 3 testing, for which the conditions are designed to simulate realistic user operations as far as possible, will likely require extensive preparation. To promote a realistic scenario for test and evaluation, the following factors need to be addressed:[1]

1. *Selection of a Test Item.* The system and its components selected for testing should represent the most current design or production configuration which incorporates the latest approved engineering changes.
2. *Selection of Test Site.* The system should be tested in the environments that will be representative of the user's environment; e.g., the arctic or tropics, desert, flat, or mountainous terrain, airborne, or ground. The test site selected should simulate these conditions as much as possible.
3. *Testing Procedures.* The achievement of test objectives usually involves the accomplishment of both operator and maintenance tasks, and the completeness of these tasks should conform to normal procedures (e.g., validated technical manuals). The recommended task sequences must be followed to ensure proper system operation.
4. *Test Personnel.* This group includes (a) the individuals who will actually operate and maintain the system throughout the test, and (b) support engineers, data recorders, analysts, and administrators who provide assistance in conducting the overall test program. Personnel selected for the first category should be representative of user requirements with respect to the recommended quantities, skill levels, and supporting training requirements.
5. *Test and Support Equipment.* The performance of system operational and maintenance tasks may require the use of ground handling equipment, support and test equipment, software, and/or a combination of these elements. Only those items that have been approved for operation should be used.
6. *Supply Support.* This includes all spares, consumables, and supporting inventories that are essential for the completion of system test and evaluation. A realistic configuration projected into a "real-world" environment is highly recommended.

---

[1]Blanchard, B.S., *Logistics Engineering and Management*, 6th Ed., Pearson Prentice Hall, Upper Saddle River, NJ, 2004, pages 184-185.

7. *Test Facilities and Resources.* The conduct of system testing may require the use of special facilities, test chambers, capital equipment, environmental controls, special instrumentation, and associated resources; e.g., heat, water, air conditioning, power, telephone. These facilities and resources must be properly identified and scheduled.

The nature of test preparation is highly dependent on the ultimate objectives of the test and evaluation. The aforementioned considerations are important to the successful achievement of such objectives.

Topic 6.1.2: Categories of Test and Evaluation

Test and evaluation during the system life cycle encompasses the *Analytical* test, *Type 1 testing*, *Type 2 testing*, *Type 3 testing,* and *Type 4 testing*. It is noted that, within the Department of Defense, Development Test and Evaluation (DT&E) basically equates to the Analytical, Type 1, and Type 2 testing, and Operational Test and Evaluation (OT&E) is equivalent to Type 3 and Type 4 testing.[2]

1. *Analytical Test.* Figure 6.1-1 shows that the first category is the *analytical* phase, which relates to certain design evaluations that can be conducted early in the system life cycle using computerized techniques including computer-aided design (CAD), computer-aided manufacturing (CAM), computer-aided support (CAS), simulation, rapid prototyping, and related approaches. With the availability of an extensive variety of models, three-dimensional databases, etc., design engineers are able to simulate human-machine interactions, equipment packaging schemes, hierarchical structures of systems, and activity/task sequences. By virtue of these capabilities, design engineers are able to do a better job of predicting, forecasting, and performing sensitivity/contingency analyses, thereby reducing future risks. Through such greatly enhanced technology, much can now be accomplished in system test and evaluation that previously could not be realized until the equipment and software became available in the latter phases of detail design and development.
2. *Type 1 Testing.* Type 1 testing concerns primarily the evaluation of system components in a laboratory environment using engineering breadboards, bench test models, service test models, rapid prototyping, and similar devices. These tests are developmental by nature and are designed primarily for the purpose of verifying performance and physical characteristics of the system components. The test models used can operate functionally but do not represent production equipment or software. Such testing is usually performed in a producer/supplier's laboratory facility by engineering technicians using jury-rigged test fixtures and engineering notes for procedures. Design concepts and technological applications are validated during this initial testing phase, and changes can be initiated with minimum cost impact.
3. *Type 2 Testing.* Type 2 testing includes formal tests and demonstrations performed during the latter stages of design and development (System Development and Demonstration Phase) when pre-production prototype equipment and software are available. Prototype equipment is similar to production equipment that will be delivered for operational use, but is not necessarily "qualified" by virtue of successful completion of environmental

[2] Blanchard, B.S., *Logistics Engineering and Management*, 6th Ed., Pearson Prentice Hall, Upper Saddle River, NJ, 2004, pages 180-183.

qualification tests (e.g., temperature cycling, shock and vibration), reliability qualification, maintainability demonstration, and supportability compatibility tests. Type 2 testing refers primarily to activities associated with the initial qualification of the system. A test program in this area may consist of a series of individual tests, tailored to the specific need, as described in the following:

a. *Environmental Qualification.* Temperature cycling, shock and vibration, humidity, sand and dust, salt spray, acoustic noise, explosive proofing, and electromagnetic interference.

b. *Reliability Qualification.* Sequential testing, life testing, environmental stress screening (ESS), and test, analyze, and fix (TAAF).

c. *Maintainability Demonstration.* Verification of maintenance tasks, task times and sequences, maintenance personnel quantities and skill levels, degree of testability and diagnostic provisions, prime equipment-test equipment interfaces, maintenance procedures, and maintenance facilities.

d. *Support Equipment Compatibility.* Verification of the compatibility among the prime equipment, test and support equipment, and ground handling equipment.

e. *Technical Data Verification.* The verification (and validation) of operating procedures, maintenance procedures, and supporting data.

f. *Personnel Test and Evaluation.* Verification to ensure compatibility between the human and equipment, personnel quantities and skill levels required, and training requirements.

g. *Software Compatibility.* Verification that software meets the system requirements, that software and hardware are compatible, and that the appropriate quality provisions have been incorporated. This test includes computer software unit (CSU) and computer software configuration item (CSCI) testing.

h. *Logistics Validation.* Validation of various logistics processes such as procurement, materials handling, transportation, warehousing and distribution, and information (i.e., the factors discussed in Topics 2.5.1, 2.5.2, 2.5.3, and 2.5.4, as applicable).

Another aspect of Type 2 testing is production sampling tests, employed when multiple quantities of an item are being produced. Although the system and its components might have successfully passed the initial qualification tests, it is necessary to ensure that the same level of quality has been maintained throughout the production process. The process can be dynamic by nature because conditions may change and there is no guarantee that the characteristics built into the design will be retained throughout production. Accordingly, a sample of products/components, determined by a percentage of the total production, may be selected for qualification tests to be conducted on a recurring basis. The results are measured and evaluated in terms of whether improvement or degradation has occurred.

4. *Type 3 Testing.* Type 3 testing entails the accomplishment of formal tests at designated field sites by user personnel over an extended period of time. These tests are typically conducted after initial system qualification and prior to completion of production/construction (Production and Deployment Phase). Operating personnel, operational test and support equipment, operational spares, applicable computer software, and validated operating and maintenance procedures are used. This test marks the first time all elements of the system (i.e., prime equipment, software, and the elements of support) are operated and evaluated on an integrated basis. A series of simulated operational exercises are usually conducted and the system evaluated in terms of such attributes as performance, effectiveness, as well as the compatibility between the prime mission-oriented segments of the system and the elements of support. Although Type 3 testing does not totally represent a fully operational situation, the tests can be designed to provide an effective approximation.

5. *Type 4 Testing.* Type 4 testing, conducted during system utilization (Operations and Support Phase), includes formal tests that are often conducted to acquire specific information relative to a certain area of operation or support. The purpose is to gain further insight about the system in the user environment, or of the user operations in the field. It may be feasible to vary the mission profile or the system utilization rate to determine the impact on the overall system effectiveness, or it may be advisable to evaluate several alternative maintenance support policies to ascertain if system operational availability can be improved. Type 4 testing should be performed at one or more user operational sites, in a realistic environment, by operator and maintenance personnel, and with support through normal logistics and maintenance capability. This is the first time the true capability of the system is defined.

Topic 6.2: Supportability Test and Evaluation

The goal in *design for supportability* is (1) to verify that various system elements are designed for supportability and (2) to ensure that sufficient logistics support is provided for the system in a timely and cost-effective manner throughout the life cycle. Specific quantitative and qualitative goals are established during conceptual design, analyses are performed during detail design and development, a fixed design configuration is assumed based on such analyses, a prototype system is produced, and test and evaluation is accomplished to compare the test results with the initially specified performance requirements. The core process of supportability test and evaluation entails reliability test and demonstration, maintainability test and demonstration, human factors test and evaluation, and validation of the logistics elements in the support infrastructure.

Topic 6.2.1: Reliability Testing

Reliability testing is conducted as part of the system test and evaluation process described in Topic 6.1.2 and portrayed by Figure 6.1-1. Specifically, reliability testing is accomplished under the categories of Type 2 and Type 3 testing. The objective of reliability testing typically is to determine if the system under test meets the specified *mean time between failure* (MTBF) requirements. To accomplish this, the system is operated in a specified manner for a designated period and failures are recorded and evaluated as the test progresses. Acceptance of the system is based on the system demonstrating a minimum acceptable life. There are a number of test methods and statistical procedures designed to measure and validate system reliability, most of which assume the applicability of the exponential distribution. To impart some understanding of the approaches used, the following will address reliability sequential qualification testing, reliability acceptance testing, and life testing.[3]

1. *Reliability Sequential Qualification Testing.* The purpose of reliability qualification testing is to provide evaluation of developmental progress, as well as the assurance that specified requirements have been met prior to proceeding to the next phase (i.e., the Production and Deployment Phase of the life cycle). Initially, a predictive reliability parameter for the MTBF is established for the system, followed by allocation of design criteria. If the predictions indicate compliance, the design process progresses to the construction of a prototype or pre-production model of the system and the commencement of qualification testing. The system under test is operated in a manner reflective of the mission cycles in a realistic operational environment. During a reliability sequential test, there are three possible decisions: (1) accept the system, (2) reject the system, or (3) continue to test. Figure 6.2.1-1 illustrates a typical sequential test profile.

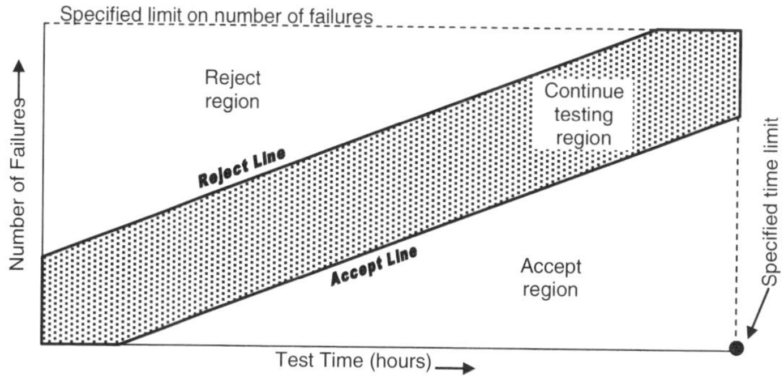

Figure 6.2.1-1. Sequential Test Plan

[3] Blanchard, B.S. and Fabrycky, W.J., *Systems Engineering and Analysis*, 3rd Ed., Upper Saddle River, NJ, 1998, pages 390-395.

Figure 6.2.1-2 portrays a sample mission profile, during which the system is operated through a series of the prescribed duty cycles under specified environmental conditions for a designated time period.

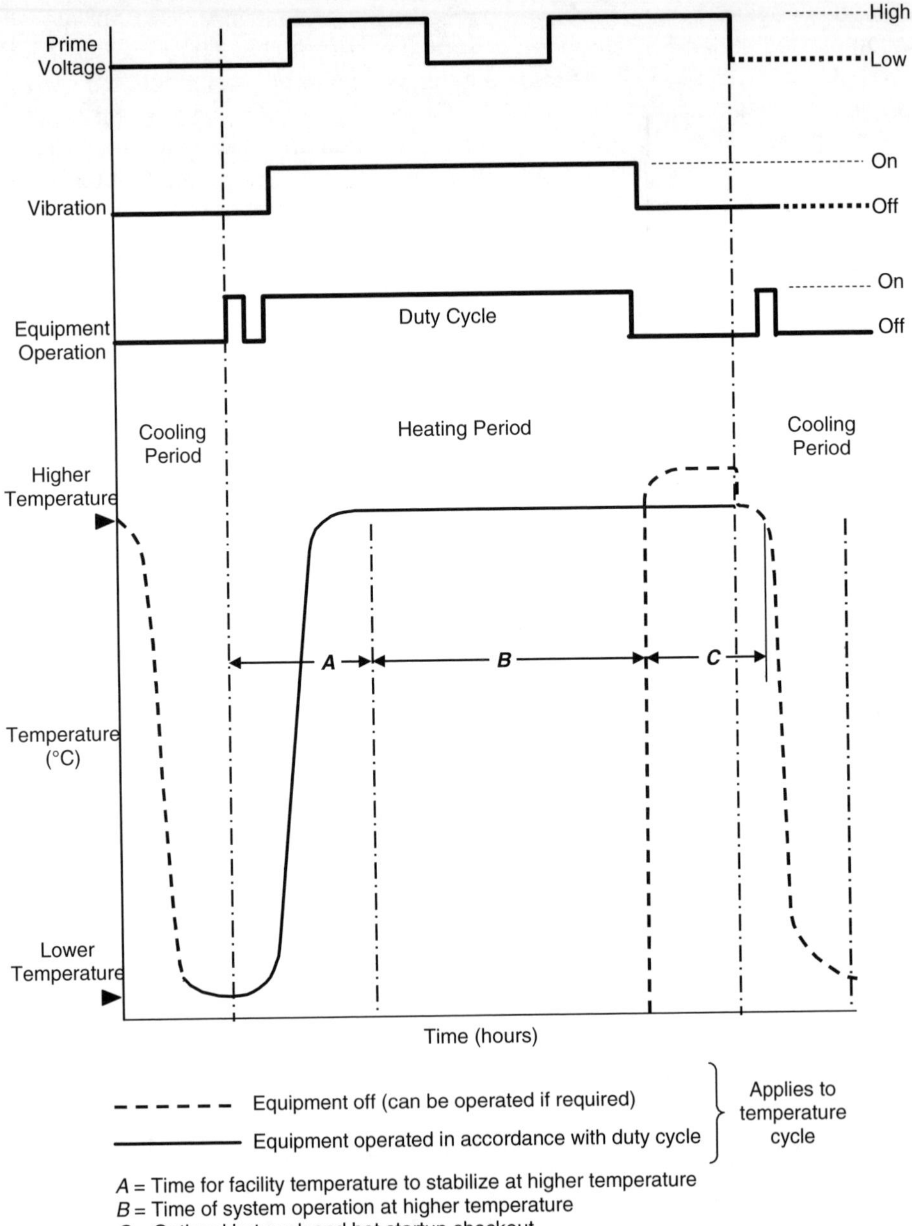

Figure 6.2.1-2. Sample Environmental Test Cycle

Referring to Figure 6.2.1-1, system operating hours accumulate along the abscissa and failures are plotted against the ordinate. If enough operating time is generated without excessive failures, a decision is made to accept the system and testing is discontinued. If there are a significant number of failures early in testing, a decision to reject will be made and the system will be considered unacceptable. A marginal condition results in a decision to continue testing until a designated point in time. Essentially, the sequential test approach allows for an early decision. If the system is highly reliable, this will be evident early in the process. If the system is not reliable, this will also be evident at an early point of time. If the system inherent reliability is marginal, the amount of test time involved could be extensive and costly. An actual example of sequential testing is described later.

2. *Risk Parameters.* Sequential test plans are influenced by the risks that the producer and user (customer) are willing to accept relative to the decisions made as a result of testing.

    a. The risks associated with test plans are defined as follows:

        i. *Producer's Risk (α).* The probability of rejection of a system when the measured MTBF is equal to or greater than the specified MTBF. This notes the probability of rejecting a system when it should be accepted, and which constitutes a risk to the system manufacturer or producer. This is also known as a Type I error.

        ii. *User's or Consumer's Risk (β).* The probability of accepting a system when the measured MTBF is less than the specified MTBF. This refers to the probability of accepting a system that should be rejected and which constitutes a risk to the user. This is known as a Type II error.

    b. The probability of an incorrect decision on the basis of test results is necessarily addressed statistically in a manner similar to hypothetical testing. An assumption is made and a test is performed to support or disprove that assumption. A null hypothesis (H_o) is established. This is a statement about a parameter such as "the true MTBF is equal to 100 hours". The alternative hypothesis (H_1) would be "the MTBF is not equal to 100 hours". When testing an item representing a sample of a total population, the question is whether or not to accept H_o. The desired result is to accept when the null hypothesis is true and reject when false, or to minimize the likelihood of making an incorrect decision. The relationship of risks in sample testing is illustrated in Table 6.2.1-1. In order to select a sequential test plan, it is first necessary to identify two values of MTBF:

        i. Specified MTBF representing the system requirement (assume that θ_0 represents this value).

ii. Minimum MTBF that is considered to be acceptable based on the results of testing (assume that θ_1 represents this value).

Table 6.2.1-1. Risks in Sample Testing – Test for MTBF=100

True State of Affairs	Accept H_0 and Reject H_1 (i.e., MTBF = 100)	Reject H_0 and Accept H_1 (i.e., MTBF $\neq$ 100)
H_0 is true (i.e., MTBF = 100)	High probability = 1-α (i.e., 0.90)	Low probability of error = α (i.e., 0.10)
H_0 is false and H_1 is true (i.e., MTBF $\neq$ 100 hours)	Low probability of error = β (i.e., 0.10)	High probability = 1-β (i.e., 0.90)

Given θ_0 and θ_1, it is necessary to next determine the values of the producer's risk (α) and consumer's risk (β). Most test plans currently in use accept risk values ranging between 5% and 25%. For example, a test where $\alpha = 0.10$ means that, on average, in 10 of 100 instances the test plan would be expected to reject items that should been accepted. Risk values are usually negotiated during the test planning phase.

Referring to Figure 6.2.1-1, the design of the sequential test plan is based on the values of θ_0, θ_1, α, and β. The *accept line* slope is derived from the following expression:

$$t_1 = \frac{\ln(\theta_0 / \theta_1)}{(1/\theta_1) - (1/\theta_0)}(r) - \frac{\ln[\beta / (1 - \alpha)]}{(1/\theta_1) - (1/\theta_0)} \tag{6.1}$$

where r is the number of failures (the accept line is plotted for the assumed value of r). The ratio of θ_0 to θ_1 is known as the *discrimination ratio,* a standard test parameter. The *reject line* slope is based on Equation (6.2).

$$t_2 = \frac{\ln(\theta_0 / \theta_1)}{(1/\theta_1) - (1/\theta_0)}(r) - \frac{\ln[(1 - \beta)/ \alpha]}{(1/\theta_1) - (1/\theta_0)} \tag{6.2}$$

Determination of the specified time limit in Figure 6.2.1-1 is generally based on a multiple of θ_1.

The example in Figure 6.2.1-3 illustrates one of a number of different sequential test plans used for reliability testing[4]. In this example, the established α and β decision risks are 10% (each), the discrimination ratio is

[4] Reference MIL-HDBK-781A, 1 April 1996.

1.5:1, and the accept-reject criteria are as specified and portrayed by the figure.

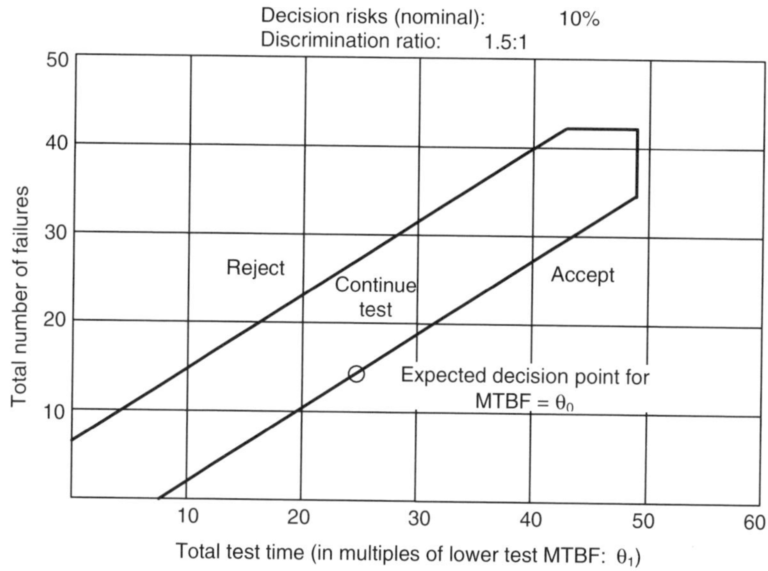

Number of Failures	Total Test Time* Reject (equal to less)	Total Test Time* Accept (equal or more)	Number of Failures	Total Test Time* Reject (equal or less)	Total Test Time* Accept (equal or more)
0	N/A	6.60	21	18.92	32.15
1	N/A	7.82	22	20.13	33.36
2	N/A	9.03	23	21.35	34.58
3	N/A	10.25	24	22.56	35.79
4	N/A	11.46	25	23.78	37.01
5	N/A	12.68	26	24.99	38.22
6	0.68	13.91	27	26.21	39.44
7	1.89	15.12	28	27.44	40.67
8	3.11	16.34	29	28.65	41.88
9	4.32	17.55	30	29.85	43.10
10	5.54	18.77	31	31.08	44.31
11	6.75	19.98	32	32.30	45.53
12	7.97	21.20	33	33.51	46.74
13	9.18	22.41	34	34.73	47.96
14	10.40	23.63	35	35.94	49.17
15	11.61	24.84	36	37.16	49.50
16	12.83	26.06	37	38.37	49.50
17	14.06	27.29	38	39.59	49.50
18	15.27	28.50	39	40.82	49.50
19	16.49	29.72	40	42.03	49.50
20	17.70	30.93	41	49.50	N/A

*Total test time is total unit hours of equipment on test and is expressed in multiples of the lower test MTBF.

Figure 6.2.1-3. Accept-Reject Criteria for a Sample Test Plan

Another approach is illustrated in Figure 6.2.1-4, representing actual experience in testing a specific system (hypothetically designated as System *ABC*). Referring to the figure, the specified MTBF for the system is 400 hours, and the maximum designated test time used for the sequential test plan is 4,000 hours, or a multiple of ten times the specified MTBF.

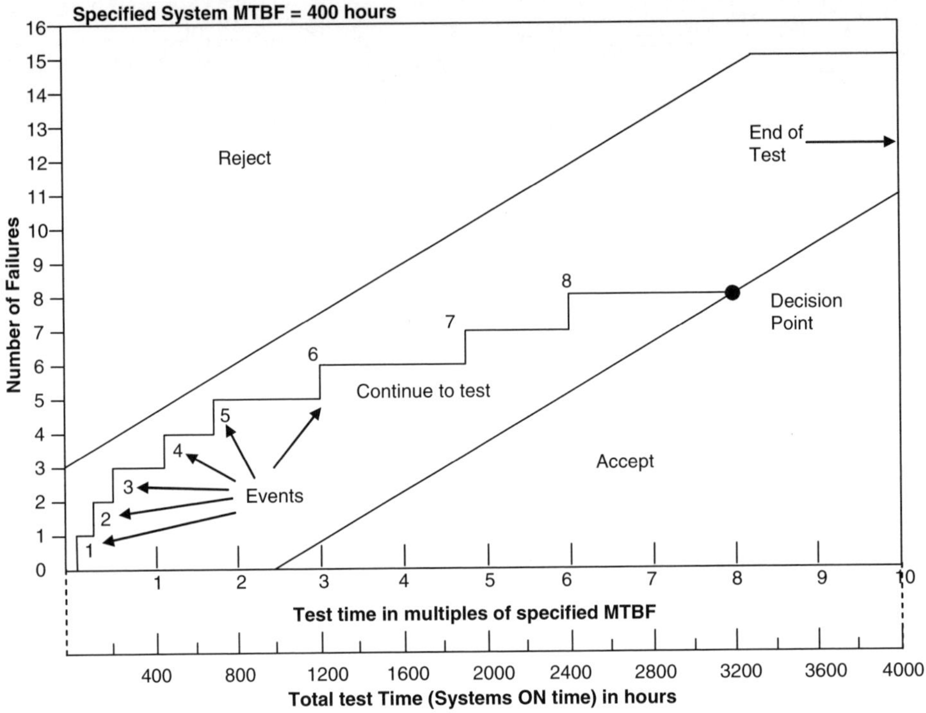

Figure 6.2.1-4. System ABC Reliability Test Plan

The test approach involves the selection of a designated quantity of systems (equipments), operating the system under prescribed performance conditions over an extended period of time, and monitoring the system for failure.

As failures, noted as events, occur, the affected system is repaired using the appropriate corrective maintenance actions, after which it is returned to full operational status for continuation of testing. An analysis of each event should determine the cause of failure, and trends may be established if more than one failure is traceable to the same cause. The latter is commonly referred to as a pattern failure and, in such cases, an engineering design change may be initiated to preclude the recurrence of failures of the same type in future testing and, eventually, operational use.

By inspection of Figure 6.2.1-4, it is observed that the system is accepted after 3,200 hours of testing, and eight events (failures) are recorded. Thus, the specified system MTBF of 400 hours is achieved. In this instance, the test sample is six units of the equipment. If a single model would be designated for testing with a minimum of 4,000 hours test time required, the length of the test program (assuming continuous resting) would obviously extend beyond six months. A test of this length might not be feasible in terms of the overall program schedule and delivery of qualified equipment. On the other hand, the use of two or more models will result in a shorter test program. The implementation of accelerated test conditions may also shorten the test period by accelerating the rate at which failures occur. These efforts should be evaluated, and an optimum balance established between a smaller sample size with longer test times and a larger sample with shorter test times.

3. *Reliability Acceptance Testing.* As previously indicated for Type 2 testing, reliability testing may be performed as part of qualification testing prior to starting full-scale production and during full-scale production on a sampling basis. To determine the effects of the production process on system reliability, it may be feasible to select a sample number of equipments from each production lot and test them in the same manner as previously described. The sample may be based on a percentage of the total equipments spread over the entire production period, or a set number of equipment(s) selected during a specified calendar time period (e.g., three items of equipment per month throughout the production phase). The selected equipment is tested and an assessed MTBF is derived from the test data. This value is compared against the specified MTBF and the measured value determined from earlier qualification testing. Growth MTBF trends (or negative trends) may be determined by plotting the resultant values as testing progresses.

4. *Reliability Life Testing.* The two basic forms of life testing are:

    a. Life tests based on a fixed-test time.

    b. Life tests based on the occurrence of a predetermined number of failures.

    In the first approach to life testing (based on time), a fixed test time is computed and a specified number of failures is predetermined. The system is accepted if the actual number of failures at the end of the scheduled test time is equal to or less than the predetermined quantity of failures.

    In the second approach to life testing (based on failures), a test plan is developed that specifies a predetermined number of failures and a computed test time based on an expected system failure rate. Testing continues until the specified quantity of failures occurs. The system is accepted if the test time is equal to or is greater than the computed time at the point of last failure.

Topic 6.2.2: Maintainability Test and Demonstration

Maintainability Demonstration generally is accomplished as part of Type 2 testing during the latter part of detail design to verify that qualitative and quantitative maintainability requirements have been achieved. The process should be accomplished in an environment that simulates as much as possible the operational and maintenance environment contemplated for the system. The maintainability demonstration may vary depending on system performance specifications and mission objectives. Two representative approaches described in the following are *demonstration method 1* and *demonstration method 2*.[5,6]

1. *Demonstration Method 1.* This method follows a sequential test approach similar to that for the reliability test described in Topic 6.2.1. Two different sequential test plans are used to demonstrate $\overline{M}_{ct}$ and M_{max} (for corrective maintenance). The test plans assume that the underlying distribution of corrective maintenance task times is log-normal. This method allows for a decision when the maintainability of the equipment under test is either far above or far below the specified values of $\overline{M}_{ct}$ and M_{max}. Maintainability testing involves the following steps:
    a. A failure is induced in the equipment without knowledge of the test team.
    b. The maintenance technician will be called to operationally check out the equipment. During the procedure, a symptom of malfunction should be detected.
    c. After detection, the maintenance technician, following the prescribed maintenance procedures with proper test and support equipment, should proceed to accomplish the necessary corrective maintenance actions as portrayed in Figure 2.2.1-1.
    d. While the technician is accomplishing the corrective maintenance tasks, a test recorder collects data on task sequences, areas of difficulty, and task times, and notes the adequacies and inadequacies of logistics support.

   The maintainability test cycle is performed *n* times, where *n* is the selected sample size. For sequential testing there could be as many as 100 demonstrations, assuming that a "continue-to-test" decision is part of the test. It is, therefore, noted that a sample size of 100 demonstrations should be planned. The task sample should be representative, with the selected tasks apportioned according to the anticipated percentage contribution of the

---

[5] Blanchard, B.S. and Fabrycky, W.J., *Systems Engineering and Analysis*, 3rd Ed., Upper Saddle River, NJ, 1998, pages 448-454.

[6] MIL-HDBK-470A, 4 August 1997.

individual tasks to the total maintenance requirements. Those items with high failure rates will fail more frequently, require more frequent maintenance and the consumption of more logistics resources, and should logically appear in the demonstration to a greater extent than items requiring maintenance less frequently.

The task selection is accomplished by proportionately distributing the 100 tasks among the major functional elements of the system. The 100 tasks may be allocated as illustrated in Table 6.2.2-1, which is based on a system of three units.

Table 6.2.2-1. Corrective Maintenance Task Allocation

(1) Item	(2) Quantity (Q)	(3) Failures/Item Per 1,000 Hours (λ)	(4) Total Failures ($Q\lambda$)	(5) % Contribution	(6) Allocated Maintenance Tasks for Demonstration
Unit A	1	0.48	0.48	21	21
Unit B	1	1.71	1.71	76	76
Unit C	1	0.06	0.06	3	3
		Totals	2.25	100	100

The table lists the elements of the system and their associated failure rates. The percentage contribution of each unit to the total anticipated corrective maintenance for the system (column 5) is calculated as follows:

$$\text{Item Percentage (\%) Contribution} = \frac{Q\lambda}{\Sigma Q\lambda} \times 100 \qquad (6.3)$$

This formula is the basis for proportionately allocating the tasks. As a further explanation, the 21 tasks designated for Unit A can be allocated to assemblies within that unit. Through a random process, one of the components within an assembly will be selected as the item where a failure will be induced by a specified method.

With the tasks identified and listed in random order, the demonstration proceeds with the first task, second task, etc. The criteria for accept-reject decisions are graphically depicted in Figure 6.2.2-1. Tasks times (M_{ct_i}) are measured and compared with the specified $\overline{M}_{ct}$ and M_{max} values. When the demonstrated time exceeds the specified value, an event is noted along the ordinate of the graph and the problem areas documented. Testing continues until the event line enters either the reject region or accept region.

An example of the demonstration test score sheet is provided as Table 6.2.2-2. The accept-reject numbers support the decision lines plotted for Test Plan A, shown in Figure 6.2.2-1. In this case, 29 tasks were completed before an accept decision was reached. The sequential test requires that both the $\overline{M}_{ct}$ criteria and M_{max} criteria be met before the equipment is fully acceptable. M_{max} may be based on either the 90th or the 95th percentile, depending on the system requirement and test plan. If one test plan is completed with the line crossing into the accept region, testing will continue until a decision is made in the other test plan.

Table 6.2.2-2. Demonstration Scoring Sheet

Maint. Task No.	Task Time M_{ct_i} (Min.)	Cum. No. M_{ct_i} > $\overline{M}_{ct}$	Accept When Cum. ≤	Reject When Cum. >
1	12	0	-	-
2	6	0	-	-
3	18	0	-	-
4	32	1	-	-
5	19	1	-	5
6	27	1	-	6
7	108	2	-	6
8	6	2	-	6
9	14	2	-	7
10	47	3	-	7
11	28	3	-	7
12	19	3	0	7
13	4	3	0	8
14	24	3	0	8
15	78	4	1	8
.	.	.	.	.
.	.	.	.	.
.	.	.	.	.
24	20	4	3	11
25	127	5	4	11
26	21	5	4	12
27	13	5	4	12
28	28	5	4	12
29	8	5	5	12
Accept for $\overline{M}_{ct}$				

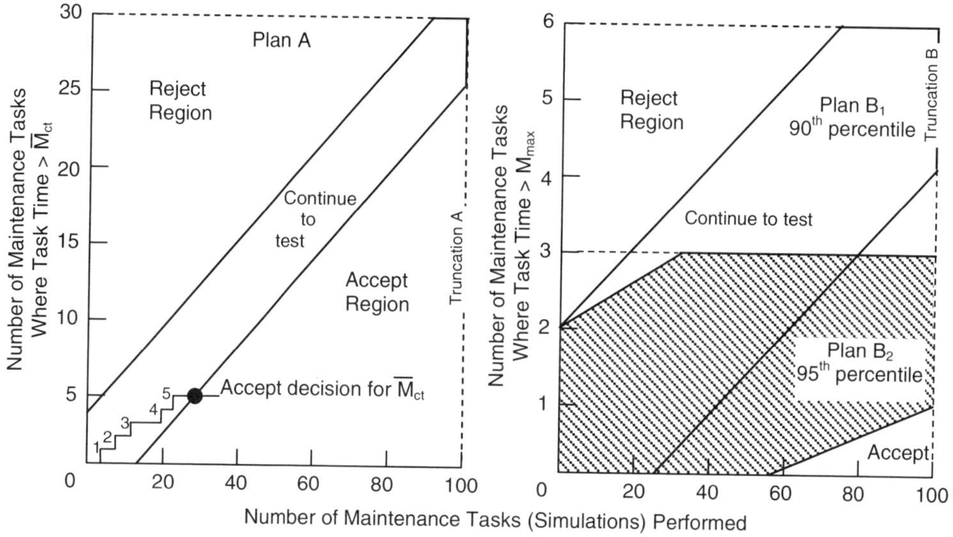

Figure 6.2.2-1. Representation of Maintainability Demonstration Plan

Referring to Figure 6.2.2-1, the criteria for sequential testing stipulates that the minimum number of tasks possible for a decision for $\overline{M}_{ct}$ is 12 (Test Plan A). For M_{max} at the 90th percentile, the least number of tasks possible is 26 (Test Plan B_1), while the figure notes a minimum of 57 for the 95th percentile (Test Plan B_2). It is obvious that, if the maintainability indicated is acceptable, demonstrating the complete sample of 100 tasks may not be necessary. As soon as the plot crosses the accept decision line, the system is deemed acceptable for $\overline{M}_{ct}$, while as soon as it crosses the reject line, it is deemed unacceptable. If 29 or fewer tasks exceed the specified $\overline{M}_{ct}$ value when the test reaches Truncation A (100 tasks), the system is deemed acceptable. The comparable parameters for M_{max} at Truncation B (100 tasks) are five or fewer for the 90th percentile, and two or fewer for the 95th percentile.

2. *Demonstration Method 2.* This method applies to the demonstration of $\overline{M}_{ct}$, $\overline{M}_{pt}$, and $\overline{M}$. The underlying distribution is not based on a specific statistical assumption, and the sample size includes 50 corrective maintenance tasks for $\overline{M}_{ct}$ and 50 preventive maintenance tasks for $\overline{M}_{pt}$. $\overline{M}$ is determined analytically from the test results for $\overline{M}_{ct}$ and $\overline{M}_{pt}$. M_{max} can also be

determined if the underlying distribution is assumed to be log-normal. This method offers the advantage of a fixed sample size. The sampling selection method and performance of maintenance tasks is similar to that described for demonstration method 1. Illustration of this method is best accomplished by means of the following example. In this example, it is assumed that the system to be demonstrated is designed to meet the following requirements.

$\overline{M}$ = 75 Minutes $\overline{M}_{ct}$ = 65 Minutes

$\overline{M}_{pt}$ = 110 Minutes M_{max} = 120 Minutes

Producer's Risk (a) = 20%

Table 6.2.2-3. Maintenance Test Time Data

Demonstration Task Number	Observed Time M_{ct_i}	$M_{ct_i} - \overline{M}_{ct}$ $(M_{ct_i} - 62)$	$(M_{ct_i} - \overline{M}_{ct})^2$
1	58	-4	16
2	72	+10	100
3	32	-30	900
.	.	.	.
.	.	.	.
.	.	.	.
50	48	-14	196
Total	3,105	-	15,016

This test is accomplished and the data collected are in Table 6.2.2-3. The determination of $\overline{M}_{ct}$ (upper confidence limit) is based on the expression:

$$\text{Upper Limit} = M_{ct} + Z\left[\frac{\sigma}{\sqrt{n_c}}\right] \tag{6.4}$$

where

$$\overline{M}_{ct} = \sum_{i=1}^{n_i} \frac{M_{ct_i}}{n_c} = \frac{3{,}105}{50} = 62.1$$

and

z = 0.84 (from Appendix E *Normal Distribution Tables*, for a based consumer's confidence factor of 80% [1.00 - α = 1.00 - 0.20 = 0.8], z = 0.84.

then

$$\sigma = \sqrt{\sum_{i=1}^{n_i} \frac{(M_{ct_i} - \overline{M}_{ct})^2}{n_c - 1}} = \sqrt{\frac{15,016}{49}} = 17.5 \tag{6.5}$$

where n_c is the corrective maintenance sample size of 50

$$\text{Upper Limit} = 62 + (0.84)\left[\frac{17.5}{\sqrt{50}}\right] = 64.08 \text{ minutes}$$

The completed $\overline{M}_{ct}$ statistic is compared to the corresponding accept-reject criteria, which is to accept if

$$\overline{M}_{ct} + Z\left[\frac{\sigma}{\sqrt{n_c}}\right] \leq \overline{M}_{ct} \text{ (Specified)} \tag{6.6}$$

and reject if

$$\overline{M}_{ct} + Z\left[\frac{\sigma}{\sqrt{n_c}}\right] > \overline{M}_{ct} \text{ (Specified)} \tag{6.7}$$

In this example, the demonstration test data show that the 64.08 minutes (the upper limit derived by test) is less than the specified requirement of 65 minutes. The system has, therefore, passed the $\overline{M}_{ct}$ test and is accepted.

For preventive maintenance the same approach is used. Fifty (50) preventive maintenance tasks are demonstrated and M_{pt_i} task times are recorded. The sample mean preventive maintenance time is computed as shown following Equation 6.8.

$$\overline{M}_{pt} = \sum_{i=1}^{n_i} \frac{M_{pt_i}}{n_p} \tag{6.8}$$

The accept-reject criteria are the same as set forth in Equations 6.5 and 6.6, except that the preventive maintenance ($\overline{M}_{pt}$) attributes instead of corrective maintenance ($\overline{M}_{ct}$) are applicable. Accordingly, accept if

$$\overline{M}_{pt} + Z\left[\frac{\sigma}{\sqrt{n_p}}\right] \leq 110 \text{ minutes}$$

and reject if

$$\overline{M}_{pt} + Z\left[\frac{\sigma}{\sqrt{n_c}}\right] > 110 \text{ minutes}$$

Given the test values for M_{ct_i} and $\overline{M}_{pt}$, the calculated mean maintenance time is:

$$\overline{M} = \frac{(\lambda)(\overline{M}_{ct}) + (f_{pt})(\overline{M}_{pt})}{\lambda + f_{pt}} \tag{6.9}$$

where

- λ = Corrective maintenance rate or the expected number of corrective maintenance (repair) tasks occurring in a designated period
- f_{pt} = Preventive maintenance rate or the expected number of preventive maintenance tasks occurring during the same period

Using the test data, the calculated value of $\overline{M}$ should be equal to or less than 75 minutes, which complies with the specified design requirements.

The M_{max} is calculated using the following algorithm:

$$M_{max} = \text{antilog}\left[\overline{\log M_{ct}} + Z\sigma_{\log M_{ct_i}}\right] \tag{6.10}$$

where:

$$\sigma_{\log M_{ct_i}} = \frac{\sum_{i=1}^{N} (\log M_{ct_i})^2 - \left(\sum_{i=1}^{N} \log M_{ct_i}\right)^2 / N}{N - 1}$$

The calculated M_{max} value should be equal to or less than 120 minutes for acceptance. If all the demonstrated values are equal to better than the specified values, following the previously defined criteria, the system is accepted. If not, retest and/or redesign may be required.

Topic 6.2.3: Human Factors Test and Evaluation

The evaluation of human factors serves to ascertain if the design of the system provides for optimum interface between the human and other elements of the system. Of special import is the critical interrelationship between human factors and considerations for safety. Given the requirement for human resources, it is necessary to ensure that such resources are utilized in the most efficient manner possible. The evaluation of human factors is an iterative process, as portrayed by Figure 6.2.3-1, which parallels the continual evolution of system design, testing, and improvement modifications to the system pursuant to evaluation of the test results.

The *performance error analysis* embodies the methodologies and techniques frequently employed for human factors testing and evaluation. This type of human performance evaluation is normally accomplished during *Type 3* testing and *Type 4* testing, as described in Topic 6.1.2, and in light of the applicable human (ergonomic) measures detailed in Topic 2.3. For example, it must be confirmed that human capabilities and attributes of the operators and maintainers can accommodate the system specifications relating to *maintainability* (task times, design configuration location of critical system components for accessibility, etc.). Another typical consideration is the validation and verification of the accuracy and completeness of the task instructions and procedures detailed in the operator and maintenance manuals. The validity of the operator and maintenance manuals has great impact on the effectiveness of the training programs developed for the personnel selected to support the system.

1. *Performance Error Analysis.* An error occurs when a human action exceeds a designated limit of acceptability for which the parameters of acceptable performance have been specified. Errors may be categorized as errors of omission, when a human fails to perform a specified task, and errors of commission, when a task is performed incorrectly (i.e., selection, sequence, or time errors). The following illustrate possible causes of error:[7]
   a. Inadequate work space and work layout–poor workstation design with respect to seating, available space, activity sequences, and accessibility to system elements.
   b. Inadequate design of facilities, equipment, and control panels for human factors–inadequate displays and readout devices, poor layout of controls, and lack of proper labeling.
   c. Poor environmental conditions–inadequate lighting, high or low temperatures, and high noise level.

[7] Blanchard, B.S., *Logistics Engineering and Management*, 6th Ed., Pearson Prentice Hall, Upper Saddle River, NJ, 2004, pages 289- 291.

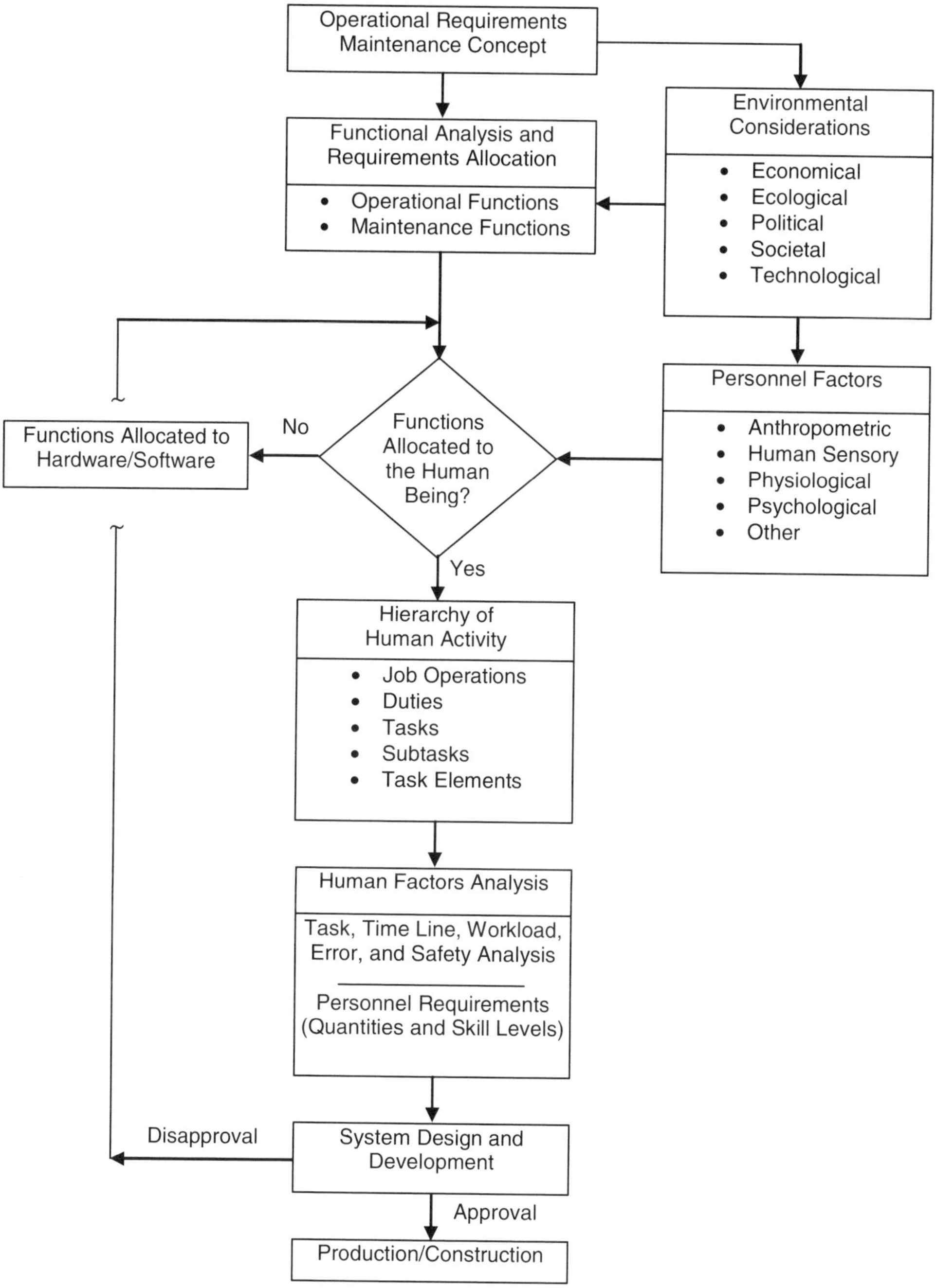

Figure 6.2.3-1. Human Factors Requirements

d. Inadequate training, job aids, and procedures–lack of proper training, and poorly written operating and maintenance procedures.

e. Poor supervision–lack of communications, no feedback, and lack of adequate planning resulting in overtime.

The performance error evaluation can be accomplished in conjunction with the *operator task analysis* (OTA), *maintenance task analysis* (MTA), during development of *operational sequence diagrams* (OSDs), and physically as part of the system test and evaluation efforts discussed in Topic 6.1.1 and Topic 6.1.2.[8] The ultimate goal is to select a human with the appropriate skills and training, simulate the operation of the system by undertaking a series of functions/tasks, and measure the number of errors that occur in the process. An Ishikawa cause-and-effect diagram, as shown by Figure 5.2.2-5, can aid in isolating the causes of the errors (e.g., how the error was introduced into the system). Additionally, it is essential to note the impacts of the error on other elements of the system and on the system as an entity.

2. *Interrelationship with the FMECA and SHA*. It is important that the performance error analysis be closely integrated with the *Failure Modes, Effects, and Criticality Analysis (FMECA)*, as detailed in Topic 5.2.2, and the *Safety/Hazard Analysis (SHA)* process as described in Topic 5.2.4.

[8] Refer to Blanchard, B.S., *Logistics Engineering and Management*, 6th Ed., Pearson Prentice Hall, Upper Saddle River, NJ, 2004, p 221 and pages 288-290.

Topic 6.2.4: Validation of Logistics Support Elements

With the system and its logistics and maintenance support infrastructure designed, the next step is that of *validation*; that is, the verification that the resulting system configuration will meet the initially specified performance requirements (i.e., the specified Performance Based Logistics (PBL) requirements). The validation process is accomplished within the context of Topic 6.1.2. With reference to Figure 6.1-1, the objective is to acquire some degree of confidence, as early as possible, that the requirements have been met. It is imperative that all elements of logistics resources specified for support of a system be verified so as to ensure the supportability of the system after distribution to the market or operational deployment. The validation can be accomplished to a certain extent through selective application of analytical methods (e.g., simulation), Type 1 testing, Type 2 testing, etc. A major part of the validation effort dedicated to logistics support can be accomplished only through Type 2, Type 3, and Type 4 testing.

As part of Type 2 testing, with the availability of prototype equipment, software, and other major elements of the system, the following tests can be conducted to validate the selected system elements:[9]

1. *Reliability Qualification Testing.* Various elements of the system are subjected to an environment characteristic of what will be later experienced by the ultimate user and are "operated" through a series of mission scenarios by personnel in the quantities and with the skills recommended for system operation during the utilization phase. (Topic 6.2.1 provides further details relevant to reliability testing.)
2. *Maintainability Demonstration.* The purpose here is to test the selected system in a true operational environment (as if the system were installed and operating in a realistic user scenario, sheltered or outside, mobile or fixed, in the mountains or in the desert, with the appropriate weather conditions, recommended personnel quantities and skills, operational procedures, and maintenance data, etc.). During the test, different maintenance tasks and task sequences are simulated using approved maintenance procedures, maintenance times are recorded, and the resources required to accomplish the tasks are noted; that is, personnel quantities and skills, tools and test equipment, spares and repair parts, packaging and transportation requirements, etc. (Topic 6.2.2 provides additional details pertinent to maintainability demonstration.)
3. *Personnel Test and Evaluation.* This testing process serves to verify the adequacy of personnel requirements recommended for system operation and maintenance support; that is, the appropriate quantities, skill levels, and

[9] Refer to Blanchard, B.S., *Logistics Engineering and Management*, 6th Ed., Pearson Prentice Hall, Upper Saddle River, NJ, 2004, pages 310-312.

classifications. Personnel requirements are initially recommended pursuant to an operator task analysis (OTA) for "operators" and the maintenance task analysis (MTA) for "maintainers".[10] Selected personnel for the test, considered to be representative of those in the user's organization, are identified, are given the appropriate recommended training, and then scheduled to perform certain designated operational and/or maintenance tasks using approved procedures. Task sequences are verified, areas of difficulty are noted along with personnel-induced faults, and the procedures used are validated. Although some of these requirements can be verified through reliability testing and maintainability demonstration, there may be some additional tasks (especially highly complex tasks) that should be evaluated.

4. *Test and Support Equipment Compatibility.* It may be necessary to ensure that significant items of test and support equipment will effectively accomplish the function(s) intended. In response, special tests will be conducted to verify compatibility between the test equipment and the element(s) of the system that it is intended to support. This may be particularly relevant for electronic–intensive equipment and large test stations that are designed to support corrective and/or preventive maintenance activities in the shop. From the results of such testing, one should be able to verify test equipment performance requirements, accuracies, reliability, and required interfaces (if any). Although some of these requirements may be verified through reliability testing and maintainability demonstration, there may be some additional tasks that need to be evaluated, particularly for test procedures that are rather complex.
5. *Logistics Validation.* Within the spectrum of Type 2 testing, one can certainly evaluate selected activities (and task sequences) pertaining to purchasing and material flow, packaging and transportation, and warehousing functions. Although a true evaluation of the overall logistics and support infrastructure cannot be accomplished at this stage, one certainly can acquire a good start in evaluating the purchasing process, the various means for transportation and handling, the processing of selected items for warehousing, the effectiveness of the data/information capability, etc.

With reference to Figure 6.1-1, system evaluation continues through Type 3 and Type 4 testing. Obviously, as one continues, the validation process becomes more valid. For instance, the first opportunity for evaluation of the overall logistics and support infrastructure as an integrated entity does not occur until the initiation of Type 3 activity, and then only a limited number of logistics activities may be accomplished. On the other hand, a complete validation of the entire logistics and support infrastructure may be completed only as a part of Type 4 testing, when contractor/

[10] Refer to Blanchard, B.S., *Logistics Engineering and Management*, 6th Ed., Pearson Prentice Hall, Upper Saddle River, NJ, 2004, p 221 and pages 288-289.

customer activities are integrated and involve a number of different user operational sites (refer to Topic 6.1.2, Categories of Test and Evaluation).

Finally, and as an integral part of the system test and evaluation process, there is the ongoing activity of data collection, analysis, and the subsequent recommendations for corrective action (in the event of not meeting a specific requirement) and/or for the purposes of product and/or process improvement. Refer to Table 6.2.4-1 for a delineation of the applications of data to the elements of the logistics and support infrastructure.[11] The system evaluation and corrective action loop are illustrated by Figure 6.2.4-1, and the requirements for system modification are discussed in Topic 6.3.

Table 6.2.4-1. Logistics and Support Data Applications

1. *General Operational and Support Factors*
   (a) Evaluation of mission requirements and performance measures.
   (b) Verification of system utilization (modes of operation and operating time).
   (c) Verification of cost/system effectiveness, operational availability, dependability, reliability, maintainability.
   (d) Evaluation of levels and location of maintenance.
   (e) Evaluation of functions/tasks by maintenance level and location.
   (f) Verification of repair level policies.
   (g) Verification of frequency distributions for unscheduled maintenance actions and repair times.
2. *Test and Support Equipment*
   (a) Verification of support equipment type and quantity by maintenance level and location.
   (b) Verification of support equipment availability.
   (c) Verification of support equipment utilization (frequency of use, location, percent of time utilized, flexibility of use).
   (d) Evaluation of maintenance requirements for support equipment (scheduled and unscheduled maintenance, downtime, logistic resource requirements).
3. *Supply Support (Spares/Repair Parts)*
   (a) Verification of spare/repair part types and quantities by maintenance level and location.
   (b) Evaluation of supply responsiveness (i.e., is a spare available when needed?).
   (c) Verification of item replacement rates, condemnation rates, attrition rates.
   (d) Verification of inventory turnaround and supply pipeline times.
   (e) Evaluation of maintenance requirements for shelf items.
   (f) Evaluation of spare/repair part replacement and inventory policies.
   (g) Identification of shortage risks.

[11] Refer to Blanchard, B.S., *Logistics Engineering and Management*, 6th Ed., Pearson Prentice Hall, Upper Saddle River, NJ, 2004, p 186.

Table 6.2.4-1. Logistics and Support Data Applications (Cont'd)

4. *Personnel and Training*
   - (a) Verification of personnel quantities and skills quantities by maintenance level and location.
   - (b) Verification of elapsed times and labor-hour expenditures by personnel skill level.
   - (c) Evaluation of personnel skill mixes.
   - (d) Evaluation of personnel training policies.
   - (e) Verification of training equipment and data requirements.
5. *Transportation and Handling*
   - (a) Verification of transportation and handling equipment type and quantities by maintenance level and location.
   - (b) Verification of availability and utilization of transportation and handling equipment.
   - (c) Evaluation of delivery response times.
   - (d) Evaluation of transportation and handling procedures.
6. *Facilities*
   - (a) Verification of facility adequacy and utilization (operation, maintenance, and training facilities).
   - (b) Evaluation of logistic resource requirements for support of operation, maintenance, and training facilities.
7. *Technical Data*
   - (a) Verification of adequacy of data coverage (level, accuracy, and method of presenting information) in operating and maintenance manuals.
   - (b) Verification of adequacy of field data, collection, analysis, and corrective action system.
8. *Software*
   - (a) Verification of adequacy of software (level, accuracy, detail, comprehensiveness).
   - (b) Verification of compatibility of software with other system elements.

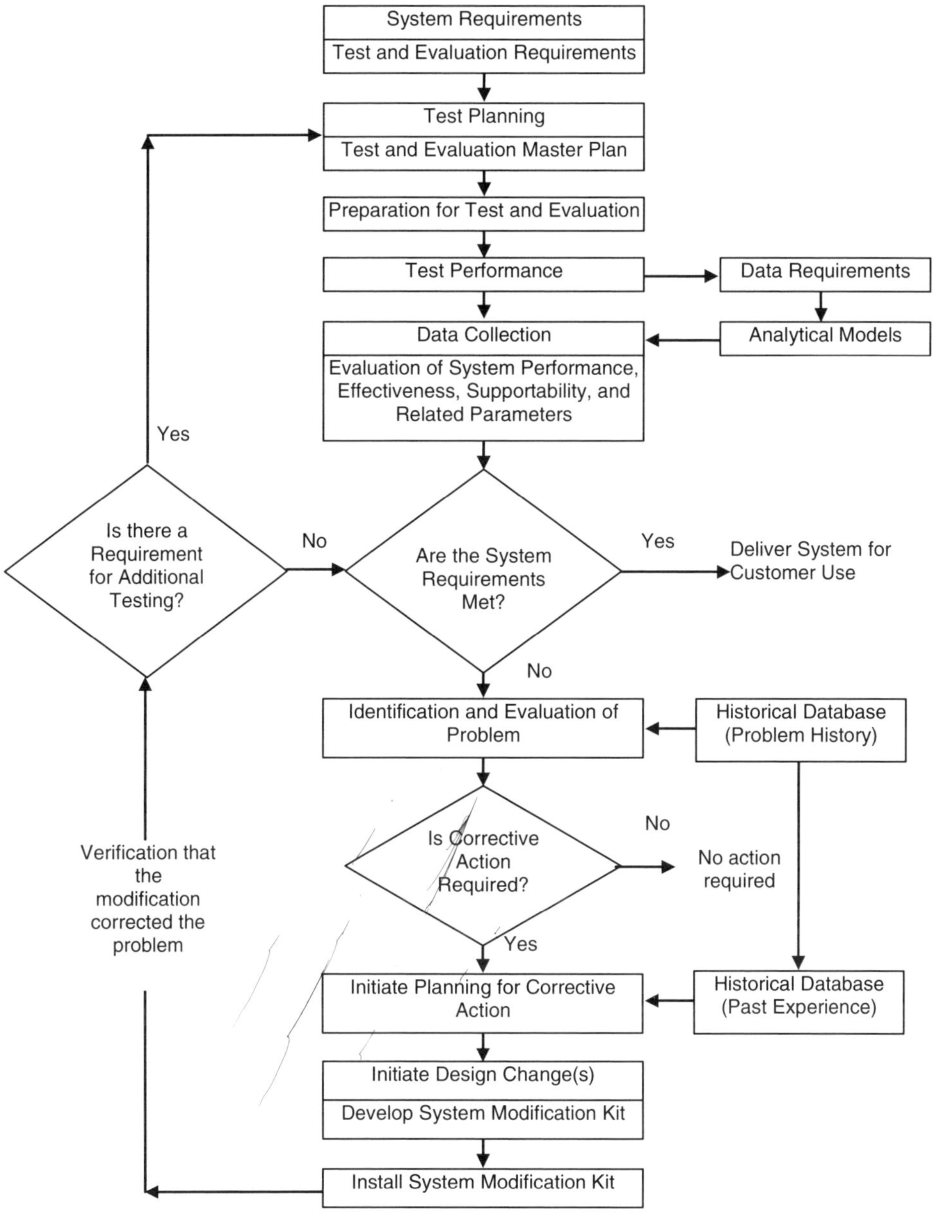

Figure 6.2.4-1. System Evaluation and Corrective Action Loop

Topic 6.3: Design Changes and System Modification(s)

A change to the configuration of a system may result from testing and could lead to redesign of a prime equipment item, a software modification, the revision of a production process, or a combination thereof. When a change (frequently cited as an *engineering change*) occurs in a maintenance and/or operating procedure, in the prime equipment, or in an element of logistics support, such system alteration will affect many other elements of the system. Procedural changes will likely affect personnel and training requirements and further dictate a change in the technical data (equipment operating and/or maintenance instructions). Hardware changes could affect software, spare/repair parts listings, support and test equipment, technical data and training documentation. Each change must be thoroughly evaluated in terms of its impact on other elements of the system prior to a decision on whether or not to incorporate the change. The feasibility of incorporating a change will be governed by the extensiveness of the change, its impact on the system's ability to accomplish its mission, the point in the life cycle when the change can be incorporated, and the cost of implementation of the change. A certain amount of evaluation and planning are necessary to make a rational determination of the feasibility of the change. If a change is to be incorporated, the necessary change control procedures must be enforced. The various components of the system (including all the elements of logistics support) must track and reflect technical integrity; otherwise, there could be a high probability of inadequate logistics support and unnecessary waste of resources. The change concepts and concerns in the production/construction phase are equally significant during the operational phase. It is emphasized that a change involving an operational system is more critical, since changes in a late stage in the life cycle will usually be more costly. The necessary discipline governing system design changes is accomplished through the process of *configuration management (CM).* Configuration management serves to ensure validation of the existing configuration baseline description as constituted by currently effective and applicable drawings, specifications, technical manuals, parts lists, computer programs, support and test equipment requirements, packaging and handling instructions, facility requirements, etc., and to enable traceability to previous baseline configurations of the same system.[12]

[12] Blanchard, B.S., *Logistics Engineering and Management*, 6th Ed., Pearson Prentice Hall, Upper Saddle River, NJ, 2004, pages 313-315.

The CM process constitutes the functions of *configuration identification*, *configuration control*, *configuration status accounting*, and *configuration audits*. A critical, correlative process is *engineering change installation.*[13]

1. *Configuration Identification.* Configuration identification is based on the family of specifications and drawings that describes the system or configuration item (CI) during the design/development cycle, becoming more precise as the design progresses toward production. This identification serves throughout the CI's life cycle as a record and description of its functional and physical characteristics. Configuration identification documents technically describe the CI and serve as a basis for change control, status accounting, and auditing. This also permits logical preparation of the technical, administrative, and management documents (e.g., work breakdown structure, technical reports, parts lists) that enhance logistics support of the system.
2. *Configuration Control.* Configuration control provides the vehicles for initiating, evaluating, approving or disapproving, releasing and directing the implementation of engineering change proposals (ECPs). This is achieved by means of documentation encompassing engineering design changes, deviations to product specifications before production, waivers to specifications for in-production or completed products pending delivery, formal changes in product specifications dictated by approved product design changes, and notices of revision to product managers of other systems which incorporate the same components affected by approved changes. Figure 6.3-1 portrays the system change control process.[14]
3. *Configuration Status Accounting.* Configuration status accounting (CSA) is the engineering management information system that provides traceability of configuration baselines and changes thereto, and facilitates the effective implementation of approved changes.
4. *Configuration Audits.* An integral part of the configuration management process is the comparison of the CI's functional performance and physical characteristics with those spelled out in the product specification and current configuration identification. Such comparison is described as a configuration audit. The audit is a formal examination to ascertain compliance with the specifications and other pertinent technical requirements and to verify that a product and its support resources conform to the established supportability documentation. Two types of audits are performed: the *functional configuration audit* and the *physical configuration audit*. Both the functional

[13] For an amplified description of the configuration management process, refer to Langford, J.W., *Logistics Principles and Applications*, McGraw-Hill, New York, NY,1995, Ch. 26, pages 529-554.

[14] The process depicted by Figure 6.3-1 applies to *major (Class I)* Engineering Change Proposals, as defined by MIL-STD-973.

and physical configurations of the items that constitute individual systems are audited as part of the configuration management process.

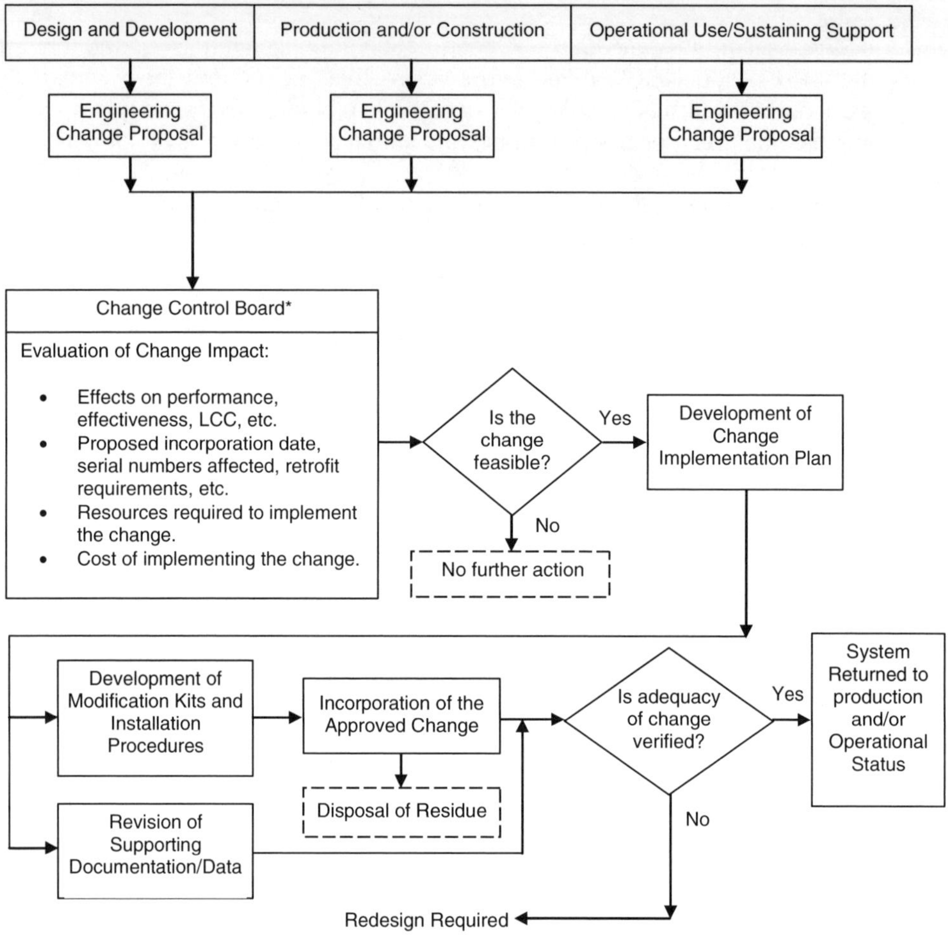

*This activity is know as the "Configuration Control Board"

Figure 6.3-1. System Change Control Process

5. *Engineering Change Installation.* The system change is implemented pursuant to approval of the change proposal by the appropriate CM authority and issuance of a directive ordering execution of the alteration by a designated change installation agent. The engineering change order (ECO) is the vehicle for execution of an approved change proposal. Engineering change orders are executed through installation on the system using engineering change kits. Each engineering change kit consists of a publications package and all parts, materials, and special tools required to accomplish and support the change. The engineering change must be validated by use of an evaluation change kit, which is normally one of the first

production kits. The evaluation kit is the basis for certification of the adequacy and completeness of the design, descriptive documentation, installation instructions, and physical composition of the change kit. It is noted that, if the change is included only in certain equipments (i.e., equipment item serial numbers 20 and up) and not retrofitted on earlier models (i.e., equipment items with serial numbers 1 through 19), each of the equipment configuration groups will require a different mix of logistics support. Such situations invariably pose challenges to logistics managers.

It is underscored that the efficacy of the configuration management (CM) process is key to life cycle cost-effectiveness of the system.

For More Information:

1. Blanchard, B.S., *Logistics Engineering and Management*, 6th Ed., Pearson Prentice Hall, Upper Saddle River, NJ, 2004.
2. Blanchard B.S., *System Engineering Management*, 3rd Ed., John Wiley & Sons, Hoboken, NJ, 2004.
3. Blanchard, B.S. and W.J. Fabrycky, *Systems Engineering and Analysis*, 3rd Ed., Prentice Hall, Upper Saddle River, N J, 1998.
4. Langford, J.W., *Logistics Principles and Applications*, McGraw-Hill, New York, NY, 1995.
5. MIL-HDBK-781A, *Handbook for Reliability Test Methods, Plans, and Environments for Engineering Development, Qualification and Production*, Department of Defense, Washington, DC, 1 April 1996.
6. MIL-HDBK-470A, *Designing and Developing Maintainable Products and Systems*, Department of Defense, Washington, DC, 4 August 1997.
7. MIL-HDBK-217F (Notice 2), *Reliability Prediction for Electronic Equipment*, Department of Defense, Washington, DC, 28 February 1995.
8. "Designing and Assessing Supportability in DOD Weapon Systems: A Guide to Increased Reliability and Reduced Logistics Footprint," prepared by the Office of Secretary of Defense, October 24, 2003.
9. O'Connor, P.D.T., D. Newton, and R. Bromley, *Practical Reliability Engineering*, Ed., John Wiley and Sons, Hoboken, NJ, 1996.
10. MIL-HDBK-502, "Department of Defense Handbook on Acquisition Logistics," Department of Defense, Washington, DC, May 1997.
11. "Reliability Toolkit: Commercial Practices Edition," Reliability Analysis Center, 201 Mill Street, Rome, NY.
12. "Maintainability Toolkit: A Practical Guide for Designing and Developing Maintainable Products and Systems," Reliability Analysis Center, 201 Mill Street Rome, NY.

13. "Quality Toolkit: A Practical Guide for Developing and Implementing a Quality Program for Products and Services," Reliability Analysis Center, 201 Mill Street, Rome, NY.

Section 7: Supportability Requirements In Production/Construction

Insight

The earlier sections of this Toolkit dealt with the design, development, testing, and evaluation of a system. The system may take the form of a configuration whose elements can be reproduced in multiple quantities through a production and manufacturing process, or a one-of-a-kind entity that is to be constructed at some designated location. While there are many logistics and related activities associated with both a production capability and the construction of a single item, the emphasis herein is primarily oriented to the former; i.e., a production or manufacturing plant.

This page intentionally left blank

Topic 7.1: Logistics and Supportability Requirements

Logistics in the production/construction stage (i.e., Production and Deployment Phase) includes three major facets of activity:

1. The aspects of logistics support in the production process itself, illustrated in Figure 7.1-1 and reflected by the activities shown in Figure 1.3-3. Production logistics support involves all supply chain and related activities to include the purchasing and physical supply of items to the producer/manufacturer, the internal flow of materials within the producer's plant operation, and the physical distribution of finished products from the manufacturer to the warehouse and/or directly to the user's operational site.[1]

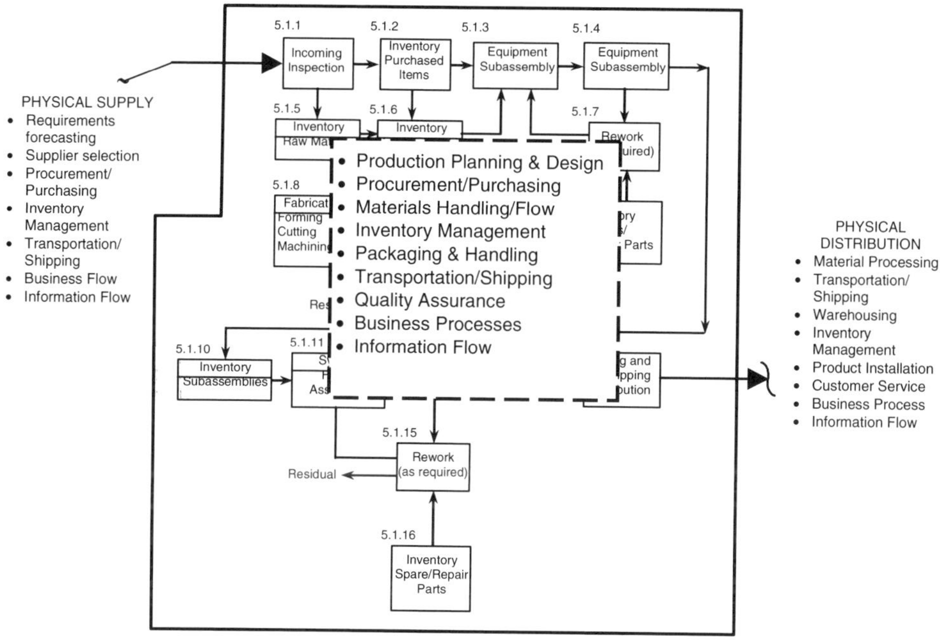

Figure 7.1-1. Logistics Requirements in Production/ Manufacturing

2. The aspects of logistics necessary for the sustaining maintenance and support of the system during its period of utilization and throughout its planned life cycle, as illustrated in Figure 1.3-4. This facet involves the

[1] This activity constitutes the spectrum of logistics covered in many of the supply chain management and business logistics textbooks today, and through the activities sponsored by the Council of Logistics Management (CLM) and others. Refer to Section A in the Bibliography in Appendix *A*

provisioning, procurement, manufacture, and distribution of the elements of support that have been identified through the supportability analysis (SA), and as reflected in Figure 1.3.5; i.e., spares/repair parts, test and support equipment, maintenance software, technical data, transportation and handling equipment, and so on.

3. The various activities associated with the sustaining maintenance and support of the production/manufacturing capability throughout the period when it is being utilized to produce the various elements of the system and its elements of support. This basically includes all of those maintenance activities covered within the spectrum of *Total Productive Maintenance (TPM)*, and as described in Section 2.8.2.[2]

In addressing the subject of "production," one must deal with the entire flow of materials, commencing with the initial procurement of components from the supplier(s) and extending through the delivery of the finished product to the ultimate user, as conveyed in Figures 1.3-3 and 7.1-1.

[2] While the elements associated with *Total Productive Maintenance (TPM)* are not usually considered to be included in the spectrum of "logistics" per se, the supportability characteristics that are built-into and inherent within the design of the equipment and processes utilized in the overall manufacturing capability are of major interest, particularly since the lack of an effective production process (i.e., a low *OEE*) can have a negative impact on the elements of the system being produced. Additionally, the maintenance support requirements for the production/ manufacturing process are comparable to those required for the system during its period of use by the customer (user).

Topic 7.2: Development and Implementation of a Logistics Infrastructure

Referring to Figure 1.6.1-2, four different "life cycles" are presented, with the second one representing the production/manufacturing life cycle. While dependent on the requirements based on the design of the prime system (and its elements), one can view this capability as a "system" in itself. Given this perspective, there are various phases of activity to include: (1) determination of the initial design requirements for the production capability; (2) design and development of the production capability as required; (3) operation of the production capability to produce the products desired; (4) sustaining maintenance and support of the production capability throughout its period of utilization; and (5) the retirement and material recycling/disposal activity when there is no longer any use for the production capability. In other words, one is dealing with the same basic phases of activity as required for the overall system.

Of particular emphasis herein is the design, development, and implementation of a total logistics and maintenance support infrastructure for the production/ manufacturing capability. This will, of course, include those requirements necessary to fulfill the objectives pertaining to the three areas of logistics activity identified in Topic 7.1. More specifically, this includes: (1) the implementation of a comprehensive supply chain (SC) network and supply chain management (SCM) capability to support all production operations; and (2) the implementation of a total productive maintenance (TPM) capability to support the production process for its life cycle (as necessary).

The objective is to address these two basic areas (i.e., the supply chain network and the TPM capability) on an "integrated" basis, and to consider these requirements early in the design stages of the production/manufacturing capability. PBL factors may be specified for such as the procurement process, materials handling, transportation, warehousing, and for the maintenance and support of the factory (e.g., a specific *OEE* design goal).[3] These factors should be specified as design objectives, considered in the day-to-day design process, and so on. Basically, the same process as described for the overall system in the earlier sections of this Toolkit should be followed.[4]

[3] The appropriate PBL factors for the various aspects of business logistics and for the supply chain (SC) may include any combination of those quantitative metrics described in Topic 2.5, and the maintenance aspect may include the specification of TPM factors as described in Topic 2.8.

[4] Rather than being redundant, it is assumed that one can follow the same basic approach and apply the same principles and concepts, with regard to the production/manufacturing capability, as are applicable for the design, development, operations, and sustaining maintenance support of the prime system. Thus, this section is somewhat abbreviated.

Topic 7.3: Assessment of the Logistics Infrastructure

The assessment and validation of the logistics and maintenance support infrastructure (and its elements) for the production/manufacturing capability may be accomplished through: (1) the implementation of an on-going data collection, analysis, feedback, and reporting system which will provide a continuing assessment of the logistics and maintenance support capability as the production/manufacturing process evolves; and/or (2) the conductance of selected tests, similar to Type 3 tests described in Section 6, with the objective of acquiring specific information pertaining to the effectiveness of the logistics and maintenance support capability as an entity, or any one of its elements. In the event of problems, the design change and system modification process described in Topic 6.3 may be implemented as necessary.

For More Information:

1. Appendix *A*, Section *A*, Items 1, 3, 5, 9, 11, 12, and 13.
2. Blanchard, B.S., *Logistics Engineering and Management*, 6th Ed., Pearson Prentice Hall, Upper Saddle River, NJ, Chapter 7, pages 319-349.

Section 8: Supportability Requirements in the System Operations, Sustaining Support, and Retirement Phases

Insight

The requirements for logistics and the maintenance support infrastructure continue to be rather intensive after the Production and Deployment Phase and throughout the System Operations and Support and the System Retirement Phases. In Sections 1 through 6, the emphasis was on the establishment of logistics requirements and the *design for supportability.* Section 7 briefly addressed the implementation of logistics requirements during the Production and Deployment Phase, and the objective of this Section is to highlight some of the sustaining logistics and supportability requirements as the system is being utilized by the customer in an operational environment and, subsequently, as the system is retired and material is processed for recycling or disposal. Basically, the logistics and maintenance support infrastructure has been designed, developed, installed, and is being utilized in support of the overall system mission requirements, and the role herein is that of system *enhancement* and *assessment.*

This page intentionally left blank

Topic 8.1: Sustaining Logistics and Maintenance Support Requirements

Logistics and supportability activities during the System Operations and Support and System Retirement Phases basically include the following:

1. Customer service functions to facilitate the transition from the time when the system and its components are initially installed at or delivered to the user's operational site(s) to the point when the system is fully operational and performing its intended mission and the user is able to operate the system in a satisfactory manner. This may include personnel on-site (on-the-job) training (operator and maintainer), servicing activities involving the upgrading of system elements after delivery, and follow-on interim contractor logistics support (CLS) for as long as required.
2. Sustaining system maintenance and support functions including the accomplishment of all required scheduled and unscheduled maintenance requirements through the period of system utilization. This phase will, of course, require the appropriate mix of maintenance personnel, test and support equipment, spares/repair parts and associated inventories, maintenance software, maintenance facilities, transportation, technical data, and information; i.e., the elements of logistics covering the activities illustrated in Figure 1.3-4.
3. Producer/contractor support to assist in the periodic upgrading of systems (i.e., new technology insertions or enhancements) through the installation of modification kits and the subsequent verification testing to ensure that the system has been returned to full operational status (refer to the process described in Topic 6.3). This activity may be accomplished by the original equipment manufacturer (OEM), a third-party logistics organization, user personnel, and/or various combinations of these. Additionally, there may be some associated training requirements, modification of the elements of the logistics and maintenance support infrastructure (e.g., test equipment, spares/repair parts, software, technical data), and follow-on interim contractor logistics support (CLS).
4. Producer/contractor/supplier supporting activities in the processing of obsolete items out of the operational inventory and to the location where the applicable materials may be recycled for other uses or, if not reusable, where they can be effectively and efficiently processed for disposal (refer to Topic 1.6.9). This can be an ongoing activity, as it will be necessary to process obsolete components resulting from system modifications throughout the System Operations and Support Phase.
5. Producer/contractor/user support necessary in the formal retirement and phase-out of the overall system. This phase may involve the disassembly of

the system and its elements, restoring the local environment as required, and the follow-on processing elements of materials for recycling and/or disposal.

The initial requirements for logistics and maintenance support during these phases stem from the development of the maintenance concept and the establishment of PBL factors during the Concept Refinement Phase (refer to Topics 3.2.3, 3.2.4, and 3.2.5); these requirements are updated on an iterative and continuing basis through the supportability analysis (SA); the results are conveyed in the detailed maintenance plan prepared during the System Development and Demonstration Phase; and these requirements are further clarified in the Integrated Logistics Support Plan (ILSP), or equivalent. Based on these sources, a plan is developed and implemented that will implement the five activity-areas mentioned above (as applicable).[1]

[1] Referring to Section 5, the basic logistics resource requirements (e.g., maintenance personnel, spares/repair parts, test and support equipment, transportation, etc.) for the system during the Operations and Support Phase are, of course, identified through the supportability analysis (SA). The same principles and concepts, applied in accomplishing the SA for this phase, can be extended to cover the required logistics resources for the System Retirement Phase.

Topic 8.2: Assessment of the Logistics and Maintenance Support Infrastructure

With the logistics and maintenance support infrastructure in-place, the next step is that of *assessment* and to verify that this capability is indeed fulfilling the objectives as planned. Hence, it is necessary to develop and implement an effective communications link for system evaluation, feedback, and corrective action as necessary. More specifically, this link will include the implementation of a good data collection, analysis, evaluation, and reporting capability. This, then, may lead to the need for system modification(s) for the purposes of corrective action and/or improvement.

Topic 8.2.1: Data Collection, Analysis, and Reporting

The purposes for establishing a good data collection and information feedback capability are to:

1. Provide field data, on a continuing basis, to evaluate and assess the performance and effectiveness of operations, maintenance, and the logistics support capability for the system in the field. The system manager needs to know exactly how the system is doing and whether the requirements are being met, and needs to know soon.
2. Provide historical data (covering existing systems in the field) that are applicable and can be utilized in the design and development of new systems having a similar function and nature. Engineering growth and potential in the future certainly depends on the ability to capture experiences from the past and subsequently be able to apply the results in terms of "what to do" and "what not to do" in new designs.

Relative to the first item, developing such a data collection capability is very critical as the *right type* of data must be available to make a good assessment of the specific system being addressed. The requirements for the design of the data forms must be consistent with the initially-specified operational requirements, the maintenance concept, and the prioritized TPM/PBL factors for the system being evaluated. In other words, if there are specific quantitative requirements for the system (e.g., *R*, MTBM, Logistics Effectiveness, Cost/MA, etc.), then the right type of information must be acquired from the field. This, in turn, will influence the particular format to be utilized in the collection of field data.

While the specific format may vary considerably, an example of one approach is presented in Tables 8.2.1-1 and 8.2.1-2. In the development of an overall data format, it is important to collect both *success data* and *maintenance data.* Success data constitute essential information covering system operations and utilization on a day-to-day basis, and the information gathered should be comparable to the factors identified in Table 8.2.1-1. Maintenance data cover each event involving scheduled and unscheduled maintenance. The events are recorded and referenced in the system operational information reports, and the factors recorded in each instance are comparable to those illustrated in Table 8.2.1-2. In the development of data forms, the following objectives should apply:

1. The data collection forms should be simple to understand and complete, as the task of recording the data may be accomplished under adverse environmental conditions by a variety of personnel with different skill levels. If the forms are difficult to understand and complete, the desired information will not be acquired.

2. The factors specified on each form must be clear and concise, and not require a lot of interpretation and manipulation to obtain. The right type of data must be collected.
3. The factors specified must have "meaning" in terms of application. The usefulness of each factor, and its relationship to a specified quantitative TPM/PBL requirement, must be verified.

These considerations are extremely important and cannot be over-emphasized. All of the analytical methods, prediction techniques, models, and so on, discussed earlier have little meaning without the proper input data. Our ability to evaluate alternatives and make assessments depends on the availability of good historical data, and the source of such stems from the data information and feedback capability developed at this stage.

Table 8.2.1-1. System Operational Information Report (Success Data)

1. Report identification information
   a. Report number
   b. Report date
   c. Name of individual preparing a report
2. System nomenclature, part number, manufacturer, and serial number
3. Description of system operation by date
   a. Mission type
   b. Mission profile
   c. Mission duration
4. Equipment utilization by date
   a. Operating time
   b. Cycles of operation
   c. Etc.
5. Description of the following needed for system operation:
   a. Personnel
   b. Transportation and handling equipment
   c. Facilities
6. Recording of maintenance events by date and time (reference maintenance event reports)

Table 8.2.1-2. Maintenance Event Report (Maintenance Data)

1. *Administrative data*
   (a) Event report number and date, and name of individual preparing report.
   (b) Work order number.
   (c) Work area and time of work (month, day, hour).
   (d) Activity (organization) identification.
2. *System factors*
   (a) Equipment part number and manufacturer.
   (b) Equipment serial number.
   (c) System operating time when event occurred (when discovered).
   (d) Segment of mission when event occurred.
   (e) Description of event (describe symptoms of failure for unscheduled actions).
3. *Maintenance factors*
   (a) Maintenance requirement (repair, calibration, servicing, etc.).
   (b) Description of maintenance tasks.
   (c) Maintenance downtime (MDT).
   (d) Active maintenance time (M_{ct_i} and M_{pt_i}).
   (e) Maintenance delays (time awaiting spare parts, delays for test equipment, work stoppage, awaiting personnel assistance, delay for weather, etc.).
4. *Logistics factors*
   (a) Start and stop times for each maintenance technician by skill level.
   (b) Technical manual or maintenance procedure used (procedure number, paragraph, date, comments on procedure adequacy).
   (c) Test and support equipment used (item nomenclature, part number, manufacturer, serial number, time of item usage, operating time on test equipment when used).
   (d) Description of facilities used.
   (e) Description of replacement parts (type and quantity).
      i. Nomenclature, part number, manufacturer, serial number, and operating time on replaced item – describe disposal of failed part.
      ii. Nomenclature, part number, manufacturer, serial number, and operating time on installed item.
5. *Other Information*
   Include any additional data considered appropriate and related to the maintenance event.

Topic 8.2.2: Evaluation of Logistics Elements

Referring to Topic 6.2.4, Section 6, a listing of the logistics elements requiring some form of assessment and validation is included in Table 6.2.4-1. While an initial assessment is accomplished through a combination of Types 3 and 4 testing, a final overall assessment is required to ensure that the proper level of support is being provided throughout the System Operations and Support Phase. This, of course, is facilitated through the implementation of the data collection, analysis, and reporting capability discussed earlier. As the field data are collected and analyzed, it is important that each of the elements of the logistics and maintenance support infrastructure be evaluated in terms of adequacy and compliance with the initially-specified TPM/PBL requirements for the system.

The system evaluation process is continuous and is illustrated in Figure 8.2.2-1. Referring to the figure, data are collected, the data are analyzed, and the results are reviewed, evaluated, and reported. Problem areas are identified and reviewed in terms of the necessity for corrective action. Corrective action may be taken in response to a system deficiency (i.e., if the system fails to meet a specified requirement), or may be taken to improve system performance, effectiveness, and/or logistics support. If corrective action is required, the necessary planning and implementation steps are a prerequisite to ensure complete compatibility of all elements of the system throughout the change process.[2]

[2] The evaluation of the different logistics elements, and the specific information desired from a data collection, analysis, and reporting capability, is discussed in Blanchard, B.S., *Logistics Engineering and Management*, 6th Ed., Prentice Hall, NJ, 2004, pages 358-367.

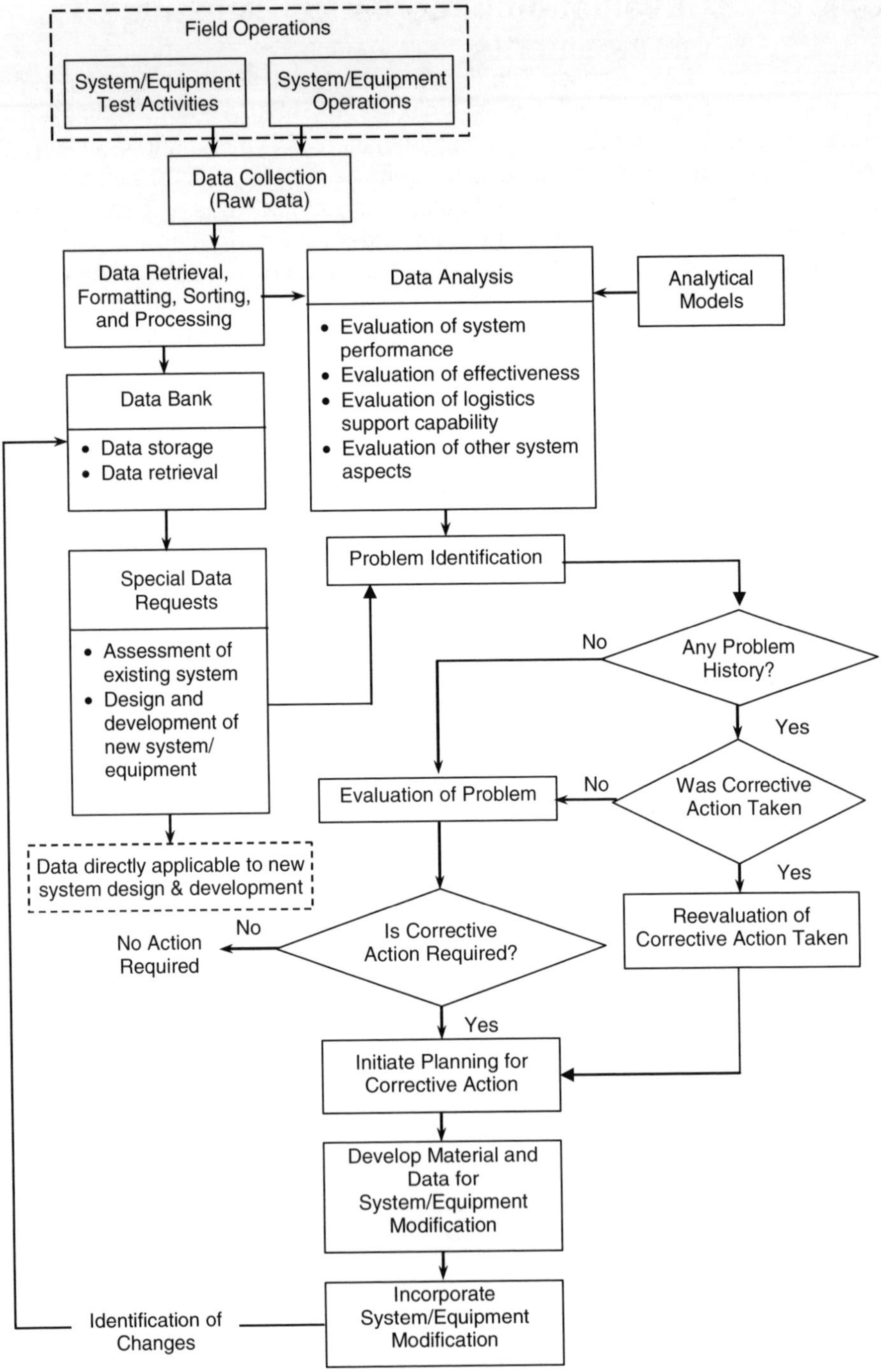

Figure 8.2.2 -1. System Evaluation and Corrective-Action Loop

Topic 8.3: System Modifications and Technology Enhancements

System changes may occur as a result of some deficiency or in the interest of "upgrading" the system through some form of "technology enhancement." When a change occurs in a procedure, in the prime equipment or software, and/or in an element of logistics support, the change in most instances will affect many other elements of the system. Procedural changes will affect personnel and training requirements and necessitate a change in technical data (equipment operating and/or maintenance instructions). Hardware changes will affect software, spare/repair parts, test and support equipment, technical data, and training requirements. Software changes will have an impact on hardware, test and support equipment, technical data, computer resources, and training requirements. Each proposed change must be evaluated in terms of its impact on other elements of the system prior to a decision on whether or not to incorporate the change. The feasibility of incorporating a change will depend on the extensiveness of the change, its impact on the system's ability to accomplish its mission, the time in the life cycle when the change can be incorporated, its impact on total system life-cycle cost, and the cost of change implementation.

If the change is to be incorporated, the necessary change control procedures must be implemented. The various components of the system (including all elements of the logistics and maintenance support infrastructure) must track! Otherwise, the probability of inadequate logistics support and unnecessary waste will be high. The configuration control and the change cycle are discussed in Topic 6.3. The concepts and concerns in the Production and Deployment Phase are equally applicable and appropriate in the System Operations and Support Phase. In fact, a change involving an "operational" system is more critical since changes in this late stage in the life cycle will usually be more costly.

Topic 8.4: System Retirement and Material Recycling/ Disposal

Referring to Topic 1.4 and Figure 1.6.1-2, the term *retirement* pertains to the final phase in the system life cycle when the requirement for the system and/or any of its components no longer exists. As one evolves through the System Operations and Support Phase, and as system changes are incorporated, the residual material (i.e., that which is no longer needed) must be processed for recycling or disposal. Additionally, after the system (as an entity) ceases to have a useful purpose in the operational inventory, it and all of its elements must be processed for recycling or disposal.

At the system level, the specific requirements for the System Retirement and Material Recycling/Disposal Phase can be determined by following the same process that was applied in developing the requirements for the System Operations and Support Phase and for the logistics and maintenance support infrastructure during the period of system utilization. Referring to Topic 3.2.6, the functional analysis can be extended to cover the major functions (activities) applicable in the process of system retirement and material phase-out. These functions can be broken down into sub-functions and to the level necessary for task definition and to provide the required visibility. Each functional entity can then be evaluated, tradeoffs are conducted, and the resources (e.g., equipment, software, facilities, personnel, data, etc.) required to accomplish the function are identified (refer to Figure 3.2.6-4). The process continues through an extension of the supportability analysis (SA) discussed in Section 5. It is important that this extension of the SA be accomplished during the System Development and Demonstration Phase, as the logistics and supportability requirements may turn out to be more extensive than initially anticipated. For example, it may be necessary to design and construct an expensive new facility for the decomposition of materials so that the environment will not be impacted in any way. Thus, it is important to provide some early visibility in this area.

Given the established requirements for this last phase of the life cycle, it is important to continue with the ongoing *assessment* in a manner similar to that discussed in Topic 8.2. While the implementation of any system-level changes at this late stage in the life cycle may not be feasible, the data collected and the subsequent assessment will certainly have some historical value for the design and development of new systems.

For More Information:

1. Blanchard, B.S., *Logistics Engineering and Management*, 6th Ed., Pearson Prentice Hall, Upper Saddle River, NJ, 2004.
2. Blanchard, B.S., *System Engineering Management*, 3rd Ed., John Wiley & Sons, Hoboken, NJ, 2004.

Section 9: Logistics and Supportability Management

Insight

Throughout the earlier sections of this Toolkit, the material presented has primarily emphasized the "technical" aspects of supportability and the design for such. To successfully accomplish the objectives, stated throughout this handbook, one must also address the subject from a *management* perspective. An organization can promote the best technical approach in the world; however, the implementation of such will not be successful unless the leadership and the proper organizational "environment" have been established to allow it to happen! The principles and concepts of *supportability* must be implemented from the top-down. The customer/producer/supplier management must both understand and believe in these principles and concepts, and the organizational structure must be such as to promote the implementation of good communications, a good technical interchange, and a participative-approach in the day-to-day decision-making processes. In other words, the successful implementation of supportability requirements depends on a highly integrated technical and management approach from the beginning.

The objective of this section is to highlight some of the key management issues associated with the implementation of supportability requirements in the design and development of new systems (and in the re-engineering of existing systems). The material presented herein stems primarily from:

1. Figure 1.7-1: Logistics and supportability throughout the system life cycle (the identification of specific tasks for each phase of the system life cycle).
2. Figure 4.3-1: Supportability requirements in the system life cycle (the layout of major supportability activities in a schedule-oriented format).

This page intentionally left blank

Topic 9.1: Supportability Program Requirements

Referring to Topic 1.7.1, the basic **program requirements** for supportability are initially established during the Concept Refinement Phase. The needs analysis is accomplished and the requirements for logistics and the maintenance support infrastructure are developed. This development process continues through the definition of system operational requirements, the maintenance concept, and the identification and prioritization of TPM/PBL factors. These requirements may be specified via several means, with the "technical" requirements being included in the *System Specification (Type "A")* and the overall "program" requirements included first in the *System Engineering Management Plan (SEMP)* which feeds into the *Integrated Logistic Support Plan (ILSP).* These requirement documents, and their interrelationships, are identified in Figure 9.1-1.[1]

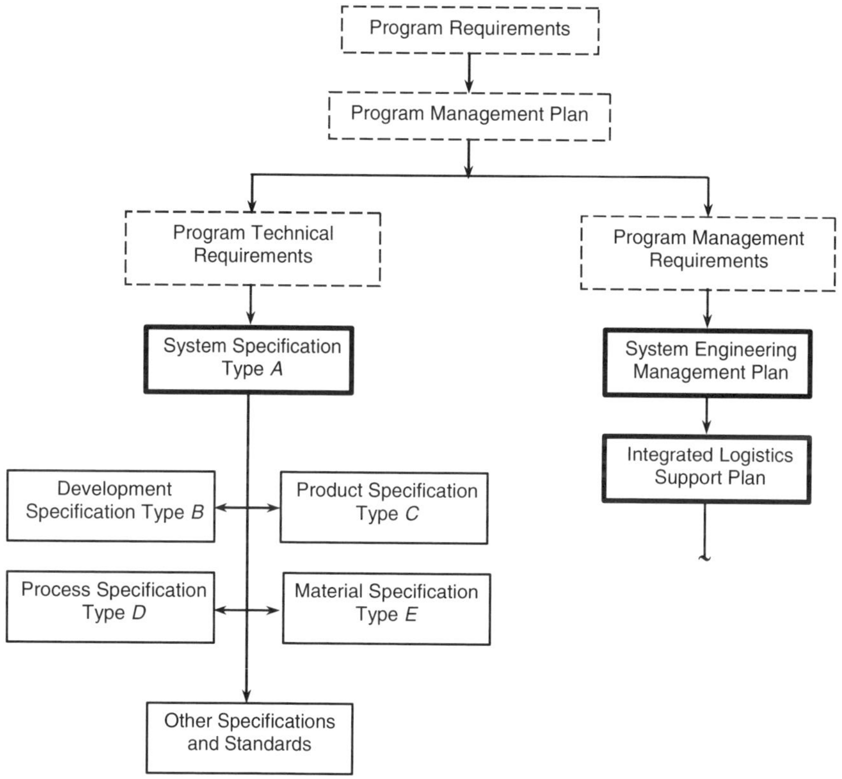

Figure 9.1-1. Supportability Program Requirements

[1] It is assumed herein that the prime logistics planning document will be identified as the ILSP. Depending on the size of the system, the nature and objectives of the program, the organizational priorities, etc., different titles may be applied to documents with the same objectives.

Topic 9.2: Supportability Program Planning

Planning a supportability program is an essential first step toward meeting the objective of developing and fielding a supportable system. The program plan must address supportability requirements, logistics, and supportability planning.

Topic 9.2.1: Development of Specifications

Referring to Topic 3.2.8, the specification of supportability requirements starts with those applicable at the system-level and included in the *System Specification (Type "A").* An example outline for this top-level specification is presented in Figure 9.2.1-1. These requirements are then allocated to the various elements of the system and included in the applicable lower-level specifications; i.e., *Development (Type "B"), Product (Type "C"), Process (Type "D"),* and *Material (Type "E") Specifications,* as applicable and shown in Figure 3.2.8-1.

System Specification

- 1.0 Scope
- 2.0 Applicable Documents
- 3.0 Requirements
  - 3.1 General Description of System Architecture
    - 3.1.1 System Operational Requirements
    - 3.1.2 Maintenance Concept
    - 3.1.3 Technical Performance Measures (TPMs)
    - 3.1.4 Functional Analysis (System Level)
    - 3.1.5 Allocation of Requirements
    - 3.1.6 Functional Interfaces
    - 3.1.7 Environmental Requirements
  - 3.2 System Characteristics
    - 3.2.1 Performance Characteristics
    - 3.2.2 Physical Characteristics
    - 3.2.3 Effectiveness Requirements
    - 3.2.4 Reliability
    - 3.2.5 Maintainability
    - 3.2.6 Usability (Human Factors)
    - 3.2.7 Supportability
    - 3.2.8 Transportability/Mobility
    - 3.2.9 Producibility
    - 3.2.10 Disposability
  - 3.3 Design and Construction
    - 3.3.1 CAD/CAM/CAS Requirements
    - 3.3.2 Materials, Processes, and Parts
    - 3.3.3 Hardware
    - 3.3.4 Software
    - 3.3.5 Electromagnetic Radiation
    - 3.3.6 Interchangeability
    - 3.3.7 Flexibility/Robustness
    - 3.3.8 Workmanship
    - 3.3.9 Safety
    - 3.3.10 Security
  - 3.4 Design Data and Database Requirements
  - 3.5 Logistics
    - 3.5.1 Supply Chain Requirements
    - 3.5.2 Spares, Repair Parts, and Inventory Requirements
    - 3.5.3 Test and Support Equipment
    - 3.5.4 Personnel and Training
    - 3.5.5 Packaging, Handling, Storage, and Transportation
    - 3.5.6 Facilities and Utilities
    - 3.5.7 Technical Data/Information
    - 3.5.8 Computer Resources (Software)
  - 3.6 Interoperability
  - 3.7 Affordability
- 4.0 Test and Evaluation
- 5.0 Maintenance and Support (Life Cycle)
- 6.0 Quality Assurance
- 7.0 Customer Service

Figure 9.2.1-1. System Specification (Example)

Topic 9.2.2: Development of Plans

The initial planning requirements for the implementation of a logistics and supportability program evolve from the *System Engineering Management Plan (SEMP)* which, for many programs, represents the top technical planning document.[2] Within the contents of the SEMP are sections covering the system engineering process, technical program planning, engineering specialty integration, program technology requirements, risk management, and so on. One of the key areas under "engineering specialty integration" is logistics and supportability engineering. From the SEMP, the specific requirements for logistics and the maintenance support infrastructure are expanded and included in an *Integrated Logistic Support Plan (ILSP)*. An example outline for an ILSP is shown in Figure 9.2.2-1.

Integrated Logistics Support Plan (ILSP)

1.0 Introduction
- 1.1 Purpose and Scope
- 1.2 System Description and Background
- 1.3 Logistics Acquisition Strategy
- 1.4 Reference and Specifications

2.0 System Characteristics
- 2.1 System Operational Requirements
- 2.2 System Maintenance Concept
- 2.3 Functional Analysis and Allocation
- 2.4 Logistics Research Results
- 2.5 Feasibility Studies
- 2.6 Performance Based Logistics (PBL) Requirements

3.0 ILS Planning and Management
- 3.1 Program Requirements
- 3.2 ILS Organizational Structure
- 3.3 Supplier/ Subcontractor Organization(s)
- 3.4 Organizational Interfaces
- 3.5 Work Breakdown Structure
- 3.6 Task Schedules and Major Milestones
- 3.7 Cost Projections
- 3.8 Program Reviews, Evaluation, and Control
- 3.9 Technical Communications (Reports and Documentation)
- 3.10 Risk Management

Figure 9.2.2-1. Integrated Logistics Support Plan (Example)

[2] A comprehensive discussion of the SEMP and its contents is included in Blanchard, B.S., *System Engineering Management*, 3rd Ed., John Wiley & Sons, Hoboken, NJ, 2004, Section 6.2, pages 254-300.

Integrated Logistics Support Plan (ILSP)
4.0 ILS Element Plans 4.1 Supportability Analysis Plan 4.2 Reliability and Maintainability Plan 4.3 Supply Support Plan 4.4 Test and Support Equipment Plan 4.5 Personnel and Training Plan 4.6 Technical Data and Information Systems Plan 4.7 Package, Handling, Storage, and Transportation Plan 4.8 Facilities Plan 4.9 Computer Resources Plan 4.10 Distribution and Customer Support Plan 4.11 Post-production Support Plan 4.12 System Retirement Plan

Figure 9.2.2-1. Integrated Logistics Support Plan (Example) (Cont'd)

Referring to the figure, the ILSP should include the usual introduction and a statement of program objectives (Section 1.0); a description of system characteristics which provides a technical baseline for the establishment of planning and management requirements (Section 2.0); the more traditional planning and management functions to include an identification and listing of program tasks, a work breakdown structure (WBS), task schedules and timelines, cost projections, program review and reporting requirements, risk management, program management and control activities, etc. (Section 3.0); and a group of individual ILS element plans (Section 4.0). Within the scope of Section 4.0, the various element plans may include (for example) a *Supportability Analysis Plan (SAP), Supply Support Plan, Test and Support Equipment Plan, Personnel and Training Plan,* and so on. It is also important to include a *Reliability Plan* (refer to Figure 4.2.1-1) and a *Maintainability Plan* (refer to Figure 4.2.2-1), or a good cross-reference to each, to ensure a close integration across the board. Figure 9.2.2-2 presents an example of what might be included as element plans.

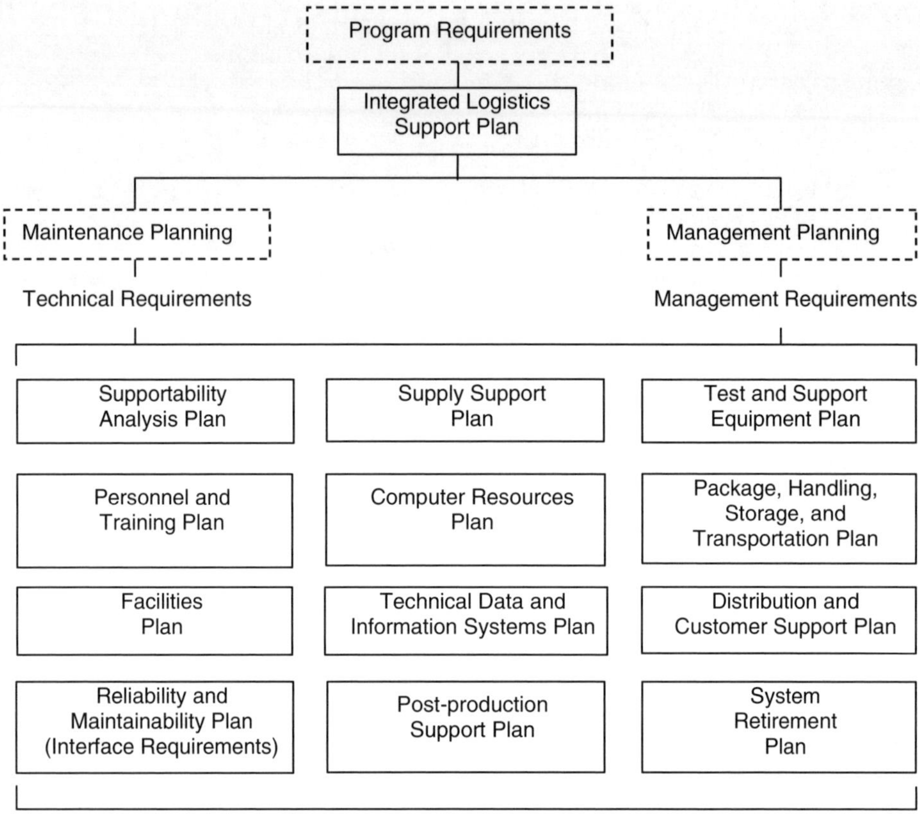

Figure 9.2.2-2. Individual Support Plans

Referring to Figure 4.3-1, the ILSP should be developed during the latter part of the Concept Refinement Phase. It should include all of the tasks essential in accomplishing the *design for supportability* objectives described throughout this Toolkit. This not only includes the initial establishment of PBL requirements, the accomplishment of the necessary design integration and review tasks, the implementation of supportability analysis (SA) activities, and the requirements for supportability validation through system test and evaluation. The ILSP must directly support the SEMP, and must be compatible with the program requirements described in Section 4 as applicable (refer to Figures 4.2.1-1, 4.2.2-1, 4.2.3-1, and 4.3-1).[3]

[3] A more in-depth discussion of the ILSP and a detailed description of what might be included in the different ILS element plans included in Blanchard, B.S., *Logistics Engineering and Management*, 6th Ed., Pearson Prentice Hall, Upper Saddle River, NJ, 2004, Section 9.3, pages 379-387. For the more traditional management tasks, reference to a good text on Project Management is recommended. Refer to Appendix *A*, Section K.

Topic 9.3: Organization for Logistics and Supportability

To address the subject of *organization for logistics and supportability* properly, one needs to understand the environment in which logistics and supportability-related tasks are performed. Although this environment may vary somewhat depending on the size of project and the stage of design and development, the discussion herein (and associated figures) is primarily directed to a large project operation characteristic for the acquisition of many large-scale systems. By assuming a large project approach, we hope to provide a better understanding of logistics (and supportability) in a somewhat complex environment. The reader must, of course, adapt to establish his/her program requirements.

Topic 9.3.1: Customer, Producer, and Supplier Requirements

For a relatively large project, logistics activity may appear at several levels, as shown in Figure 9.3.1-1. The customer (or consumer) may establish a logistics organization to accomplish the basic supportability-related tasks identified in Figures 1.7-1 and 4.3-1, or these tasks may be relegated to the producer (or prime contractor) through some form of a contractual structure. The question is -- *who is responsible and has the authority to perform the logistics supportability functions as defined?*

In some instances, the customer may assume full responsibility for the overall design and integration of the system and its elements. Top system-level design, preparation of the system specification, preparation of the ILSP, and the completion of the major supportability tasks in Figures 1.7-1 and 4.3-1 are accomplished by the customer's organization. Individual producers, subcontractors, and components suppliers perform supporting tasks as necessary.

In other cases, although the customer provides overall program guidance in terms of issuing a general statement of work (SOW) or an equivalent contractual document, the producer (prime contractor) is held responsible for system design and integration. In this situation, the definition and expansion of system requirements (and the detailed development of the applicable PBL factors) is accomplished by the producer, to include the definition of logistics and supportability requirements and the preparation of the ILSP, with supporting tasks being accomplished by individual suppliers as required. In other words, although both the customer and the producer may have established logistics organizations, the basic responsibility (and authority) for fulfilling the objectives described throughout this Toolkit lies within the producer's organization. This is the "model" that will serve as the basis for much of the discussion from hereon.

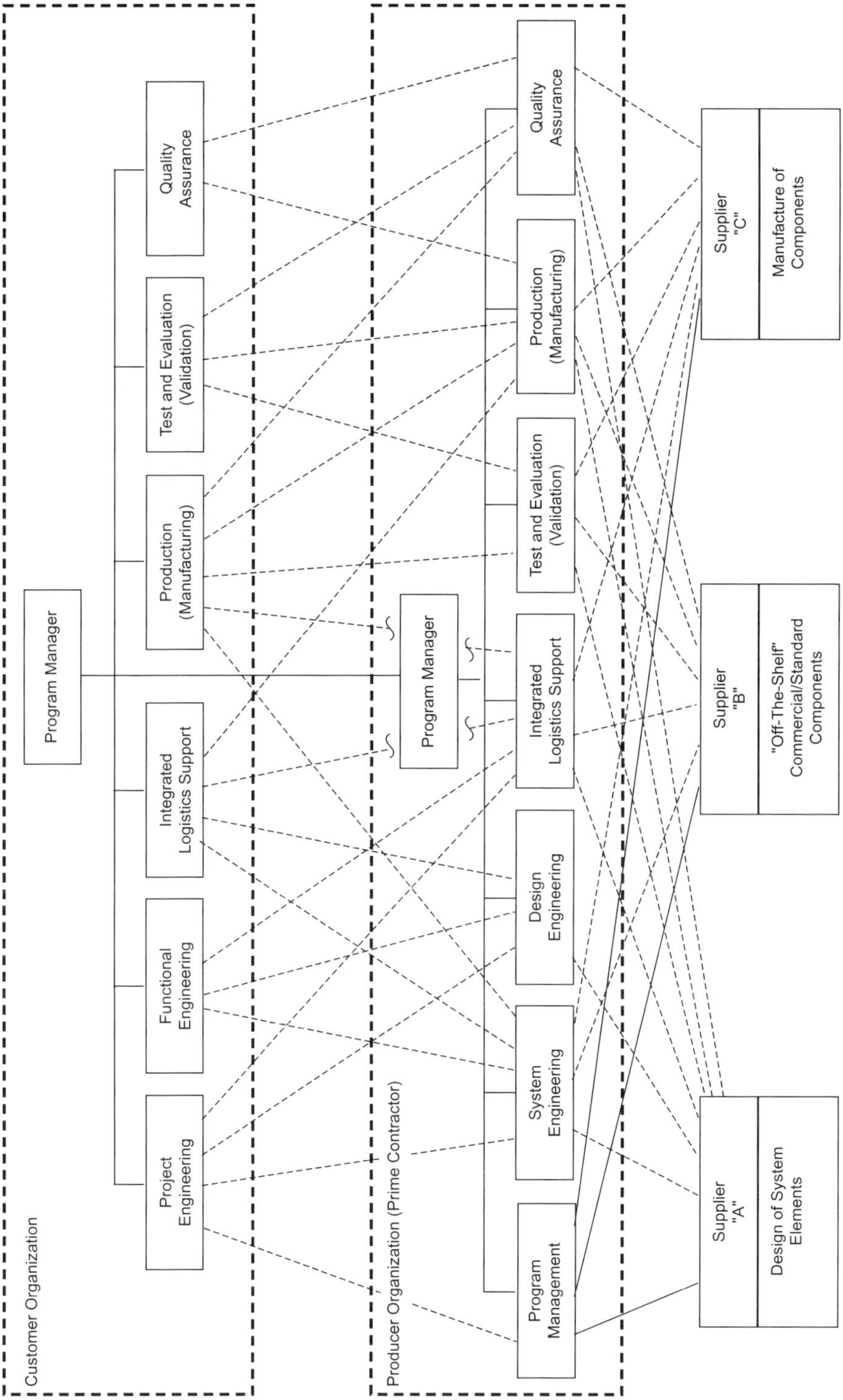

Figure 9.3.1-1. Customer/Producer/Supplier Interfaces

As Figure 9.3.1-1 illustrates, basic logistics and supportability requirements, stated in one form or another, must evolve from the *top*; i.e., must be specified by the customer. Otherwise, there is no guarantee that the requirements discussed throughout this Toolkit will be implemented. Further, and as shown in the figure, an extensive amount of communications must exist between the logistics organization and other organizations within the project. Basic system-level design and the establishment of requirements evolves as a result of a *team* approach and, for example, while supportability requirements may initially stem from the ILS organization they need to be properly integrated into the overall System Engineering requirements. In other words, the interfaces are many, and the successful accomplishment of logistics and supportability tasks requires an in-depth understanding of the responsibilities and tasks performed by these many and varied organizations. The communication requirements are not only extensive within the producer's organization, but must be established both upward from producer to customer and downward from producer to supplier.[4]

[4] Referring to Figure 9.3.1-1, the more *formal* contractually-related communications are reflected by the solid lines, whereas the day-to-day *informal* communications that need to occur are represented by the dash lines.

Topic 9.3.2: Producer Organization and Functions

For most large-scale projects, the producer (or prime contractor) will undertake the bulk of activities associated with the design and development of a new system. The customer will specify system-level and program requirements through the preparation of a request-for-proposal (RFP) or an invitation-for-bid (IFB), and various industrial firms will respond in terms of a formal proposal. As there may be a number of responding proposals, a formal competition is initiated, individual proposals are reviewed and evaluated, a selection is made, and contract negotiations take place. The successful producer (contractor) then proceeds with the project.

In addressing program requirements, it is essential that the successful producer have access to all information and data leading up to the requirements specified in the technical portions of the RFP/IFB. In some instances, the RFP will include a system specification covering the technical aspects of system design and development, along with a SOW directed toward program tasks and the management aspects of the program. As part of this total package for some programs, the customer may include an ILSP, along with the required PBL factors as an input to the design effort. In any event, the transition process from customer to producer is one of the most critical steps in the implementation of a program. It is at this stage that the necessary communications must be established.

Given that system-level requirements are properly defined and that a producer (prime contractor) has been selected to design and develop a new system, the next step is to discuss the subject of logistics (and supportability) in the context of the producer's organization. Organizational structures will vary from the pure *functional* approach to *project* or *product line*, *matrix*, combined *functional (staff)/project*, and so on. These organizational patterns are highlighted in the subsections immediately following.

1. *Functional organizational structure.* The primary building block for most organizational patterns is the functional structure reflected in Figure 9.3.2-1. This approach involves the grouping of specialties or disciplines into separately identifiable entities. The intent is to perform similar activities within one organizational element. For example, all engineering would be the responsibility of one executive, all logistics activities would be the responsibility of another executive, and so on.

   As with any organizational structure, there are advantages and disadvantages. The President (or General Manager) has under his/her control all of the functional entities necessary to design and develop, produce, deliver, and support a system. Each department maintains a strong concentration of technical expertise, so that a project can benefit from the most advanced technology in the field. Additionally, levels of authority and responsibility are clearly defined, communication channels are well structured, and the necessary controls over budgets and costs can be easily established. In general, this type of organizational structure is well suited for a large single project.

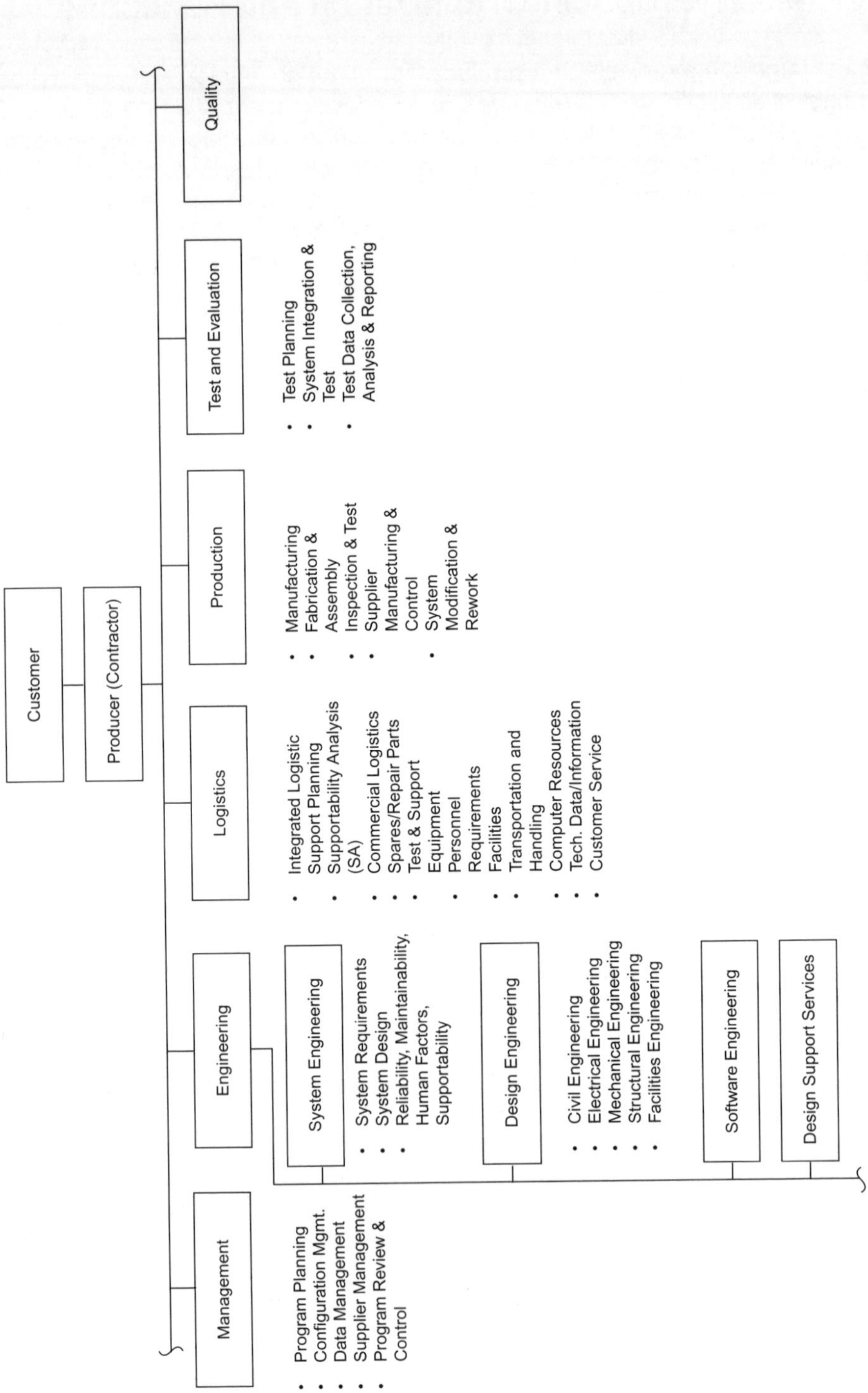

Figure 9.3.2-1. Producer Functional Organization (Example)

Conversely, the pure functional structure may not be as appropriate for large multi-project firms or agencies. Where there are many different projects, each competing for special attention and the necessary resources, there are some disadvantages. The main problem is that there is no central authority or individual responsible for the total system project. As a result, the integration of activities and the crossing of functional lines often become difficult. Conflicts occur as each functional activity struggles for power and resources, and decisions are often made on the basis of what is best for the functional group rather than what is best for the project. The "customer focus" is often lacking. Further, the decision-making processes are sometimes slow and tedious because all communications must be channeled through upper-management. Projects may fall behind and suffer in the classical functional organization structure.

2. *Project organizational structure.* As industrial firms grow and there are more systems (products) being developed, it is often convenient to establish a number of different projects, as illustrated in Figure 9.3.2-2. A project organization is one that is solely responsive to the planning, design and development, production, and support of a single system or a large product. It is time-limited and directly oriented to the life cycle of the system, and the commitment of personnel and materials is purely for the purposes of accomplishing tasks unique to that system. Each project will contain its own engineering function, logistics capability, production operations, and so on. The Project Manager has the authority and responsibility for all aspects of the project, whether it is success or failure. The project structure is extremely "customer-focused."

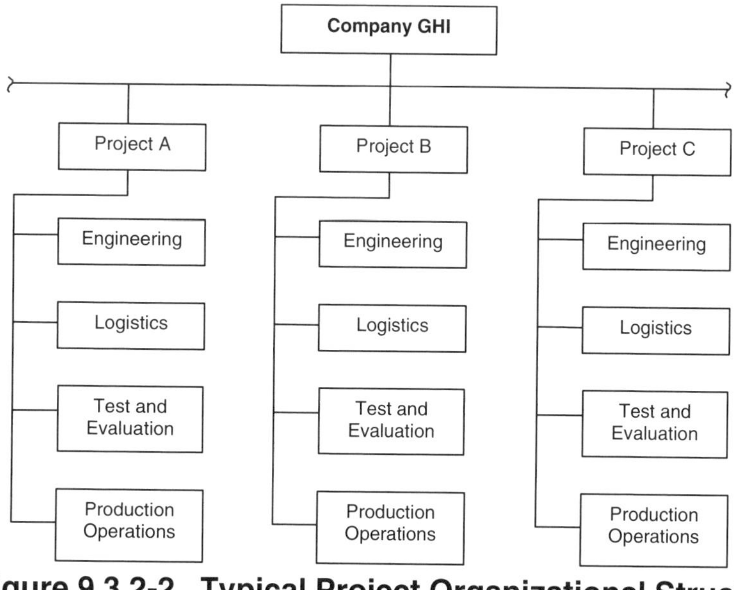

Figure 9.3.2-2. Typical Project Organizational Structure

In the case of the pure project organization, the lines of authority and responsibility are clearly defined, and there is little question as to priorities. On the other hand, there is potential for the duplication of activities within a firm, which could prove to be quite costly. This, in turn, may result in the reduction of the resources necessary to develop and maintain a high-quality level of effort for any specific project because of having to spread those resources that are available across too many other organizational entities. Decentralization in key areas often results in the "weakening" of a firm's capability.

3. *Matrix organizational structure.* The matrix organization structure attempts to combine the advantages of the pure functional organization and the pure project organization. In the functional organization, technology is emphasized while project-oriented tasks, schedules, and time constraints are often sacrificed. For the pure project, technology tends to suffer, since there is no single and independent group for its planning and development. Matrix management is an attempt to acquire the greatest amount of technology consistent with project schedules, time and cost constraints, and related customer requirements. Figure 9.3.2-3 presents a typical matrix organizational structure. Project responsibilities extend horizontally, while functional responsibilities are vertical.

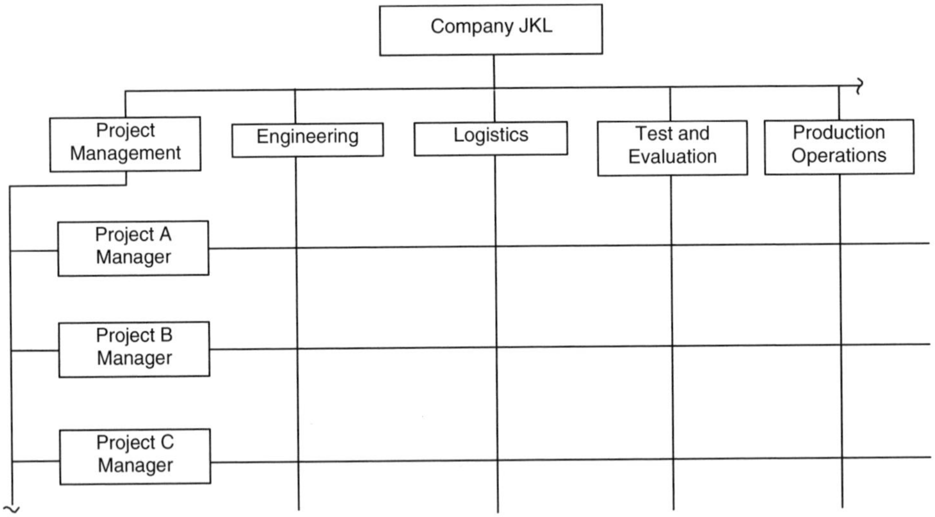

Figure 9.3.2-3. Typical Matrix Organizational Structure

Referring to the figure, each Project Manager reports to a Vice President (or equivalent) and has the overall responsibility and is accountable for project success. At the same time, the functional departments are responsible for maintaining technical excellence and for ensuring that all available technical information and support is exchanged between the projects. The functional managers, who also report to a Vice President, are responsible for ensuring that their personnel are highly trained and knowledgeable in their respective fields of endeavor.

To ensure success in the implementation of matrix management, a highly cooperative and mutually supportive environment must be created within the company. Managers and workers alike must be committed to the objectives of matrix management. A few key point are noted as follows:

a. Good communication channels (vertical and horizontal) must be established to allow for a free and continuing flow of information between projects and the functional departments.

b. Both project and functional managers should participate in the initial establishment of company-wide and program-oriented objectives.

c. In the event of conflict, a quick and effective method for resolution must be established. A procedure must be developed with the participation and commitment from both project and functional managers.

d. For personnel representing the technical functions and assigned to a project, both the Project Manager and the Functional Department Manager should agree to the duration of assignment, the tasks to be accomplished, and the basis upon which the individual(s) involved will be evaluated.

The matrix structure provides the best of several worlds; that is, it is a composite of the pure project approach and the traditional functional approach. The main advantage pertains to the capability of providing a mix of technology and project-related activities. At the same time, a major disadvantage relates to the conflicts that arise on a continuing basis as a result of a power struggle between project and functional managers, changes in priorities, and so on.

The preceding material provides an overview of the major characteristics of the functional, project, and matrix organizational structures. The advantages and disadvantages of each need to be assessed in arriving at a preferred organization for logistics. Within the broad and interdisciplinary field of logistics there are many activities for which a centralized integrated functional organization approach is preferred, and there are other areas for which a strong project organization appears to be desirable. For instance, in the processing of supportability analysis (SA) data and in the production of technical manuals, it is often more efficient to integrate these activities within a functional organization in support of all projects. This allows data (and experience) from one project to be used to help fulfill the requirements for another. In the publications area, the combining of equipment and associated resources provides for a better overall production capability.

Conversely, for activities that are directly related to the design and development of a particular system (e.g., development of design criteria, design participation and the accomplishment of trade-off analyses, design review and evaluation), it is more appropriate to assume a project organizational approach. The design-related tasks of reliability, maintainability, human factors, supportability, and those SA tasks that directly support the day-to-day design process can be accomplished more effectively through a pure project structure. In fact, it is often appropriate to physically locate those individuals who are assigned to design-oriented tasks in the same area with the responsible system/product designers. The objective is to promote the "team" approach to design and to enhance the communications among its members.

Given the wide variety of functions within the overall logistics domain, it is often feasible to establish an organizational configuration similar to the structure in Figure 9.3.2-4. In this instance, a large proportion of the logistics activities are accomplished through a functionally-oriented logistics department; however, design-related tasks are accomplished for each project, and many additional tasks require support from marketing, purchasing and contracts, test and evaluation, production operations, and quality assurance. Some of the major activity interfaces are identified in Table 9.3.2-1.

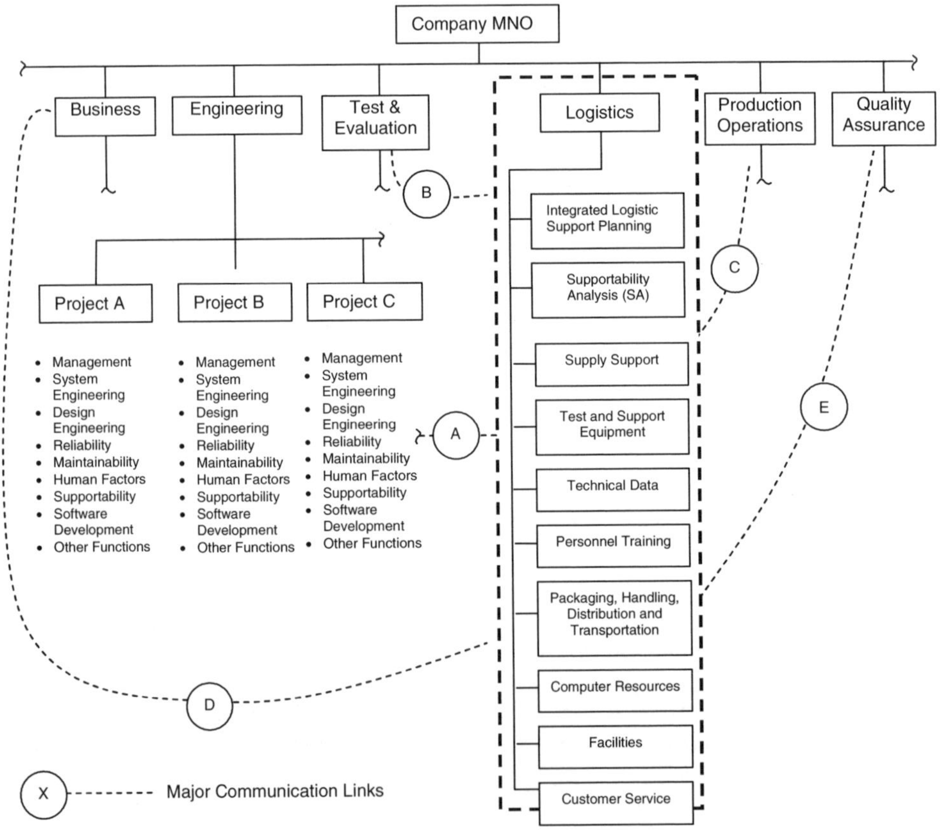

Figure 9.3.2-4. Combined Project-Staff Organizational Structure

Table 9.3.2-1. Major Logistics Activity Interfaces

Communication Channel (Figure 9.3.2-4)	Supporting Organization (Interface Requirements)
A	Project Activities: 1. *Project Management:* to determine specific logistics requirements for a given project (i.e., tasks, schedules, cost, etc.), and to establish continuing liaison and close communications to ensure successful completion of project tasks. 2. *System Engineering*: to provide initial design input and to acquire documentation from feasibility analyses, system operational requirements, the maintenance concept, functional analyses, and the allocation of requirements; to provide assistance in the preparation of the system specification and subordinate specifications as applicable; and to participate in formal design reviews. 3. *Design Engineering:* to provide initial design criteria for supportability; to provide day-to-day assistance in design; to acquire design information/data for supportability analyses, review and evaluation; and to initiate recommendations for improvement/corrective action as feasible. 4. *Reliability Engineering:* to establish and maintain close liaison in the initial determination of reliability requirements for a given program and in the subsequent completion of reliability tasks; and to acquire reliability information/data for the supportability analysis. 5. *Maintainability Engineering:* to establish and maintain close liaison in the initial determination of maintainability requirements for a given program and in the subsequent completion of maintainability tasks; and to acquire maintainability information/data for the supportability analysis. 6. *Human Factors Engineering:* to establish and maintain close liaison in the initial determination of human factors requirements for a given program and in the subsequent completion of human factors tasks; and to acquire human factors information/data for the supportability analysis. 7. *Supportability Engineering:* to determine supportability requirements for a given program and to accomplish logistics engineering functions in the design. 8. *Software Development:* to provide initial design criteria for software development requirements as they pertain to system maintenance and support, and to ensure the ultimate compatibility between system software and computer resources for logistics.
B	*Test and Evaluation*: to establish initial requirements for supportability test and evaluation; to directly support system test and evaluation activities; and to review and evaluate tests and evaluation results for supportability assessment.

Table 9.3.2-1. Major Logistics Activity Interfaces (Cont'd)

Communication Channel (Figure 9.3.2-4)	Supporting Organization (Interface Requirements)
C	*Production Operations:* (1) to assist in establishing the initial requirements for and to maintain ongoing support relative to the flow of materials in the manufacture, assembly and test of prime mission-oriented equipment; (2) to establish the initial requirements for and to provide ongoing support in the manufacture, assembly and test of spare/repair parts, special test equipment, etc.; and (3) to provide assistance relative to the packaging, handling, distribution, warehousing and transportation of system components to the consumer.
D	Business Operations: 1. *Marketing and Sales:* to acquire and sustain the necessary communications with the customer. Supplemental information pertaining to customer requirements, system operational and maintenance support requirements, changes in requirements, outside competition, etc., is needed. This is above and beyond the formal "contractual" channel of communications. 2. *Accounting*: to acquire both budgetary and actual cost data in support of economic analysis efforts (e.g., life-cycle cost analyses). 3. *Purchasing:* to assist in the identification, evaluation, and selection of component suppliers (with regard to supportability implications) and in the provision and procurement of components. 4. *Human Resources (Personnel):* to solicit assistance in the initial recruiting and hiring of qualified logistics personnel, and in the subsequent training and maintenance of personnel skills; to conduct training programs in logistics concepts, objectives, and the implementation of program requirements. 5. *Contract Management:* to keep abreast of contractual requirements (of a technical nature) between customer and the contractor; to ensure the appropriate contract structure and relationships are established and maintained with suppliers in meeting the logistics needs for a given program.
E	*Quality Assurance:* to establish initial quality requirements for the elements of logistic support, and to maintain quality standards in the acquisition of these elements through supplier activity; to provide support for the company wide total quality management (TQM) program.

As Figure 9.3.2-4 illustrates, there may be a variety of approaches within the same company. One or two large projects may exist along with numerous smaller projects. The large projects will tend to support an organizational structure similar to that presented in Figure 9.3.2-2, whereas the smaller projects will probably follow a matrix organization as shown in Figure 9.3.2-3. While the larger projects can afford to support significant numbers of personnel on a full-time basis, the smaller projects may be able to support only a select number of individuals on a part-time basis. The specific requirements are dictated through the generation of tasks by the project organization; that is, the Project Manager initiates a request for assistance, with the task(s) being completed within the functional organization.

Project size will vary not only with the type and nature of the system being developed, but with the specific stage of development. A large-scale system in the early stages of conceptual design (i.e., the Concept Refinement Phase) may be represented by a small project structure, selectively utilizing the resources from the various functional organizations as required (refer to Figure 9.3.2-4). As system development progresses into preliminary system design and detail design and development (i.e., Technology Development and System Development and Demonstration Phases), the organizational structure may shift somewhat, replicating the configuration in Figure 9.3.2-2. In other words, the characteristics and structure of organizations are often highly *dynamic* by nature. The organizational structure must be adapted to the needs of the project(s) at the time, and these needs may shift from one phase to the next.

Although there are differences from one program to the next, the logistics and supportability tasks identified in Figures 1.7-1 and 4.3-1 are assumed as being somewhat typical. These tasks cover the major areas of activity described throughout the earlier sections of this Toolkit, and their application must be properly tailored to the particular program need. It is not intended to imply that the logistics organization does everything. Further, the successful completion of these tasks is not meant to convey a tremendous level of effort. Many of these tasks are iterative in nature, starting with a minimal amount of activity during the Concept Refinement Phase and progressively expanding to include more in-depth coverage of the entire system and its elements during the System Development and Demonstration Phase. The emphasis is on the process and on maintaining continuity throughout.

Figure 9.3.2-4 and Table 9.3.2-1 depict the numerous interfaces that exist within a traditional company organization. The resources required to fulfill logistics (and supportability) objectives are located throughout many different departments (e.g., marketing, purchasing, contract management, business processes, engineering, production operations, transportation and distribution (shipping), test and evaluation, quality assurance). The ultimate success in meeting logistics objectives is, of course, highly dependent on managerial support from the top-down. The CEO (President or General Manager), Vice-President of Engineering, Vice-President of Production Operations, Project *B* Manager, and other high-level managers must each *understand* and *believe in* the principles, concepts, and objectives as presented herein. If the Logistics Manager is to be successful, these higher-level managers must be directly supportive 100% of the time. There will be numerous occasions in which individual design engineers or middle-level managers will go off on their own, making decisions that will conflict with logistics objectives. When this occurs, the Logistics Manager must have the necessary support to ensure that actions are taken to get things back on track.

Topic 9.3.3: Supplier Organization and Functions

The term *supplier*, as defined herein, refers to a broad category of organizations that provide system components to the producer (i.e., the prime contractor). These components may range from a large element of the system (e.g., a facility, an intermediate-level maintenance shop full of test equipment, Unit *B* of System *XYZ*) to a small non-repairable item (e.g., resistor, fastener, bracket, hose, cable, battery). In some instances, the components may be newly developed and require some detail design activity. The supplier will design and produce the component in the quantities desired. In other cases, the supplier will serve as a manufacturing source for the component. The supplier will produce the desired quantities from a given data set. In other words, the component design has been completed (whether by the same firm or some other), and the basic service being provided is "production." A third scenario may involve the supplier as being an inventory source for one or more common and standard off-the-shelf (COTS) items. There are no design or production activities, just distribution and materials handling functions. As one can see, the supplier roles can vary significantly.

In addressing the category of "supplier," one often finds a layering effect, as illustrated in Figure 9.3.3-1. There are a wide variety of suppliers with varying organizational patterns. Although many are functionally oriented, the practices and procedures of each will be different. With regard to logistics, it is unlikely that a supplier organization will include a department, group, or section identified as such; however, the functions described previously must be accomplished as applicable. Although the overall responsibilities for logistics (and the design for supportability) are assigned to the producer (prime contractor), there must be some identifiable organizational element responsible for the *technical integration* of those tasks assigned to the supplier. It is important that logistics concepts and objectives be established and understood from the beginning, and the contribution of each supplier is essential to the realization of these objectives.

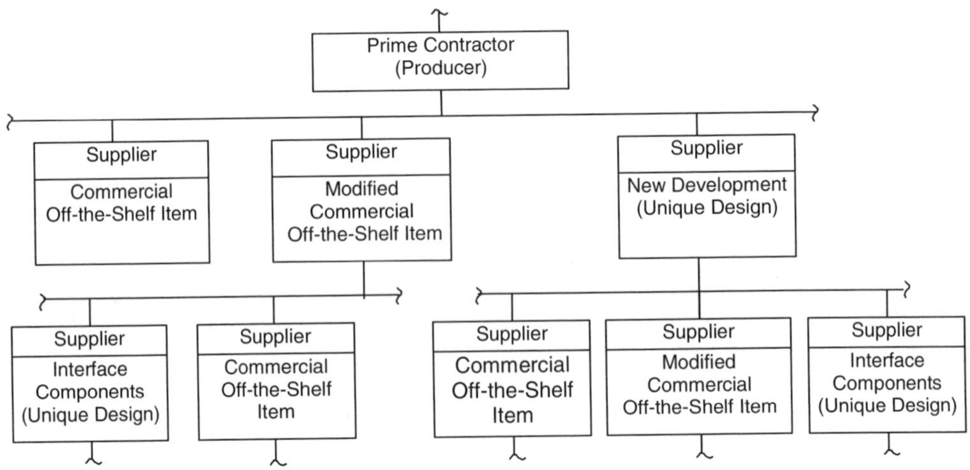

Figure 9.3.3-1. Layering of Suppliers

Of significance, in working with suppliers, is the proper specification of the requirements for each. There must be a top-down "traceability" of supportability and related requirements from the system-level and down to the subsystem and lower-level elements of the system. Referring to Topic 3.2.8 (and Figure 3.2.8-1), these requirements must first be properly stated, by the producer (prime contractor), in the appropriate specification (Type *B, C, D,* or *E*). Subsequently, there must be an on-going liaison and "teaming" relationship between the supplier and the producer. This is particularly important with suppliers who are located throughout the world, and where there must be a mutual understanding of both the technical and management objectives associated with the development of a given system.

Topic 9.3.4: Staffing the Organization

The interdisciplinary nature of logistics is such that a broad spectrum of personnel backgrounds and skills is required for the staffing of a logistics organization. There are planning and management functions, detail design engineering functions, analysis functions, technical writing functions, provisioning and procurement functions, transportation and distribution functions, warehousing functions, maintenance and customer service functions, and so on. These various functions are dispersed throughout the system life cycle and may differ from one program phase to another, and the logistics manager must be particularly sensitive to these non-homogeneous characteristics in the recruiting, hiring, placement, and supervision of personnel. While there certainly are different personnel classifications depending on the organizational approach, Figure 9.3.4-1 is presented to convey an example of how these various personnel skills can be applied during the early phases of the system life cycle. In the interest of simplicity, four basic categories have been assumed for discussion purposes.[5]

Referring to the figure, these categories include:

1. *Logistics (supportability) requirements and planning.* A person who has an in-depth understanding of logistics functions throughout the system life cycle, who can analyze technical requirements and define design criteria for supportability, and who is aware of the interrelationships among the elements of the logistics and maintenance support infrastructure and with the other facets of the program. This individual would be responsible for providing input to the System Specification (Type "A") and the System Engineering Management Plan (SEMP), and for preparing the Integrated Logistics Support Plan (ILSP).
2. *System/product design support.* A design-oriented individual, preferably with some design experience, who is knowledgeable about current design methods (e.g., CAD, CAM, CAS) and has an in-depth knowledge of customer (user) operations, organization, facilities, skills, and the system operational and maintenance support environment in general. This person must have an engineering background, be able to converse fluently with the design engineer (on the latter's terms), and must be able to translate system requirements and field experience into specific supportability and related design criteria and subsequent requirements.
3. *Supportability analysis (SA).* A person with a combination of skills in reliability and maintainability analyses, maintenance task analysis, statistical analysis, logistics modeling, computer-aided support (CAS) requirements,

[5] For a more detailed coverage of personnel and staffing requirements, refer to Blanchard, B.S., *Logistics Engineering and Management*, 6th Ed., Pearson Prentice Hall, Upper Saddle River, NJ, 2004, Section 9.6.5, pages 425-428.

technical documentation, provisioning and procurement, and data processing methods. This person must be capable of reviewing and evaluating design data, identifying future maintenance tasks and the specific resource requirements for logistics (e.g., spare/repair parts, test and support equipment, facilities, personnel and training, technical data and information, etc.), developing logistics management information (LMI), and processing supportability analysis (SA) data. This individual must be thoroughly familiar with the customer (user) environment, levels of maintenance and support, user operating and maintenance procedures, organizational structure and personnel skills, and should have had some actual hands-on logistics and maintenance experience in the field.

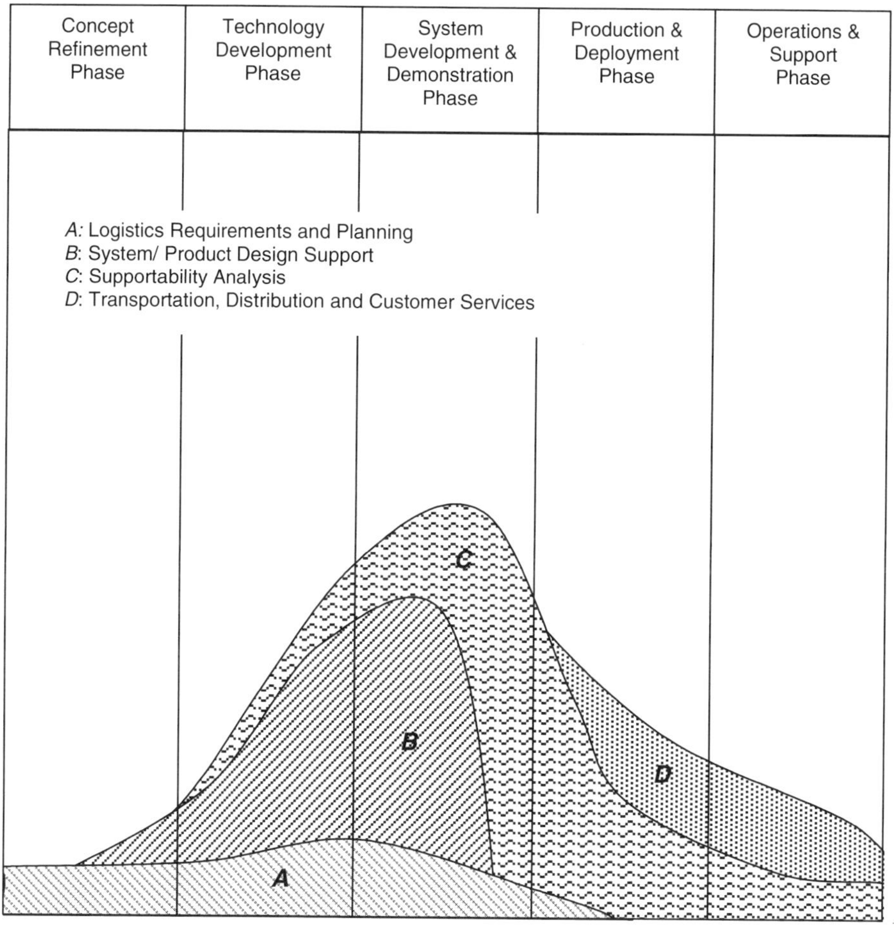

Figure 9.3.4-1. Assigned Logistics Personnel Loading

4. *Transportation, distribution, and customer service.* A person knowledgeable of the different facets of the overall supply chain (SC), supply chain

management (SCM), material flow, and product support in the field. The supply chain includes the initial procurement of materials, material flow throughout the production stage, packaging, transportation and distribution of products from the producer, warehousing, the final distribution and installation of the system and its elements at the user's site, the training of customer personnel and the sustaining customer service as required. Personnel in this area should be conversant in business practices, financial analysis, operations research methods, principles of inventory control and management, statistics, and logistics modeling. The other major thrust within this category is product support, or the sustaining maintenance and support of the system/product during its period of utilization. Such support may be in the form of a technician dispatched to the field to perform maintenance on the system, or to install some modification. Persons performing this activity should have a good knowledge of the system and its components, and should know how the system is to perform its operations in the field.

In staffing an organization, the logistics manager needs to develop the appropriate mix of personnel backgrounds and skills. The staffing, of course, will require a close working relationship with the Human Resources Department (or equivalent) in establishing the initial requirements for personnel, in developing position descriptions and advertising material, in recruiting and the conducting of interviews, in the selection of qualified candidates, and in the final hiring of individuals for employment within the logistics organization.[6]

[6] In most companies/agencies, the Human Resources Department (or some organizational entity with an equivalent function) is responsible for establishing job classifications and salary structures for the recruiting and hiring of personnel, for initiating employee benefit coverage, for providing employee opportunities for education and training, and so on. It is incumbent on the logistics department manager to ensure that his/her organizational requirements are initially understood and subsequently met through recruiting, employment, and training activities.

Topic 9.3.5: Measurement and Evaluation of Organizations

An important objective in the establishment and operation of any organizational entity is the implementation of a *measurement and evaluation capability* for that organization. This is particularly important when the organization is highly interdisciplinary by nature and is very much dependent on other organizational elements to accomplish its day-to-day functions and tasks. The logistics organization (as presented in Figures 9.3.1-1 through 9.3.2-4) is very much dependent on various facets of project engineering, design engineering, production, quality assurance, the activities of various suppliers, and associated business activities in order to successfully meet the objectives of *design for supportability*, as described throughout this Toolkit. Given this, and assuming that an organization such as "logistics" must maintain its *technical competency* and gain the *respect* from the outside, it is important that the logistics organization establish some key "benchmarks" and prepare and implement an *Organizational Development Plan*, with future growth as an objective.

The term *benchmark* may be defined in different ways, depending on one's individual background and experience. *Webster's Collegiate Dictionary* (10th Edition) defines "benchmark" as *a point of reference from which measurements may be made; something that serves as a standard by which others may be measured.* With this in mind, it would be appropriate for the logistics organization to "baseline" its current organizational capability and to establish some "benchmarks" (goals and objectives) for the future. Such benchmarks need to address both the *technical* capability (or expertise) of the organization and the *effectiveness* and *efficiency* of the organization in its implementation of the required day-to-day program/project activities.[7]

With regard to a logistics organization, one could select such key areas as: *knowledge of the logistics discipline; capability to perform all logistics tasks; organizational effectiveness; organizational efficiency; ability to attract new programs/projects; supplier relationships; customer relations*; and so on. Specific measures can be established for each category from the beginning, and a system can be implemented for the "tracking" of these measures over a specified period of time (i.e., at least annually). For a similar approach, one should refer to the literature and the implementation of such tools as the *Systems Engineering Capability Maturity Model (SE-CMM),* the *System Engineering Capability Assessment Model (SECAM),* and/or the *Capability Maturity Model Integration (CMMI).* These tools/models are currently being used to aid in the assessment of many system engineering organizations and

[7] Three references for benchmarking are: (a) Camp, R.C., *Benchmarking – The Search for Industry Best Practices That Lead to Superior Performance*, ASQ Quality Press, Milwaukee, 1989; (b) Balm, G.J., *Benchmarking: A Practitioner's Guide for Becoming and Staying the Best of the Best*, QPMA Press, Schaumburg, IL, 1992; and (c) Blanchard, B.S., *System Engineering Management*, 3rd Ed., John Wiley & Sons, Hoboken, NJ, 2004, pages 381-383.

software engineering projects, and it is envisioned that this same approach can be applied in the measurement and evaluation of a logistics organization.[8]

In summary, it is extremely important that the logistics organization, independent of the structure within which it belongs (i.e., whether functional, project, matrix, or various combinations of the organizational structures described in Topics 9.3.1-9.3.3), gain the respect and viability that will lead to growth and the effectiveness in meeting the objectives and performing the functions/tasks described throughout this Toolkit.

[8] Three references are: (a) EIA/IS-731, "*Systems Engineering Capability Model (SECM),*" Government Electronics and Information Technology Association, Washington, DC, 2001; (b) SECMM-95-01, *A Systems Engineering Capability Maturity Model (SE-CMM)*, Carnegie Melon University, Software Engineering Institute (SEI), Pittsburgh, PA, 1995; and (c) Blanchard, B.S., *System Engineering Management*, 3rd Ed., John Wiley & Sons, Hoboken, NJ, 2004, pages 383-389.

Topic 9.4: Management and Control

Successful program implementation includes the day-to-day managerial functions associated with the influencing, guiding, and supervising of personnel in the accomplishment of the organizational activities identified in the Integrated Logistics Support Plan (ILSP). This implementation requires a continuing awareness, on the part of management, of the specifics pertaining to the tasks being performed and of whether the results are fulfilling the program objectives in an effective manner. The results desired include those factors identified through the definition of system operational requirements, the maintenance concept, the identification and prioritization of TPM/PBL factors, and the functional analysis and allocation process (refer to Topic 3.2).

The "controlling" aspect pertains to: (1) the ongoing monitoring of program activities in order to assess the status of the tasks being performed at designated times in the program and the results of those tasks in terms of the desired characteristics and metrics pertaining to the system being developed or re-engineered, and (2) the initiation of corrective action as required to overcome any noted deficiencies in the event that program objectives are not being met. Periodic program management reviews are usually conducted throughout the system design and development process. The nature and frequency of these reviews will vary with each program and will be a function of design complexity, the number and location of organizations participating in the program, the number of suppliers, and the number of problems being encountered. Systems being developed that incorporate new technologies, where the risks are usually greater, will require more frequent reviews than will other systems.

The objective of these management-oriented reviews is to determine the status of selected program tasks, to assess the degree of progress with respect to fulfilling the system program requirements, to review program schedule and cost data, to review supplier activity, and to coordinate the results of the technical design reviews (refer to Topic 4.3). These reviews are directed toward dealing with issues of a program nature and are not intended to duplicate the formal design review process, which deals with design-related problems of a technical nature. However, of common interest is the "tracking" of critical TPM/PBL factors, as illustrated in Figure 9.4-1.

The data system that includes information needed for these program reviews must be designed in such a way as to provide the right type of information content, to the appropriate locations and in the proper format, in a timely manner and at the desired frequency, with the right degree of reliability, and at the right cost. The nature and content of the data should be based on several factors, both of a technical nature (e.g., TPM/PBL factors) and of a program management nature (e.g., program schedules and cost projections and status). In addition to the more conventional type of program/ project reporting, including schedule and cost status, the applicable TPM/PBL factors must be addressed as well.

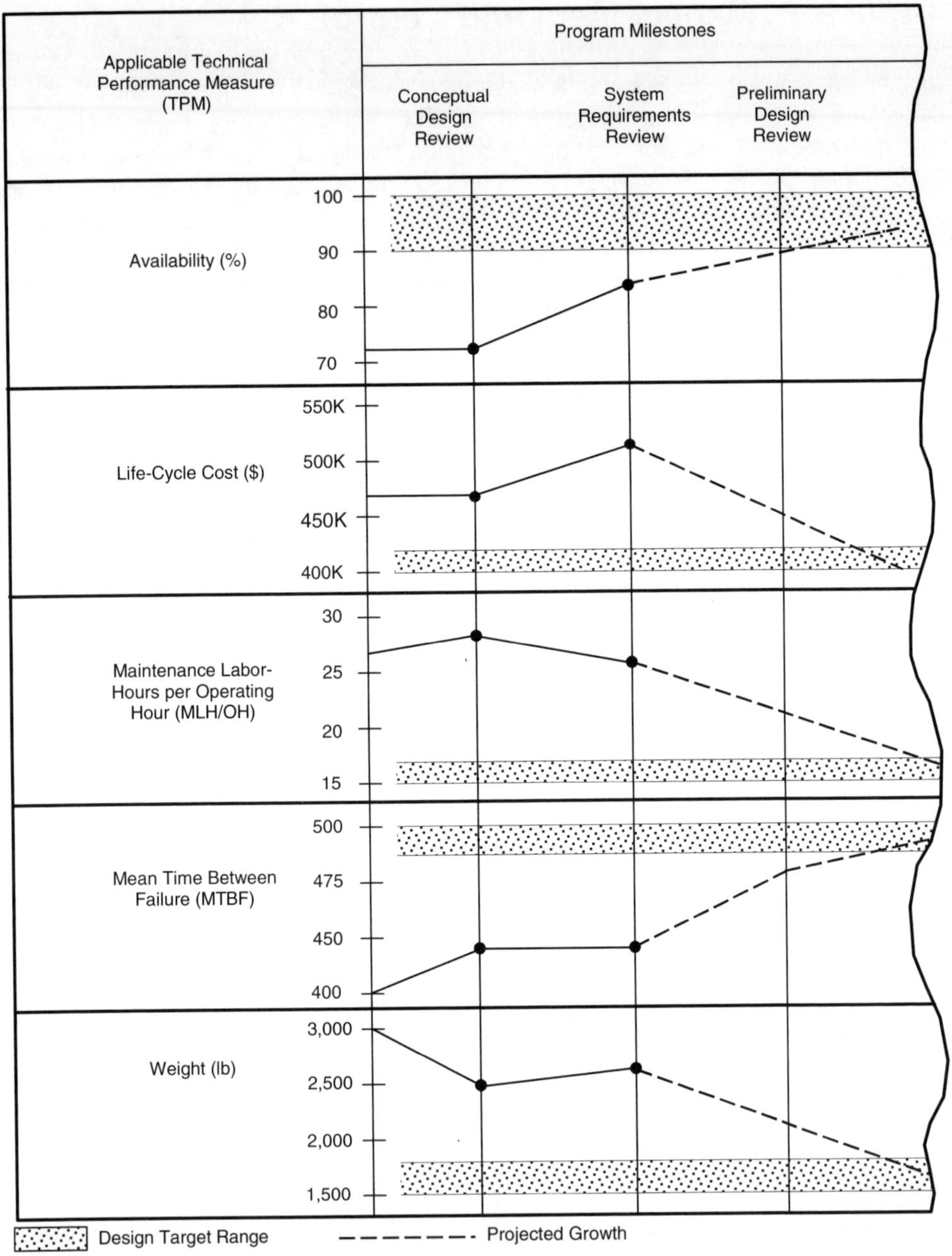

Figure 9.4-1. Performance Parameter Measurement at Significant Milestones (e.g., SRR)

In summary, the successful implementation of the objectives specified throughout this Toolkit depends on both a good *technical* approach in the design and development of systems and on establishing the proper *management* environment to facilitate its happening. The objective in the preparation and publication of this Toolkit is to address both sides of the spectrum. The authors acknowledge that a greater depth of material coverage could have been provided in any one of the topics discussed herein. However, it is hoped that what has been included provides a good overview of the subject, and will be helpful in the implementation of your day-to-day activities pertaining to the *design for supportability.*

For More Information:

1. Blanchard, B.S., *Logistics Engineering and Management*, 6th Ed., Pearson Prentice Hall, Upper Saddle River, NJ, 2004.
2. Blanchard, B.S., *System Engineering Management*, 3rd Ed., John Wiley & Sons, Hoboken, NJ, 2004.
3. Appendix A, "Selected Bibliography," Section K (Management and Supporting Areas -- Refer to Project/Program Management References).

This page intentionally left blank

Appendix A
Logistics -- Selected Bibliography

This page intentionally left blank

When addressing the subject of *logistics*, one should become familiar not only with the available literature in this field, but with some of the subject areas that are closely aligned with logistics. "Logistics," by nature, is *interdisciplinary* and acquiring knowledge in these related areas is essential if one is to progress and successfully accomplish his/her objectives in the field. With this in mind, this bibliography has been developed to cover a few selected references in each of the following areas:

A. Logistics, Logistics Engineering, Supply Chain Management, Supportability, and Integrated Logistic Support (ILS).
B. Systems, Systems Engineering, and Systems Analysis.
C. Concurrent Engineering.
D. Software and Computer-Aided Systems.
E. Reliability Engineering.
F. Maintainability Engineering and Maintenance.
G. Human Factors and Safety Engineering.
H. Production, Manufacturing, Quality, Quality Control, and Quality Assurance.
I. Operations Research and Operations Management.
J. Engineering Economy, Life-Cycle Cost Analysis, and Cost Estimation.
K. Management and Supporting Areas.

A. Logistics, Logistics Engineering, Supply Chain Management, Supportability, and Integrated Logistic Support (ILS)

1. Ackerman, K.B., *Practical Handbook of Warehousing*, 4th Ed., Chapman & Hall, 1997, New York, NY (ISBN 0412125110).
2. Arnold, T. and S. Chapman, *Introduction to Materials Management*, 4th Ed., Prentice Hall, Upper Saddle River, NJ, 2001 (ISBN 0130144908).
3. Ballou, R.H., *Business Logistics Management: Planning, Organizing, and Controlling the Supply Chain*, 4th Ed., Prentice Hall, Upper Saddle River, NJ,1998 (ISBN 0137956592).
4. Blanchard, B.S., *Logistics Engineering and Management*, 6th Ed., Prentice Hall, Upper Saddle River, NJ, 2004 (ISBN 0-13-142915-9).
5. Bowersox, D.J. and D.J. Closs, *Logistical Management: The Integrated Supply Chain Process*, McGraw Hill, New York, NY, 1996 (ISBN 0070068836).
6. Christopher, M., *Logistics and Supply Chain Management*, 2nd Ed., Prentice Hall, Upper Saddle River, NJ, 1998 (ISBN 0273630490).
7. Council of Logistics Management (CLM), *Logistics Comment*, CLM Newsletter, 2805 Butterfield Rd., Suite 200, Oak Brook, IL 60523.
8. Council of Logistics Management (CLM), *Annual Conference Proceedings*, CLM, 2805 Butterfield Rd., Suite 200, Oak Brook, IL 60523.
9. Council of Logistics Management (CLM), *Journal of Business Logistics*, CLM, 2805 Butterfield Rd., Suite 200, Oak Brook, IL 60523.

10. Copacino, W.C., *Supply Chain Management: The Basics and Beyond*, The St. Lucie Press/APICS Series on Resource Management, Boca Raton, FL, 1997.
11. Coyle, J.J., E.J. Bardi, and R.A. Novack, *Transportation*, 5th Ed., South-Western Publisher, St. Paul, MN, 2000 (ISBN 0538881801).
12. Coyle, J.J., E.J. Bardi, and C.J. Langley, *The Management of Business Logistics*, 7th Ed., South-Western, Mason, OH, 2003 (ISBN 032007515).
13. DoD, *Designing and Assessing Supportability in DoD Weapon Systems: A Guide to Increased Reliability and Reduced Logistics Footprint*, Department of Defense, Washington, DC, 2003.
14. Frazelle, E.H., *Supply Chain Strategy: The Logistics of Supply Chain Management*, McGraw-Hill, New York, NY, 2002 (ISBN 0-07-137599-6).
15. Handfield, R.B. and E.Z. Nichols, *Introduction to Supply Chain Management*, Prentice Hall, Upper Saddle River, NJ, 1998 (ISBN 0136216161).
16. Jones, J.V., *Integrated Logistics Support Handbook*, Special Reprint Ed., McGraw-Hill, New York, NY, 1998 (ISBN 0070331391).
17. Lambert, D.M., J.R. Stock, and L.M. Ellram, *Fundamentals of Logistics Management*, McGraw-Hill, New York, NY, 1998 (ISBN 0256141177).
18. Langford, J.W., *Logistics: Principles and Applications*, McGraw-Hill, New York, NY, 1995 (ISBN 007036415X).
19. MIL-HDBK-59A, Military Handbook, *Computer-Aided Acquisition Logistic Support (CALS) Implementation Guide*, Department of Defense, Washington, DC (latest edition).
20. MIL-HDBK-502, Department of Defense Handbook, *Acquisition Logistics*, Department of Defense, Washington, DC (latest edition).
21. MIL-PRF-49506, Performance Specification, *Logistics Management Information*, Department of Defense, Washington, DC (latest edition).
22. MIL-STD-1840A, Military Standard, *Automated Interchange of Technical Information*, Department of Defense, Washington, DC (latest edition).
23. Ministry of Defence, Defence Standard 00-60, *Integrated Logistic Support*, Kentigern House, 65 Brown St., Glasgow G2 8EX, United Kingdom (latest ed.), 1996 (ISBN 0899309852).
24. Reed Business Information, *Supply Chain Management Review*, Reed Elsevier, Inc., 360 Park Ave. South, New York, NY (bi-monthly publication).
25. SOLE-The International Society of Logistics, *Logistics Spectrum*, Quarterly Journal, SOLE, 8100 Professional Place, Suite 111, Hyattsville, MD 20785 (ISSN 00245852).
26. SOLE-The International Society of Logistics, *SOLEtech*, Monthly Newsletter, SOLE, 8100 Professional Place, Suite 111, Hyattsville, MD 20785.
27. SOLE-The International Society of Logistics, *Annual Symposium Proceedings*, SOLE, 8100 Professional Place, Suite 111, Hyattsville, MD 20785.

28. Stock, J.R., *Development and Implementation of Reverse Logistics Program*, CLM Press, Oak Brook, IL, 1998.
29. Tilanus, B., *Information Systems in Logistics and Transformation*, 2nd Ed., Elsevier Science Ltd., New York, NY, 1997 (ISBN 0080430546).
30. Tompkins, J.A. and J.D. Smith, *The Warehouse Management Handbook*, Tompkins Press, Raleigh, NC, 1998.
31. U.S. Air Force, *Air Force Journal of Logistics*, Quarterly Journal, Superintendent of Documents, U.S. Government Printing Office, Washington, DC 20402.

B. Systems, Systems Engineering, and Systems Analysis

1. Andriole, S.J., *Managing System Requirements: Methods, Tools, and Cases*, Prentice Hall, Upper Saddle River, NJ, 1996 (ISBN 0070019746).
2. Blanchard, B.S., *System Engineering Management*, 3rd Ed., John Wiley & Sons, Hoboken, NJ, 2004 (ISBN 0471190861).
3. Blanchard, B.S. and W.J. Fabrycky, *Systems Engineering and Analysis*, 3rd Ed., Prentice Hall, Upper Saddle River, NJ, 1998 (ISBN 0131350471).
4. Buede, D.M., *The Engineering Design of Systems: Models and Methods*, John Wiley & Sons, Hoboken, NJ, 2000 (ISBN 0-471-28225-1).
5. Checkland, P., *Systems Thinking, Systems Practice*, John Wiley & Sons, Hoboken, NJ, 1981 (ISBN 0471279110).
6. EIA/IS-632, *Processes For Engineering A System*, Electronic Industries Association (EIA), Washington, DC (latest edition).
7. EIA/IS-731, *System Engineering Capability Maturity Model (SECM)*, Electronic Industries Association (EIA), Washington, DC (latest edition).
8. Eisner, H., *Essentials of Project and Systems Engineering Management*, John Wiley & Sons, Hoboken, NJ, 1997 (ISBN 0471148466).
9. Forrester, J.W., *Principles of Systems*, The MIT Press, Cambridge, MA, 1968.
10. Grady, J.O., *System Engineering Deployment*, CRC Press, Boca Raton, FL, 2000 (ISBN 0-8493-7839-7).
11. Grady, J.O, *System Engineering Planning and Enterprise Identity*, CRC Press, Boca Raton, FL, 1995 (ISBN 084937832X).
12. Hall, A.D., *A Methodology for Systems Engineering,* Van Nostrand, Princeton, NJ, 1962.
13. IEEE-1220, *Standard for Application and Management of the Systems Engineering Process*, Institute of Electrical and Electronics Engineers (IEEE), 345 East 47th Street, New York, NY 10017 (latest edition).
14. INCOSE-The International Council on Systems Engineering, *Systems Engineering*, Quarterly Journal, published by John Wiley & Sons, Hoboken, NJ.

15. INCOSE-The International Council on Systems Engineering, *INSIGHT*, Quarterly Newsletter, INCOSE, 2150 North 107th Street, Suite 205, Seattle, WA 98133-9009.

16. Kossiakoff, A. and W. Sweet, *Systems Engineering: Principles and Practice*, John Wiley & Sons, Hoboken, NJ, 2003 (ISBN 0471234435).

17. Maier, M.W. and E. Rechtin, *The Art of Systems Architecting*, 2nd Ed., CRC Press, Boca Raton, FL, 2000 (ISBN 0849378362).

18. Martin, J.N., *Systems Engineering Guidebook: A Process for Developing Systems and Products*, CRC Press, Boca Raton, FL, 1996 (ISBN 0849378370).

19. Pugh, S., *Total Design: Integrated Methods for Successful Product Engineering*, Addison-Wesley Publishing, Reading, MA, 1991 (ISBN 0-201-41639-5).

20. Rechtin, E., *Systems Architecting: Creating and Building Complex Systems*, Prentice Hall, Upper Saddle River, NJ, 1991 (ISBN 0138803455).

21. Sage, A.P., *Systems Engineering*, John Wiley & Sons, Hoboken, NJ, 1992 (ISBN 0471536393).

22. Sage, A.P. and W.B. Rouse (Editors), *Handbook of Systems Engineering and Management*, John Wiley & Sons, Hoboken, NJ, 1999 (ISBN 0471154059).

23. Sage, A.P. and J.E. Armstrong, *Introduction to Systems Engineering*, John Wiley & Sons, Hoboken, NJ, 2000 (ISBN 0-471-02766-9).

24. Sommerville, I. and P. Sawyer, *Requirements Engineering: A Good Practice Guide*, John Wiley & Sons, Hoboken, NJ, 1997 (ISBN 0471974447).

25. Thome, B., *Systems Engineering: Principles and Practice of Computer-Based Systems Engineering*, John Wiley & Sons, Hoboken, NJ 1992 (ISBN 0471935522).

26. Young, R.R., *Effective Requirements Practices*, Addison-Wesley Publishing, Reading, MA, 2001 (ISBN 0-201-70912-0).

C. Concurrent Engineering

1. Hartley, J.R., *Concurrent Engineering: Shortening Lead Times, Raising Quality, and Lowering Costs*, Productivity Press, Portland, OR, 1998 (ISBN 1563271893).

2. Kusiak, A. (Editor), *Concurrent Engineering: Automation, Tools, and Techniques*, John Wiley & Sons, Hoboken, NJ, 1992 (ISBN 0471554928).

3. Miller, L.C.G., *Concurrent Engineering Design: Integrating the Best Practices for Process Improvement*, Society of Manufacturing, New York, NY, 1993 (ISBN 0872634337).

4. Prasad, B., *Concurrent Engineering Fundamentals: Integrated Product and Process Organization*, Prentice Hall, Upper Saddle River, NJ, 1996 (ISBN 0131474634).

5. Shina, S.G. (Editor), *Successful Implementation of Concurrent Engineering Products and Processes*, John Wiley & Sons, Hoboken, NJ, 1997 (ISBN 0471285102).

D. Software and Computer-Aided Systems

1. Bass, L., P. Clements, R. Kazman, and K. Bass, *Software Architecture in Practice*, Addison-Wesley, Reading, MA, 1998 (ISBN 0201199300).
2. Boehm, B.W., *Software Engineering Economics*, Prentice Hall, Upper Saddle River, NJ, 1981 (ISBN 0138221227).
3. *Crosstalk-The Journal of Defense Software Engineering*, published by the Software Technology Support Center (STSC), 00-ALC/TISE, 7278 Fourth Street, Hill AFB, UT 84056-5205 (monthly).
4. Humphrey, W.S., *A Discipline for Software Engineering*, Addison-Wesley, Reading, MA, 1995 (ISBN 0201546108).
5. IEEE/EIA 12207, Commercial Standard, *Information Technology-Software Life-Cycle Processes*, Defense Automated Printing Services, Building 4/D, 700 Robins Avenue, Philadelphia, PA 19111-5094.
6. ISO/IEC 15939, *Software Engineering-Software Measurement Process*, 2002.
7. Jones, C., *Applied Software Measurement*, 2nd Ed., McGraw-Hill, New York, NY, 1996.
8. Jones, C., *Software Assessments, Benchmarks, and Best Practices*, Addison Wesley Longman, Reading, MA, 2000.
9. Leach, R., *Introduction to Software Engineering*, CRC Press, Boca Raton, FL, 2000 (ISBN 0-8493-1445-3).
10. McConnell, S., *Software Project Survival Guide*, Microsoft Press, Redmond, WA, 1998 (ISBN 1-57231-621-7).
11. Moore, J.W., *Software Engineering Standards; A User's Road Map*, IEEE Computer Society, Los Alamitos, CA, 1998 (ISBN 0-8186-8008-3).
12. Pfleeger, S.L., *Software Engineering: Theory and Practice*, Prentice Hall, Upper Saddle River, NJ, 1998 (ISBN 013624842X).
13. Pressman, R.S., *Software Engineering: A Practitioner's Approach*, 4th Ed., McGraw-Hill, New York, NY, 1996 (ISBN 0070521824).
14. Sage, A.P. and J.D. Palmer, *Software Systems Engineering*, John Wiley & Sons, Hoboken, NJ, 1990 (ISBN 047161758X).
15. Sage, A.P., *Systems Management for Information Technology and Software Engineering*, John Wiley & Sons, Hoboken, NJ, 1995 (ISBN 0471015830).
16. Thayer, R., M. Dorfman, and S.C. Bailin (Editors), *Software Requirements Engineering*, IEEE Computer Society, Los Alamitos, CA, 1997 (ISBN 0818677384).

17. Wiegers, K.E., *Software Requirements*, Microsoft Press, Redmond, WA, 1999 (ISBN 0-7356-0631-5).

E. Reliability Engineering

1. *Annual Reliability and Maintainability Symposium*, Proceedings, sponsored by nine technical societies, Evans Associates, 804 Vickers Avenue, Durham, NC 27701-3141.
2. Barlow, R.E., *Engineering Reliability (Statistics and Applied Probability)*, Society of Industrial & Applied Mathematics, New York, NY, 1998 (ISBN 0898714052).
3. Blischke, W.R. and D.N.P. Murthy, *Reliability: Modeling, Prediction, and Optimization*, John Wiley & Sons, Hoboken, NJ, 2000.
4. Ebeling, C.E., *An Introduction to Reliability and Maintainability Engineering*, McGraw-Hill, New York, NY, 1996 (ISBN 0070188521).
5. Elsayed, E.A., *Reliability Engineering*, Addison Wesley, Reading, MA, 1996.
6. Institute of Electrical and Electronics Engineers (IEEE), *IEEE Transactions on Reliability*, published quarterly, IEEE, P.O. Box 1331, Piscataway, NJ 08854-1331 (ISSN 00189529).
7. Ireson, W.G. (Editor), *Handbook of Reliability Engineering and Management*, McGraw Hill, New York, NY, 1996 (ISBN 0070127506).
8. Kales, P., *Reliability: For Technology, Engineering, and Management*, Prentice Hall, Upper Saddle River, NJ, 1997 (ISBN 0134858220).
9. Kececioglu, D., *Reliability Engineering Handbook, Vol. 1 and Vol. 2*, Prentice Hall, Upper Saddle River, NJ, 1991 (ISBN 013772294X).
10. Knezevic, J., *Reliability, Maintainability, and Supportability: A Probabilistic Approach*, McGraw-Hill, New York, NY, 1993 (ISBN 0077076915).
11. Lewis, E.E., *Introduction to Reliability Engineering*, 2nd Ed., John Wiley & Sons, Hoboken, NJ, 1995 (ISBN 0471018333).
12. Lyu, M.R. (Editor), *Handbook of Software Reliability Engineering*, McGraw-Hill, New York, NY, 1996 (ISBN 0070394008).
13. MIL-HDBK-217F, *Reliability Prediction For Electronic Equipment*, Department of Defense, Washington, DC (latest edition).
14. Musa, J.D., *Software Reliability Engineering: More Reliable Software, Faster Development and Testing*, McGraw-Hill, New York, NY, 1998 (ISBN 0079132715).
15. O'Connor, P.D.T., D. Newton, and R. Bromley, *Practical Reliability Engineering*, 3rd Ed., John Wiley & Sons, Hoboken, NJ, 1996 (ISBN 0471957674).
16. Palady, P., *Failure Modes and Effects Analysis*, Pitman Publishing, New York, NY, 1995 (ISBN 0945456174).
17. Pecht, M. (Editor), *Product Reliability, Maintainability, and Supportability Handbook*, CRC Press, Boca Raton, FL, 1995 (ISBN 0849394570).

18. Pham, H. (Editor), *Software Reliability and Testing*, IEEE Computer Society, Los Alamitos, CA, 1995 (ISBN 0818668520).
19. Reliability Analysis Center (RAC), *Reliability Toolkit: Commercial Practices* Edition, RAC, 201 Mill Street, Rome, NY 13440-6916, 1994.

F. Maintainability Engineering and Maintenance

1. American Institute for Total Productive Maintenance (AITPM), *TPM Report*, Monthly Newsletter, Productivity, Inc., 541 NE 20th Ave., Portland, OR 97232.
2. Blanchard, B.S., D. Verma, and E.L. Peterson, *Maintainability: A Key to Effective Serviceability and Maintenance Management*, John Wiley & Sons, Hoboken, NJ, 1995 (ISBN 0471591327).
3. Dhillon, B.S., *Engineering Maintenance*, CRC Press, Boca Raton, FL, 2002 (ISBN 1-58716-142-7).
4. Hartmann, E.H., *Successfully Installing TPM in a Non-Japanese Plant: Total Productive Maintenance*, Productivity Press, Portland, OR, 1992 (ISBN 1882258002).
5. Kececioglu, D., *Maintainability, Availability, and Operational Readiness Engineering Handbook*, Prentice Hall, Upper Saddle River, NJ, 1995 (ISBN 0135736277).
6. Knezevic, J., *Systems Maintainability: Analysis, Engineering, and Management*, Chapman and Hall, London, UK, 1997 (ISBN 0412802708).
7. Maintenance Steering Group 3 Task Force, *Operator/Manufacturer Schedule Maintenance Development*, Air Transport Association (ATA) of America, Washington, DC, April 2001 (http://www.airlines.org).
8. Moubray, J., *Reliability-Centered Maintenance*, 2nd Ed., Industrial Press, Boca Raton, FL, 1997 (ISBN 0831130784).
9. Nakajima, S., *Introduction to TPM: Total Productive Maintenance*, Productivity Press, Portland, OR, 1994 (ISBN 0915299232).
10. Nakajima, S. (Editor), *TPM Development Program: Implementing Total Productive Maintenance*, Productivity Press, Portland, OR, 1989 (ISBN 0915299372).
11. Niebel, B.W., *Engineering Maintenance Management*, 2nd Ed., Marcel Dekker, New York, NY, 1994 (ISBN 0824792475).
12. Nyman, D. and J. Levitt, *Maintenance Planning, Scheduling, and Coordination*, Industrial Press, Boca Raton, FL, 2001.
13. Reliability Analysis Center (RAC), *Maintainability Toolkit: A Practical Guide for Designing and Developing Maintainable Products and Systems*, RAC, 201 Mill Street, Rome, NY 13440-6916, 2000.
14. Smith, A.M., *Reliability-Centered Maintenance*, McGraw-Hill, New York, NY, 1992 (ISBN 007059046X).

15. Willmott, P., *Total Productive Maintenance*, Butterworth-Heinemann, New York, NY, 1995 (ISBN 0750619252).

16. Wireman, T., *Developing Performance Indicators for Managing Maintenance*, Industrial Press, Boca Raton, FL, 1998 (ISBN 0831130806).

G. Human Factors and Safety Engineering

1. Bahr, N.J., *System Safety Engineering and Risk Assessment: A Practical Approach*, Taylor & Francis, New York, NY, 1997 (ISBN 1560324163).
2. Chapanis, A., *Human Factors in Systems Engineering*, John Wiley & Sons, Hoboken, NJ, 1996 (ISBN 0-471-13782-0)
3. Kroemer, K.H.E. and H.J. Kroemer, *Engineering Physiology: Basis of Human Factors/Ergonomics*, 3rd Ed., John Wiley & Sons, Hoboken, NJ, 1997 (ISBN 0471287989).
4. Roland, H.E. and B. Moriarity, *System Safety Engineering and Management*, 2nd Ed., John Wiley & Sons, Hoboken, NJ, 1990 (ISBN 0471618160).
5. Salvendy, G. (Editor), *Handbook of Human Factors and Ergonomics*, 2nd Ed., John Wiley & Sons, Hoboken, NJ, 1997 (ISBN 0471116904).
6. Sanders, M.S. and E.J. McCormick, *Human Factors in Engineering and Design*, 7th Ed., McGraw-Hill, New York, NY, 1992 (ISBN 007054901X).
7. Wickens, C.D., S. Gordon, and Y. Liu, *An Introduction to Human Factors Engineering*, Addison-Wesley, Reading, MA, 1997 (ISBN 0321012291).

H. Production, Manufacturing, Quality, Quality Control, and Quality Assurance

1. Abbott, J.C., *SPC: Practical Understanding of Capability by Implementing Statistical Process Control*, Robert Houston Smith Publishers, New York, NY, 1999.
2. Breyfogle, F.W., J.M. Cupello, and B. Meadows, *Managing Six Sigma*, John Wiley & Sons, Hoboken, NJ, 2000.
3. Cohen, L., *Quality Function Deployment: How to Make QFD Work for You*, Addison-Wesley, Reading, MA, 1995 (ISBN 0201633302).
4. Crosby, P.B., *Quality Is Free: The Art of Making Quality Certain*, Mentor Books, Reprinted 1992, New York, NY (ISBN 0451625854).
5. Demming, W.E., *Out of Crisis*, MIT Press, Cambridge, MA 1986.
6. Fasser, Y. and D. Brettner, *Management for Quality in High-Technology Enterprises*, John Wiley & Sons, Hoboken, NJ, 2002 (ISBN 0-471-20958-9).
7. Feigenbaum, A.V., *Total Quality Control*, 3rd Ed., McGraw-Hill, New York, NY, 1991 (ISBN 0070203547).

8. Fowlkes, W.Y. and C.M. Creveling, *Engineering Methods for Robust Design: Using Taguchi Methods in Technology and Product*, Addison-Wesley, Reading, MA, 1995 (ISBN 0201633671).

9. Fryman, M., *Quality and Process Improvement*, Delmar Publishers, New York, NY, 2001.

10. Gaal, A., *ISO 9001:2000 for Small Business: Implementing Process Approach in Quality Management*, St. Lucie Press, Boca Raton, FL, 2001 (ISBN 1-57444-307-0).

11. Juran, J.M. and G.A. Blanton (Editors), *Juran's Quality Control Handbook*, 5th Ed., McGraw-Hill, New York, NY, 1999.

12. Gyrna, F.M. and J.M. Juran, *Quality Planning and Analysis: From Product Development Through Use*, 3rd Ed., McGraw-Hill, New York, NY, 1993 (ISBN 0070331839).

13. Hoyle, D., *ISO 9000 Quality Systems Development Handbook: A Systems Engineering Approach*, Butterworth-Heinemann, New York, NY, 1998 (ISBN 0750625627).

14. Kidd, P.T. and W. Karwowski (Editors), *Advances in Agile Manufacturing: Integrating Technology, Organization and People*, IOS Press, New York, NY, 1995 (ISBN 9051991762).

15. Lowenthal, J.N., *Six Sigma Project Management: A Pocket Guide*, ASQ Quality Press, Milwaukee, WI, 2002.

16. Madu, C.N. (Ed.), *Handbook of Environmentally Conscious Manufacturing*, Kluwer Academic Publishers, New York, NY, 2000.

17. McMahon, C. and J. Browne, *CADCAM: Principles, Practice and Manufacturing Management*, 2nd Ed., Addison-Wesley, Reading, MA, 1998 (ISBN 0201178192).

18. Pyzdek, T., *The Six Sigma Handbook*, McGraw-Hill, New York, NY, 2001.

19. Reliability Analysis Center (RAC), *Quality Toolkit*, RAC, 201 Mill Street, Rome, NY 13440-6916 (ISBN 0-9712853-7-3).

20. Revelle, J.B., J.W. Moran, and C. Cox, *The QFD Handbook*, John Wiley & Sons, Hoboken, NJ, 1997 (ISBN 0471173819).

21. Revelle, J.B., *Manufacturing Handbook of Best Practices: An Innovation, Productivity, and Quality Focus,* St. Lucie Press, Boca Raton, FL, 2001.

22. Ross, P.J., *Taguchi Techniques for Quality Engineering: Loss Function, Orthogonal Experiments, Parameter and Tolerance Design*, 2nd Ed., McGraw-Hill, New York, NY, 1995 (ISBN 0070539588).

23. Roy, R.K., *Design of Experiments Using the Taguchi Approach*, John Wiley & Sons, Hoboken, NJ, 2001.

24. Smith, G.M., *Statistical Process Control and Quality Improvement*, 4th Ed., Prentice Hall, Upper Saddle River, NJ, 2001 (ISBN 0130255637).

25. Swamidass, P.M. (Ed.), *Innovations in Competitive Manufacturing*, American Management Association (AMACOM), Boston, MA, 2002 (ISBN 0-7923-7896-2).

26. Tompkins, J.A., Y.A. Bozer, E. Frazelle, J. Tanchoco, and J. White, *Facilities Planning*, 2nd Ed., John Wiley & Sons, Hoboken, NJ, 1996 (ISBN 0471002526).

I. Operations Research and Operations Management

1. Eppen, G.D., *Introductory Management Science*, Prentice Hall, Upper Saddle River, NJ, 1998 (ISBN 0536012024).
2. Evans, J.R., *Production/Operations Management: Quality, Performance, and Value*, 5th Ed., West/Wadsworth, St. Paul, MN, 1996 (ISBN 0314062475).
3. Hillier, F.S. and G.J. Lieberman, *Introduction to Operations Research*, 6th Ed., McGraw-Hill, New York, NY, 1995 (ISBN 0078414474).
4. Krajewski, L.J. and L.P. Ritzman, *Operations Management: Strategy and Analysis*, 5th Ed., Addison-Wesley, Reading, MA, 1998 (ISBN 0201331187).
5. Taha, H.A., *Operations Research: An Introduction*, 6th Ed., Prentice Hall, Upper Saddle River, NJ, 1996 (ISBN 0132729156).

J. Engineering Economy, Life-Cycle Cost Analysis, and Cost Estimation

1. Canada, J.R., W.G. Sullivan, and J.A. White, *Capital Investment Analysis for Engineering and Management*, 2nd Ed., Prentice Hall, Upper Saddle River, NJ, 1996 (ISBN 0133110362).
2. Fabrycky, W.J. and B.S. Blanchard, *Life-Cycle Cost and Economic Analysis,* Prentice Hall, Upper Saddle River, NJ, 1991 (ISBN 0135383234).
3. Fabrycky, W.J., G.J. Thuesen, and D. Verma, *Economic Decision Analysis*, Prentice Hall, Upper Saddle River, NJ, 1997 (ISBN 0133702499).
4. Grant, E.L., W.G. Ireson, and R.S. Leavenworth, *Principles of Engineering Economy*, 8th Ed., John Wiley & Sons, Hoboken, NJ, 1990 (ISBN 047163526X).
5. Hicks, D.T., *Activity-Based Costing: Making it Work for Small and Mid-Sized Companies*, 2nd Ed., John Wiley & Sons, Hoboken, NJ, 1998 (ISBN 0471249599).
6. Ostwald, P.F., *Engineering Cost Estimating*, 3rd Ed., Prentice Hall, Upper Saddle River, NJ, 1992 (ISBN 0132766272).
7. Stewart, R.D., *Cost Estimating*, 2nd Ed., John Wiley & Sons, Hoboken, NJ, 1991 (ISBN 0471857076).
8. Thuesen G.J. and W.J. Fabrycky, *Engineering Economy*, 9th Ed., Prentice Hall, Upper Saddle River, NJ, 2001 (ISBN 0-13-028128-X).

K. Management and Supporting Areas

1. Boxwell, R.J., *Benchmarking for Competitive Advantage*, McGraw-Hill, New York, NY, 1994 (ISBN 0070068992).
2. Brigham, E.F., D.A. Clark, and J.F. Houston, *Fundamentals of Financial Management*, Dryden Press, New York, NY, 1997 (ISBN 003024434X).
3. Camp, R.C., *Business Process Benchmarking: Finding and Implementing Best Practices*, Quality Resources, Milwaukee, WI, 1995 (ISBN 0873892968).
4. Cleland, D.I., *Project Management: Strategic Design and Implementation*, 3rd Ed., McGraw-Hill, New York, NY, 1998 (ISBN 007012020X).
5. Corbitt, R.A., *Standard Handbook of Environmental Engineering*, McGraw-Hill, New York, NY, 1999 (ISBN 0-07-013160-0).
6. DoD 5000.2-R, *Mandatory Procedures for Major Defense Acquisition Programs (MDAPS) and Major Automated Information System (MAIS) Acquisition Programs*, Office of the Secretary of Defense, The Pentagon, Washington, DC 20301, April 5, 2002.
7. Defense Acquisition University (DAU), *Program Manager*, published bi-monthly, DAU Press, Fort Belvoir, VA 22060-5565 (ISSN 0199-7114).
8. Gibson, J.L., J.H. Donnelly, J.M. Ivancevich, and R. Konopaske, *Organizations: Behavior, Structure, and Processes*, McGraw-Hill/Irwin, New York, NY, 2003 (ISBN 0-07-252409-X).
9. Gordon, J.R. and S.R. Gordon, *Information Systems: A Management Approach*, 2nd Ed., Dryden Press, New York, NY, 1998 (ISBN 0030224691).
10. Griffin, R.W., *Management*, 5th Ed., Houghton Mifflin, 1995 (ISBN 0395731100).
11. Haimes, Y.Y., *Risk Modeling, Assessment, and Management*, John Wiley & Sons, Hoboken, NJ, 1998 (ISBN 0-471-24005-2).
12. Hall, E.M., *Managing Risk*, Addison-Wesley Publisher, Reading, MA, 1998 (ISBN 0-201-25592-8).
13. Hesse, R., *Managerial Spreadsheet Modeling and Analysis*, Richard D. Irwin, New York, NY, 1996 (ISBN 0256215308).
14. Johnson, R.A., I. Miller, and J.E. Freund, *Probability and Statistics for Engineers*, 5th Ed., Prentice Hall, Upper Saddle River, NJ, 1994 (ISBN 0137214081).
15. Kerzner, H., *Project Magement: A Systems Approach to Planning, Scheduling, and Controlling*, 7th Ed., John Wiley & Sons, Hoboken, NJ, 2000 (ISBN 0471288357).
16. Koontz, H. and H. Weihrich, *Essentials of Management*, 5th Ed., McGraw-Hill, New York, NY, 1990 (ISBN 007035605X).
17. Laudon, K.C. and J.P. Laudon, *Essentials of Management Information Systems*, Prentice Hall, Upper Saddle River, NJ, 1997.

18. Lewis, J.P., *Fundamentals of Project Management*, AMACOM, Boston, MA, 1995 (ISBN 0814478352).

19. Magrab, E.B., *Integrated Product and Process Design and Development: The Product Realization Process*, CRC Press, Boca Raton, FL, 1997.

Appendix B
Logistics -- List of Web-Based Contact Points

This page intentionally left blank

The URLs shown were current and active when this publication went to press. Readers are cautioned that URLs are subject to change frequently and without notice. Therefore, it is possible that a given URL referenced herein may not function.

1. *Air Force Logistics Management Agency (AFLMA)* – provides access to the *Air Force Journal Of Logistics* and provides links to other logistics activities in the USAF. Web Site: <http://www.il.hq.af.mil/aflma>.
2. *Air Transport Association of America* – a trade association in which corporations hold membership (23 US airlines and five foreign airlines hold membership). ATA, 1301 Pennsylvania Avenue NW, Suite 1100, Washington, DC 20004 (tel: 202-626-4000; Web Site: <http://www.air-transport.org>).
3. *American Society of Transportation and Logistics (AST&L)* – a professional organization in transportation and logistics with objectives to: establish, promote, and maintain high standards of knowledge and professional training; formulate a code of ethics for the profession; advance the professional interests of members of the organization; serve as a source of information and guidance for the fields of traffic and transportation, logistics, and physical distribution management; and serve the industry as a whole by fostering professional accomplishments. This site is primarily "outbound and transportation" oriented. AST&L, 229 Peachtree Street, Suite 401, Atlanta, GA 30303 (tel: 404-524-3555; Web Site: <http://www.astl.org>).
4. *American Production and Inventory Control Society (APICS) – The Educational Society for Resource Management*: a professional organization, with both individual and corporate membership, with the objective of meeting the changing needs of business by providing broad-based education focused on integrating resources for improved productivity in the manufacturing and service sectors. APICS, 5301 Shawnee Road, Alexandria, VA 22312 (tel: 703-354-8851; Web Site: <http://www.apics.org>).
5. *Army Logistics Management College (ALMC)* – Army educational programs, research and publications, and source for the "Army Logistician" Journal. Web Site: <http://www.almc.army.mil/>.
6. *Army Logistic Support Activity – Acquisition Logistics Center (ALC)* – part of the US Army Materiel Command Logistic Support Activity (LOGSA) which conducts logistics research activities, develops, and publishes logistics specifications and handbooks (MIL-HDBK-502 – *Acquisition Logistics Handbook*; MIL-PRF-49506 – *Logistics Management Information*), and sponsors an annual *DOD Government/ Industry Supportability Engineering Exchange (GISEE) Conference.* Web Site: <http://www.logpars.army.mil>.
7. *Associazione Italiana di Logistica (AILOG)* – a professional organization dedicated to promote research, the exchange of experiences, and the collaboration among individuals practicing in the logistics field. AILOG, 19-20159 Milano, Italy (tel: +392-6671-0622; fax: +392-670-1483).

8. *Canadian Association of Logistics Management (CALM)* – a professional organization of business professionals interested in improving their logistics and/or distribution management skills. CALM, 590 Alden Road, Suite 211, Markham, Ontario L3R 8N2, Canada (tel: 905-513-7300; fax: 905-513-1248; E-Mail: <admin@calm.org>; Web Site: <http://www.calm.org>).

9. *Council of Logistics Management (CLM)* – a professional organization dedicated to provide: leadership in developing, defining, understanding, and enhancing the logistics process worldwide; a forum for the exchange of concepts and best practices among logistics professionals; research and advanced knowledge leading to enhanced customer value and supply chain performance; and education and career development in the logistics field. CLM, 2805 Butterfield Road, Suite 200, Oak Brook, IL 60523 (tel: 630-574-0985; fax: 630-574-0989; Web Site: <http://www.clm1.org>). Effective January 2005, CLM became the Council of Supply Chain Management Professionals (CSCMP), and the web site is <http://www.cscmp.org>.

10. *Defense Logistics Agency (DLA)* – provides access to Department Of Defense (DOD) logistics sites. Web Site: <http://www.dla.mil>.

11. *Defense Acquisition University (DAU)* – source for many education/training programs and supporting documentation (i.e., "Acquisition Logistics"). Source of the *Acquisition Logistics Guide* and the *Systems Engineering Management Guide.* Web Site: <http://www.dau.mil>.

12. *Deutsche Gesellschaft fur Logistik (German Society of Logistics)* – a professional organization with individual and corporate memberships dedicated to promote research and the dissemination of information in the field of logistics. GfL, Joseph-von-Fraunhofer-Str 20, Dortmund 44227, Germany (tel: +49 231 9700 120; fax: +49 231 9700-464; E-Mail: <dgflev@t-online.de>).

13. *European Federation of National Maintenance Societies (EFNMS)* – a Federation of European Nations organized with the objective of providing a link and an open forum for communications, development of standards, and the promotion of education and training in the field of maintenance. The EFNMS sponsors a Congress on an every-other-year basis, and includes a membership involving Belgium, Croatia, Denmark, Finland, France, Germany, Great Britain, Ireland, Italy, Netherlands, Norway, Portugal, Slovenia, Spain, Sweden, and Switzerland. Web Site: <http://www.ini.hr/efnms>.

14. *European Logistics Association (ELA)* – a Federation of European Nations organized with the objective of providing a link and an open forum for communications between countries, a vehicle for formulating standards, and a mechanism for providing training and education. ELA member organizations include Austria, Belgium, Denmark, Finland, France, Germany, Great Britain, Greece, Hungary, Ireland, Italy, Luxembourg, Netherlands, Norway, Poland, Portugal, Spain, Sweden, and Switzerland. ELA, Rue Archimede 5/11, B-1000 Brussels, Belgium (tel: +32 2 230-0211; Web Site: <http://www.elalog.org>).

15. *Institute of Industrial Engineers (IIE)* – a professional organization dedicated to advancing the general welfare of human kind through the resources and creative

abilities of the industrial engineering profession. Included within the IIE organizational structure is a *Transportation, Distribution and Logistics Division.* IIE, 25 Technology Park, Norcross, GA 30043 (tel: 770-449-0461; fax: 770-263-8532; Web Site: <http://www.iienet.org>).

16. *Institute of Logistics and Transport* – a professional organization dedicated to promote professional excellence and social responsibility in the fields of transport and supply-chain management. IOLT, 80 Portland Place, London WIN 4DP, United Kingdom (tel: +44(0) 207 467-9400; fax: +44(0) 207 467-9440; Web Site: <http://www.iolt.org>.uk>).

17. *International Council On Systems Engineering (INCOSE)* – an international organization dedicated to the promotion of systems engineering principles and concepts. Included is a quarterly newsletter, *Insight*, and a quarterly Journal, *Systems Engineering* INCOSE, 2150 N. 107th Street, Suite 205, Seattle, WA 98133-9009 (tel: 206-361-6607; fax: 206-367-8777; E-Mail: <incose@halcyon.com>; Web Site: <http://www.incose.org>).

18. *International Customer Service Organization (ICSA)* – a professional organization dedicated to promoting the development and awareness of the customer service profession through networking, education, and research. ICSA, 401 North Michigan Avenue, Chicago, IL 60611-4267 (tel: 312-321-6800; fax: 312-245-1084; E-Mail: icsa@sba.com>).

19. *International Society of Logistics (SOLE)* – an international organization devoted to promoting scientific, educational, and literary endeavors to enhance the art and science of logistics technology and management. Included is a "Certification Professional Logistician (CPL);" a monthly newsletter, *SOLEtech*; and a quarterly Journal, *Logistics Spectrum.* SOLE, 8100 Professional Place, Suite 211, Hyattsville, MD 20785 (tel: 301-459-8446; fax: 301-459-1522; E-Mail: SOLEHQ@erols.com>; Web Site: <http://www.sole.org>).

20. *Logistics World* – this site provides a directory of logistics resources on the Internet, or a guide to thousands of sites related to logistics, transportation, warehousing, distribution, supply, maintenance, manufacturing, management, reliability, and quality. One of its most useful features is LogLink.com, which constitutes a large and comprehensive list by categories of providers of transportation and logistics services. Available through the site is "Logistics World: Guide To Transportation And Logistics On The Internet." Web Site: <http://www.logisticsworld.com>.

21. *LOGTOOLS – Tools To Aid Implementation Of Naval Logistics* – covers a great deal of information pertaining to "Logistics Functions" and "Logistics Guidance." Web Site: <http://www.logtools.navsea.navy.mil/>.

22. *Maintenance Engineering Societies-Australia (MESA)* – an integrated group of maintenance societies throughout Australia. Web Site: <http://www.mesa.org.au>.

23. *Transportation Research Board (TRB)* – to advance knowledge concerning the nature and performance of transportation systems by stimulating research and

the disseminating the information derived from research. TRB, 21-1 Constitutional Ave. NW, Washington, DC 20418 (tel: 202-334-2936; fax: 202-334-2920; Web Site: <http://www.nas.edu/trb>).

24. *VDI-Gesellschaft Foerdertechnik Meterialfluss Logistik (VDI-FML)* – to promote the science and development in the fields of material flow, transportation, and logistics. VDI-FML, Postfach 10 11 39, 40002 Dusseldorf, Germany (tel: +49 211 6214-231; fax: +49 211 6214-155).

25. *Warehousing Education And Research Council (WERC)* – to provide education, conduct research concerning the warehousing process, and refine the art and science of managing warehouses. WERC, 1100 Jorie Blvd., Suite 170, Oak Brook, IL 60523-2243 (tel: 630-990-0001; fax: 630-990-0256; E-Mail: <tsharpe@werc.org>).

Additional Web Sites Of Interest:

1. *Acquisition Reform* – Naval Acquisition Reform. Web Site: <http://www.acq-ref.navy.mil>. Also check <http://lrc3.monmouth.army.mil/cecom/lrc/leo/eladiv/logistics/flexguide/flexguide-toc.html>. This contains the "Flexible Sustainment Guide" and coverage of "Reliability-Based Logistics (RBL)" and "Trigger-Based Asset Management (TBAM)."

2. *Annual Reliability And Maintainability Symposium (RAMS)* – Web Site: <http://www.rams.org>.

3. *Association For Field Service Management International (AFSMI)* – dedicated to furthering the knowledge, understanding, and career development of executives, managers, and professionals in the services and support industries. AFSMI, 1342 Colonial Blvd., Suite 25, Ft. Myers, FL 33907. Web Site: <http://www.afsmi.org>.

4. *Defense Modeling And Simulation Office (DMSO)* – provides the "Modeling and Simulation Information Analysis Center (MSIAC)" which, in turn, will provide near real-time analysis and research support on a quick reaction basis. Web Site: <http://www.msiac.dmso.mil>.

5. Georgia Tech – *Logistics Institute* – conducts academic programs and research in the field. Web Site: <http://www.tli.gatech.edu>.

6. *Global Logistics Village* – this site is involved largely with transport and the tracking/tracing of goods that have been transported. It is more of a hosted tracking site or provider for identified partners in a particular supply chain than a general reference source for industry. Web Site: <http://www.glovill.com>.

7. *Government-Industry Data Exchange Program (GIDEP)* – includes a reliability database covering existing systems. GIDEP Operations Center, 2300 Fifth St., Norco, CA 92860 (tel: 909-273-4677; fax: 909-273-5200; Web Site: <http://www.gidep.org>).

8. *Institute Of Transportation Engineers* – ITE Journal. Web Site: <http://www.ite.org>.

9. *Intermodal Association Of North America (IANA)* – a leading industry trade association representing the combined interests of intermodal freight transportation companies. Web Site: <http://intermodal.org>.
10. *International Society Of Parametric Analysts (ISPA)* – provides a "Parametric Estimating Handbook" useful in cost estimating for life cycle cost analyses. Web Site: <http://www.ispa-cost.org/PEIWeb/ch1.htm>. Also check: <http://www.contracts.hq.navsea.navy.mil/webdata/pceh/pceh.html>.
11. *International Warehouse Logistic Association* – linked to other associations, education courses, education and research foundation, and logistics today. Web Site: <http://logistx.dartgc.com>.
12. *Logistics Management Institute (LMI)* – non-profit corporation that provides research and consulting services for the government in logistics and related areas. Web Site: <http://www.lmi.org>.
13. *Manufacturing Net* – provides news, editorial content, search and retrieval capabilities in a wide variety of fields. Web Site: <http://www.manufacturing.net>. Also check <http://www.manufacturing.net/magazine/logistics/main.html>. This includes a magazine that provides news in the logistics world.
14. *Material Handling Industry (MHI)* – MHI, 8720 Red Oak Blvd., Suite 201, Charlotte, NC 28217 (tel: 704-676-1190; fax: 704-676-1199; E-Mail: <jnofsinger@mhia.org>). Web Site: <http://www.mhi.org>.
15. *National Association Of Purchasing Management (NAPM)* – NAPM, P.O. Box 22160, Tempe, AZ 85285-2160 (tel: 480-752-6276; fax: 480-752-7890; E-Mail: <pnovak@napm.org>).
16. *National Information Center For Reliability Engineering* – supported by the Center for Reliability Engineering, University of Maryland. College Park, MD 20742, which is a good source for educational courses, research accomplishments, and literature in the field. Web Site: <http://www.enre.umd.edu/mainnojs.html>.
17. *National Defense Industrial Association (NDIA)* – includes a Logistics Division. NDIA, 2111 Wilson Blvd., Suite 400, Arlington, VA (tel: 703-522-1820; fax: 703-522-1885; Web Site: <http://www.ndia.org>).
18. *North American Logistics Association* – the focus of this organization is on warehousing and distribution services – a joint effort of three warehousing associations. Web Site: <http://www.iwla.com> and/or <http://www.cargolog.com>.
19. *National Defense Transportation Association (NDTA)* – NDTA, 50 South Pickett Street, Suite 220, Alexandria, VA 22304-7296 (tel: 703-751-5011; fax: 703-823-8761; E-Mail: <ndta@ndtahq.com>).
20. *Reliability Analysis Center (RAC)* – an excellent source for reliability information, training programs, research results and documentation, specifications and standards, etc. Web Site: <http://www.rac.alionscience.com>.

21. *Society Of Maintenance And Reliability Professionals (SMRP)* – a non-profit society by and for practitioners in the field of maintenance and reliability. Web Site: <http://www.smrp.org>.

22. *Society Of Reliability Engineers (SRE)* – a technical reliability organization. Web Site: <http://www.sre.org>.

23. *Supply Chain Council* – a trade association in which corporations hold membership. Members of the Council work together to develop, maintain, communicate, and support the "Supply-Chain Operations Reference Model (SCOR). SCOR contains standard process definitions, standard terminology, standard metrics, supply-chain best practices, and enabling information technology when applicable. SCC, 303 Freeport Road, Pittsburgh, PA 15215 (tel: 412-781-4104; fax: 412-781-2871; Web Site: <http://www.supplychainworld.org>).

Appendix C
Exponential Distribution Table
(Values of $e^{-\lambda t}$ for Selected Values of λt)

This page intentionally left blank

If System Reliability = $e^{-\lambda t}$

where:

$$t = \text{system operating time, and } \lambda = \text{System failure rate} = \frac{1}{\text{MTBF}}$$

If the operating time t = 500 hours and MTBF = 1,000 hours, then λ = 1/,1000 = 0.001 and λt = (0.001)(500) = 0.5.

From Table C-1, for λt = 0.5, $e^{-\lambda t}$ = 0.606531, or Reliability at t = 1,000 is about 60.65%.

Table C-1. Exponential Distribution Table – Values of $e^{-\lambda t}$

λt	$e^{-\lambda t}$	λt	$e^{-\lambda t}$	λt	$e^{-\lambda t}$	λt	$e^{-\lambda t}$	λt	$e^{-\lambda t}$	λt	$e^{-\lambda t}$
0.000	1.0000	0.250	0.7788	0.500	0.6065	0.750	0.4724	1.000	0.3679	1.250	0.2865
0.010	0.9900	0.260	0.7711	0.510	0.6005	0.760	0.4677	1.010	0.3642	1.260	0.2837
0.020	0.9802	0.270	0.7634	0.520	0.5945	0.770	0.4630	1.020	0.3606	1.270	0.2808
0.030	0.9704	0.280	0.7558	0.530	0.5886	0.780	0.4584	1.030	0.3570	1.280	0.2780
0.040	0.9608	0.290	0.7483	0.540	0.5827	0.790	0.4538	1.040	0.3535	1.290	0.2753
0.050	0.9512	0.300	0.7408	0.550	0.5769	0.800	0.4493	1.050	0.3499	1.300	0.2725
0.060	0.9418	0.310	0.7334	0.560	0.5712	0.810	0.4449	1.060	0.3465	1.310	0.2698
0.070	0.9324	0.320	0.7261	0.570	0.5655	0.820	0.4404	1.070	0.3430	1.320	0.2671
0.080	0.9231	0.330	0.7189	0.580	0.5599	0.830	0.4360	1.080	0.3396	1.330	0.2645
0.090	0.9139	0.340	0.7118	0.590	0.5543	0.840	0.4317	1.090	0.3362	1.340	0.2618
0.100	0.9048	0.350	0.7047	0.600	0.5488	0.850	0.4274	1.100	0.3329	1.350	0.2592
0.110	0.8958	0.360	0.6977	0.610	0.5434	0.860	0.4232	1.110	0.3296	1.360	0.2567
0.120	0.8869	0.370	0.6907	0.620	0.5379	0.870	0.4190	1.120	0.3263	1.370	0.2541
0.130	0.8781	0.380	0.6839	0.630	0.5326	0.880	0.4148	1.130	0.3230	1.380	0.2516
0.140	0.8694	0.390	0.6771	0.640	0.5273	0.890	0.4107	1.140	0.3198	1.390	0.2491
0.150	0.8607	0.400	0.6703	0.650	0.5220	0.900	0.4066	1.150	0.3166	1.400	0.2466
0.160	0.8521	0.410	0.6637	0.660	0.5169	0.910	0.4025	1.160	0.3135	1.410	0.2441
0.170	0.8437	0.420	0.6570	0.670	0.5117	0.920	0.3985	1.170	0.3104	1.420	0.2417
0.180	0.8353	0.430	0.6505	0.680	0.5066	0.930	0.3946	1.180	0.3073	1.430	0.2393
0.190	0.8270	0.440	0.6440	0.690	0.5016	0.940	0.3906	1.190	0.3042	1.440	0.2369
0.200	0.8187	0.450	0.6376	0.700	0.4966	0.950	0.3867	1.200	0.3012	1.450	0.2346
0.210	0.8106	0.460	0.6313	0.710	0.4916	0.960	0.3829	1.210	0.2982	1.460	0.2322
0.220	0.8025	0.470	0.6250	0.720	0.4868	0.970	0.3791	1.220	0.2952	1.470	0.2299
0.230	0.7945	0.480	0.6188	0.730	0.4819	0.980	0.3753	1.230	0.2923	1.480	0.2276
0.240	0.7866	0.490	0.6126	0.740	0.4771	0.990	0.3716	1.240	0.2894	1.490	0.2254
0.250	0.7788	0.500	0.6065	0.750	0.4724	1.000	0.3679	1.250	0.2865	1.500	0.2231

Table C-1. Exponential Distribution Table – Values of $e^{-\lambda t}$ (Cont'd)

λt	$e^{-\lambda t}$	λt	$e^{-\lambda t}$	λt	$e^{-\lambda t}$	λt	$e^{-\lambda t}$	λt	$e^{-\lambda t}$	λt	$e^{-\lambda t}$
1.500	0.2231	1.850	0.1572	2.200	0.1108	2.550	0.0781	2.900	0.0550	3.250	0.0388
1.510	0.2209	1.860	0.1557	2.210	0.1097	2.560	0.0773	2.910	0.0545	3.260	0.0384
1.520	0.2187	1.870	0.1541	2.220	0.1086	2.570	0.0765	2.920	0.0539	3.270	0.0380
1.530	0.2165	1.880	0.1526	2.230	0.1075	2.580	0.0758	2.930	0.0534	3.280	0.0376
1.540	0.2144	1.890	0.1511	2.240	0.1065	2.590	0.0750	2.940	0.0529	3.290	0.0373
1.550	0.2122	1.900	0.1496	2.250	0.1054	2.600	0.0743	2.950	0.0523	3.300	0.0369
1.560	0.2101	1.910	0.1481	2.260	0.1044	2.610	0.0735	2.960	0.0518	3.310	0.0365
1.570	0.2080	1.920	0.1466	2.270	0.1033	2.620	0.0728	2.970	0.0513	3.320	0.0362
1.580	0.2060	1.930	0.1451	2.280	0.1023	2.630	0.0721	2.980	0.0508	3.330	0.0358
1.590	0.2039	1.940	0.1437	2.290	0.1013	2.640	0.0714	2.990	0.0503	3.340	0.0354
1.600	0.2019	1.950	0.1423	2.300	0.1003	2.650	0.0707	3.000	0.0498	3.350	0.0351
1.610	0.1999	1.960	0.1409	2.310	0.0993	2.660	0.0699	3.010	0.0493	3.360	0.0347
1.620	0.1979	1.970	0.1395	2.320	0.0983	2.670	0.0693	3.020	0.0488	3.370	0.0344
1.630	0.1959	1.980	0.1381	2.330	0.0973	2.680	0.0686	3.030	0.0483	3.380	0.0340
1.640	0.1940	1.990	0.1367	2.340	0.0963	2.690	0.0679	3.040	0.0478	3.390	0.0337
1.650	0.1920	2.000	0.1353	2.350	0.0954	2.700	0.0672	3.050	0.0474	3.400	0.0334
1.660	0.1901	2.010	0.1340	2.360	0.0944	2.710	0.0665	3.060	0.0469	3.410	0.0330
1.670	0.1882	2.020	0.1327	2.370	0.0935	2.720	0.0659	3.070	0.0464	3.420	0.0327
1.680	0.1864	2.030	0.1313	2.380	0.0926	2.730	0.0652	3.080	0.0460	3.430	0.0324
1.690	0.1845	2.040	0.1300	2.390	0.0916	2.740	0.0646	3.090	0.0455	3.440	0.0321
1.700	0.1827	2.050	0.1287	2.400	0.0907	2.750	0.0639	3.100	0.0450	3.450	0.0317
1.710	0.1809	2.060	0.1275	2.410	0.0898	2.760	0.0633	3.110	0.0446	3.460	0.0314
1.720	0.1791	2.070	0.1262	2.420	0.0889	2.770	0.0627	3.120	0.0442	3.470	0.0311
1.730	0.1773	2.080	0.1249	2.430	0.0880	2.780	0.0620	3.130	0.0437	3.480	0.0308
1.740	0.1755	2.090	0.1237	2.440	0.0872	2.790	0.0614	3.140	0.0433	3.490	0.0305
1.750	0.1738	2.100	0.1225	2.450	0.0863	2.800	0.0608	3.150	0.0429	3.500	0.0302
1.760	0.1720	2.110	0.1212	2.460	0.0854	2.810	0.0602	3.160	0.0424	3.510	0.0299
1.770	0.1703	2.120	0.1200	2.470	0.0846	2.820	0.0596	3.170	0.0420	3.520	0.0296
1.780	0.1686	2.130	0.1188	2.480	0.0837	2.830	0.0590	3.180	0.0416	3.530	0.0293
1.790	0.1670	2.140	0.1177	2.490	0.0829	2.840	0.0584	3.190	0.0412	3.540	0.0290
1.800	0.1653	2.150	0.1165	2.500	0.0821	2.850	0.0578	3.200	0.0408	3.550	0.0287
1.810	0.1637	2.160	0.1153	2.510	0.0813	2.860	0.0573	3.210	0.0404	3.560	0.0284
1.820	0.1620	2.170	0.1142	2.520	0.0805	2.870	0.0567	3.220	0.0400	3.570	0.0282
1.830	0.1604	2.180	0.1130	2.530	0.0797	2.880	0.0561	3.230	0.0396	3.580	0.0279
1.840	0.1588	2.190	0.1119	2.540	0.0789	2.890	0.0556	3.240	0.0392	3.590	0.0276
1.850	0.1572	2.200	0.1108	2.550	0.0781	2.900	0.0550	3.250	0.0388	3.600	0.0273

Appendix D
Selected Interest Tables

This page intentionally left blank

6% Interest Factors for Annual Compounding

n	Single Payment		Equal Payment Series				Uniform Gradient-Series Factor
	Compound-Amount Factor	Present-Worth Factor	Compound-Amount Factor	Sinking-Fund Factor	Present-Worth Factor	Capital-Recovery Factor	
	To find *F* given *P*, *F/P*, *i*, *n*	To find *P* given *F*, *F/P*, *i*, *n*	To find *F* given *A*, *F/A*, *i*, *n*	To find *A* given *F*, *A/F*, *i*, *n*	To find *P* given *A*, *P/A*, *i*, *n*	To find *A* given *P*, *A/P*, *i*, *n*	To find *A* given *G*, *A/G*, *i*, *n*
1	1.060	0.9434	1.000	1.0000	0.9434	1.0600	0.0000
2	1.124	0.8900	2.060	0.4854	1.8334	0.5454	0.4854
3	1.191	0.8396	3.184	0.3141	2.6730	0.3741	0.9612
4	1.262	0.7921	4.375	0.2286	3.4651	0.2886	1.4272
5	1.338	0.7473	5.637	0.1774	4.2124	0.2374	1.8836
6	1.419	0.7050	6.975	0.1434	4.9173	0.2034	2.3304
7	1.504	0.6651	8.394	0.1191	5.5824	0.1791	2.7676
8	1.594	0.6274	9.897	0.1010	6.2098	0.1610	3.1952
9	1.689	0.5919	11.491	0.0870	6.8017	0.1470	3.6133
10	1.791	0.5584	13.181	0.0759	7.3601	0.1359	4.0220
11	1.898	0.5268	14.972	0.0668	7.8869	0.1268	4.4213
12	2.012	0.4970	16.870	0.0593	8.3838	0.1193	4.8113
13	2.133	0.4688	18.882	0.0530	8.8527	0.1130	5.1920
14	2.261	0.4423	21.015	0.0476	9.2950	0.1076	5.5635
15	2.397	0.4173	23.276	0.0430	9.7122	0.1030	5.9260
16	2.540	0.3936	25.673	0.0390	10.1059	0.0990	6.2794
17	2.693	0.3714	28.213	0.0354	10.4773	0.0954	6.6240
18	2.854	0.3503	30.906	0.0324	10.8276	0.0924	6.9597
19	3.026	0.3305	33.760	0.0296	11.1581	0.0896	7.2867
20	3.207	0.3118	36.786	0.0272	11.4699	0.0872	7.6051
21	3.400	0.2942	39.993	0.0250	11.7641	0.0850	7.9151
22	3.604	0.2775	43.392	0.0230	12.0416	0.0830	8.2166
23	3.820	0.2618	46.996	0.0213	12.3034	0.0813	8.5099
24	4.049	0.2470	50.816	0.0197	12.5504	0.0797	8.7951
25	4.292	0.2330	54.865	0.0182	12.7834	0.0782	9.0722
26	4.549	0.2198	59.156	0.0169	13.0032	0.0769	9.3414
27	4.822	0.2074	63.706	0.0157	13.2105	0.0757	9.6029
28	5.112	0.1956	68.528	0.0146	13.4062	0.0746	9.8568
29	5.418	0.1846	73.640	0.0136	13.5907	0.0736	10.1032
30	5.743	0.1741	79.058	0.0126	13.7648	0.0726	10.3422
31	6.088	0.1643	84.802	0.0118	13.9291	0.0718	10.5740
32	6.453	0.1550	90.890	0.0110	14.0840	0.0710	10.7988
33	6.841	0.1462	97.343	0.0103	14.2302	0.0703	11.0166
34	7.251	0.1379	104.184	0.0096	14.3681	0.0696	11.2276
35	7.686	0.1301	111.435	0.0090	14.4982	0.0690	11.4319
36	8.147	0.1227	119.121	0.0084	14.6210	0.0684	11.6298
37	8.636	0.1158	127.268	0.0079	14.7368	0.0679	11.8213
38	9.154	0.1092	135.904	0.0074	14.8460	0.0674	12.0065
39	9.704	0.1031	145.058	0.0069	14.9491	0.0669	12.1857
40	10.286	0.0972	154.762	0.0065	15.0463	0.0665	12.3590
45	13.765	0.0727	212.744	0.0047	15.4558	0.0647	13.1413
50	18.420	0.0543	290.336	0.0034	15.7619	0.0634	13.7964
55	24.650	0.0406	394.172	0.0025	15.9905	0.0625	14.3411
60	32.988	0.0303	533.128	0.0019	16.1614	0.0619	14.7909
65	44.145	0.0227	719.083	0.0014	16.2891	0.0614	15.1601
70	59.076	0.0169	967.932	0.0010	16.3845	0.0610	15.4613
75	79.057	0.0126	1,300.949	0.0008	16.4558	0.0608	15.7058
80	105.796	0.0095	1,746.600	0.0006	16.5091	0.0606	15.9033
85	141.579	0.0071	2,342.982	0.0004	16.5489	0.0604	16.0620
90	189.465	0.0053	3,141.075	0.0003	16.5787	0.0603	16.1891
95	253.546	0.0039	4,209.104	0.0002	16.6009	0.0602	16.2905
100	339.302	0.0029	5,638.368	0.0002	16.6175	0.0602	16.3711

7% Interest Factors for Annual Compounding

n	Single Payment		Equal Payment Series				Uniform Gradient-Series Factor
	Compound-Amount Factor	Present-Worth Factor	Compound-Amount Factor	Sinking-Fund Factor	Present-Worth Factor	Capital-Recovery Factor	
	To find *F* given *P*, *F/P*, *i*, *n*	To find *P* given *F*, *F/P*, *i*, *n*	To find *F* given *A*, *F/A*, *i*, *n*	To find *A* given *F*, *A/F*, *i*, *n*	To find *P* given *A*, *P/A*, *i*, *n*	To find *A* given *P*, *A/P*, *i*, *n*	To find *A* given *G*, *A/G*, *i*, *n*
1	1.070	0.9346	1.000	1.0000	0.9346	1.0700	0.0000
2	1.145	0.8734	2.070	0.4831	1.8080	0.5531	0.4831
3	1.225	0.8163	3.215	0.3111	2.6243	0.3811	0.9549
4	1.311	0.7629	4.440	0.2252	3.3872	0.2952	1.4155
5	1.403	0.7130	5.751	0.1739	4.1002	0.2439	1.8650
6	1.501	0.6663	7.153	0.1398	4.7665	0.2098	2.3032
7	1.606	0.6227	8.654	0.1156	5.3893	0.1856	2.7304
8	1.718	0.5820	10.260	0.0975	5.9713	0.1675	3.1465
9	1.838	0.5439	11.978	0.0835	6.5152	0.1535	3.5517
10	1.967	0.5083	13.816	0.0724	7.0236	0.1424	3.9461
11	2.105	0.4751	15.784	0.0634	7.4987	0.1334	4.3296
12	2.252	0.4440	17.888	0.0559	7.9427	0.1259	4.7025
13	2.410	0.4150	20.141	0.0497	8.3577	0.1197	5.0648
14	2.579	0.3878	22.550	0.0443	8.7455	0.1143	5.4167
15	2.759	0.3624	25.129	0.0398	9.1079	0.1098	5.7583
16	2.952	0.3387	27.888	0.0359	9.4466	0.1059	6.0897
17	3.159	0.3166	30.840	0.0324	9.7632	0.1024	6.4110
18	3.380	0.2959	33.999	0.0294	10.0591	0.0994	6.7225
19	3.617	0.2765	37.379	0.0268	10.3356	0.0968	7.0242
20	3.870	0.2584	40.995	0.0244	10.5940	0.0944	7.3163
21	4.141	0.2415	44.865	0.0223	10.8355	0.0923	7.5990
22	4.430	0.2257	49.006	0.0204	11.0612	0.0904	7.8725
23	4.741	0.2109	53.436	0.0187	11.2722	0.0887	8.1369
24	5.072	0.1971	58.177	0.0172	11.4693	0.0872	8.3923
25	5.427	0.1842	63.249	0.0158	11.6536	0.0858	8.6391
26	5.807	0.1722	68.676	0.0146	11.8258	0.0846	8.8773
27	6.214	0.1609	74.484	0.0134	11.9867	0.0834	9.1072
28	6.649	0.1504	80.698	0.0124	12.1371	0.0824	9.3289
29	7.114	0.1406	87.347	0.0114	12.2777	0.0814	9.5427
30	7.612	0.1314	94.461	0.0106	12.4090	0.0806	9.7487
31	8.145	0.1228	102.073	0.0098	12.5318	0.0798	9.9471
32	8.715	0.1147	110.218	0.0091	12.6466	0.0791	10.1381
33	9.325	0.1072	118.933	0.0084	12.7538	0.0784	10.3219
34	9.978	0.1002	128.259	0.0078	12.8540	0.0778	10.4987
35	10.677	0.0937	138.237	0.0072	12.9477	0.0772	10.6687
36	11.424	0.0875	148.913	0.0067	13.0352	0.0767	10.8321
37	12.224	0.0818	160.337	0.0062	13.1170	0.0762	10.9891
38	13.079	0.0765	172.561	0.0058	13.1935	0.0758	11.1398
39	13.995	0.0715	185.640	0.0054	13.2649	0.0754	11.2845
40	14.974	0.0668	199.635	0.0050	13.3317	0.0750	11.4233
45	21.002	0.0476	285.749	0.0035	13.6055	0.0735	12.0360
50	29.457	0.0339	406.529	0.0025	13.8007	0.0725	12.5287
55	41.315	0.0242	575.929	0.0017	13.9399	0.0717	12.9215
60	57.946	0.0173	813.520	0.0012	14.0392	0.0712	13.2321
65	81.273	0.0123	1,146.755	0.0009	14.1099	0.0709	13.4760
70	113.989	0.0088	1,614.134	0.0006	14.1604	0.0706	13.6662
75	159.876	0.0063	2,269.657	0.0004	14.1964	0.0704	13.8136
80	224.234	0.0045	3,189.063	0.0003	14.2220	0.0703	13.9273
85	314.500	0.0032	4,478.576	0.0002	14.2403	0.0702	14.0146
90	441.103	0.0023	6,287.185	0.0002	14.2533	0.0702	14.0812
95	618.670	0.0016	8,823.854	0.0001	14.2626	0.0701	14.1319
100	867.716	0.0012	12,381.662	0.0001	14.2693	0.0701	14.1703

8% Interest Factors for Annual Compounding

n	Single Payment		Equal Payment Series				Uniform Gradient-Series Factor
	Compound-Amount Factor	Present-Worth Factor	Compound-Amount Factor	Sinking-Fund Factor	Present-Worth Factor	Capital-Recovery Factor	
	To find *F* given *P*, *F/P, i, n*	To find *P* given *F*, *F/P, i, n*	To find *F* given *A*, *F/A, i, n*	To find *A* given *F*, *A/F, i, n*	To find *P* given *A*, *P/A, i, n*	To find *A* given *P*, *A/P, i, n*	To find *A* given *G*, *A/G, i, n*
1	1.080	0.9259	1.000	1.0000	0.9259	1.0800	0.0000
2	1.166	0.8573	2.080	0.4808	1.7833	0.5608	0.4808
3	1.260	0.7938	3.246	0.3080	2.5771	0.3880	0.9487
4	1.360	0.7350	4.506	0.2219	3.3121	0.3019	1.4040
5	1.469	0.6806	5.867	0.1705	3.9927	0.2505	1.8465
6	1.587	0.6302	7.336	0.1363	4.6229	0.2163	2.2763
7	1.714	0.5835	8.923	0.1121	5.2064	0.1921	2.6937
8	1.851	0.5403	10.637	0.0940	5.7466	0.1740	3.0985
9	1.999	0.5002	12.488	0.0801	6.2469	0.1601	3.4910
10	2.159	0.4632	14.487	0.0690	6.7101	0.1490	3.8713
11	2.332	0.4289	16.645	0.0601	7.1390	0.1401	4.2395
12	2.518	0.3971	18.977	0.0527	7.5361	0.1327	4.5957
13	2.720	0.3677	21.495	0.0465	7.9038	0.1265	4.9402
14	2.937	0.3405	24.215	0.0413	8.2442	0.1213	5.2731
15	3.172	0.3152	27.152	0.0368	8.5595	0.1168	5.5945
16	3.426	0.2919	30.324	0.0330	8.8514	0.1130	5.9046
17	3.700	0.2703	33.750	0.0296	9.1216	0.1096	6.2037
18	3.996	0.2502	37.450	0.0267	9.3719	0.1067	6.4920
19	4.316	0.2317	41.446	0.0241	9.6036	0.1041	6.7697
20	4.661	0.2145	45.762	0.0219	9.8181	0.1019	7.0369
21	5.034	0.1987	50.423	0.0198	10.0168	0.0998	7.2940
22	5.437	0.1839	55.457	0.0180	10.2007	0.0980	7.5412
23	5.871	0.1703	60.893	0.0164	10.3711	0.0964	7.7786
24	6.341	0.1577	66.765	0.0150	10.5288	0.0950	8.0066
25	6.848	0.1460	73.106	0.0137	10.6748	0.0937	8.2254
26	7.396	0.1352	79.954	0.0125	10.8100	0.0925	8.4352
27	7.988	0.1252	87.351	0.0114	10.9352	0.0914	8.6363
28	8.627	0.1159	95.339	0.0105	11.0511	0.0905	8.8289
29	9.317	0.1073	103.966	0.0096	11.1584	0.0896	9.0133
30	10.063	0.0994	113.283	0.0088	11.2578	0.0888	9.1897
31	10.868	0.0920	123.346	0.0081	11.3498	0.0881	9.3584
32	11.737	0.0852	134.214	0.0075	11.4350	0.0875	9.5197
33	12.676	0.0789	145.951	0.0069	11.5139	0.0869	9.6737
34	13.690	0.0730	158.627	0.0063	11.5869	0.0863	9.8208
35	14.785	0.0676	172.317	0.0058	11.6546	0.0858	9.9611
36	15.968	0.0626	187.102	0.0053	11.7172	0.0853	10.0949
37	17.246	0.0580	203.070	0.0049	11.7752	0.0849	10.2225
38	18.625	0.0537	220.316	0.0045	11.8289	0.0845	10.3440
39	20.115	0.0497	238.941	0.0042	11.8786	0.0842	10.4597
40	21.725	0.0460	259.057	0.0039	11.9246	0.0839	10.5699
45	31.920	0.0313	386.506	0.0026	12.1084	0.0826	11.0447
50	46.902	0.0213	573.770	0.0017	12.2335	0.0817	11.4107
55	68.914	0.0145	848.923	0.0012	12.3186	0.0812	11.6902
60	101.257	0.0099	1,253.213	0.0008	12.3766	0.0808	11.9015
65	148.780	0.0067	1,847.248	0.0005	12.4160	0.0805	12.0602
70	218.606	0.0046	2,720.080	0.0004	12.4428	0.0804	12.1783
75	321.205	0.0031	4,002.557	0.0002	12.4611	0.0802	12.2658
80	471.955	0.0021	5,886.935	0.0002	12.4735	0.0802	12.3301
85	693.456	0.0014	8,655.706	0.0001	12.4820	0.0801	12.3772
90	1018.915	0.0010	12,723.939	0.0001	12.4877	0.0801	12.4116
95	1497.121	0.0007	18,701.507	0.0001	12.4917	0.0801	12.4365
100	2199.761	0.0005	27,484.516	0.0000	12.4943	0.0800	12.4545

9% Interest Factors for Annual Compounding

n	Single Payment		Equal Payment Series				Uniform Gradient-Series Factor
	Compound-Amount Factor	Present-Worth Factor	Compound-Amount Factor	Sinking-Fund Factor	Present-Worth Factor	Capital-Recovery Factor	
	To find *F* given *P*, *F/P, i, n*	To find *P* given *F*, *F/P, i, n*	To find *F* given *A*, *F/A, i, n*	To find *A* given *F*, *A/F, i, n*	To find *P* given *A*, *P/A, i, n*	To find *A* given *P*, *A/P, i, n*	To find *A* given *G*, *A/G, i, n*
1	1.090	0.9174	1.000	1.0000	0.9174	1.0900	0.0000
2	1.188	0.8417	2.090	0.4785	1.7591	0.5685	0.4785
3	1.295	0.7722	3.278	0.3051	2.5313	0.3951	0.9426
4	1.412	0.7084	4.573	0.2187	3.2397	0.3087	1.3925
5	1.539	0.6499	5.985	0.1671	3.8897	0.2571	1.8282
6	1.677	0.5963	7.523	0.1329	4.4859	0.2229	2.2498
7	1.828	0.5470	9.200	0.1087	5.0330	0.1987	2.6574
8	1.993	0.5019	11.028	0.0907	5.5348	0.1807	3.0512
9	2.172	0.4604	13.021	0.0768	5.9952	0.1668	3.4312
10	2.367	0.4224	15.193	0.0658	6.4177	0.1558	3.7978
11	2.580	0.3875	17.560	0.0569	6.8052	0.1469	4.1510
12	2.813	0.3555	20.141	0.0497	7.1607	0.1397	4.4910
13	3.066	0.3262	22.953	0.0436	7.4869	0.1336	4.8182
14	3.342	0.2992	26.019	0.0384	7.7862	0.1284	5.1326
15	3.642	0.2745	29.361	0.0341	8.0607	0.1241	5.4346
16	3.970	0.2519	33.003	0.0303	8.3126	0.1203	5.7245
17	4.328	0.2311	36.974	0.0270	8.5436	0.1170	6.0024
18	4.717	0.2120	41.301	0.0242	8.7556	0.1142	6.2687
19	5.142	0.1945	46.018	0.0217	8.9501	0.1117	6.5236
20	5.604	0.1784	51.160	0.0195	9.1285	0.1095	6.7674
21	6.109	0.1637	56.765	0.0176	9.2922	0.1076	7.0006
22	6.659	0.1502	62.873	0.0159	9.4424	0.1059	7.2232
23	7.258	0.1378	69.532	0.0144	9.5802	0.1044	7.4357
24	7.911	0.1264	76.790	0.0130	9.7066	0.1030	7.6384
25	8.623	0.1160	84.701	0.0118	9.8226	0.1018	7.8316
26	9.399	0.1064	93.324	0.0107	9.9290	0.1007	8.0156
27	10.245	0.0976	102.723	0.0097	10.0266	0.0997	8.1906
28	11.167	0.0895	112.968	0.0089	10.1161	0.0989	8.3571
29	12.172	0.0822	124.135	0.0081	10.1983	0.0981	8.5154
30	13.268	0.0754	136.308	0.0073	10.2737	0.0973	8.6657
31	14.462	0.0691	149.575	0.0067	10.3428	0.0967	8.8083
32	15.763	0.0634	164.037	0.0061	10.4062	0.0961	8.9436
33	17.182	0.0582	179.800	0.0056	10.4644	0.0956	9.0718
34	18.728	0.0534	196.982	0.0051	10.5178	0.0951	9.1933
35	20.414	0.0490	215.711	0.0046	10.5668	0.0946	9.3083
36	22.251	0.0449	236.125	0.0042	10.6118	0.0942	9.4171
37	24.254	0.0412	258.376	0.0039	10.6530	0.0939	9.5200
38	26.437	0.0378	282.630	0.0035	10.6908	0.0935	9.6172
39	28.816	0.0347	309.066	0.0032	10.7255	0.0932	9.7090
40	31.409	0.0318	337.882	0.0030	10.7574	0.0930	9.7957
45	48.327	0.0207	525.859	0.0019	10.8812	0.0919	10.1603
50	74.358	0.0134	815.084	0.0012	10.9617	0.0912	10.4295
55	114.408	0.0087	1,260.092	0.0008	11.0140	0.0908	10.6261
60	176.031	0.0057	1,944.792	0.0005	11.0480	0.0905	10.7683
65	270.846	0.0037	2,998.288	0.0003	11.0701	0.0903	10.8702
70	416.730	0.0024	4,619.223	0.0002	11.0844	0.0902	10.9427
75	641.191	0.0016	7,113.232	0.0001	11.0938	0.0901	10.9940
80	986.552	0.0010	10,950.574	0.0001	11.0998	0.0901	11.0299
85	1517.932	0.0007	16,854.800	0.0001	11.1038	0.0901	11.0551
90	2335.527	0.0004	25,939.184	0.0000	11.1064	0.0900	11.0726
95	3593.497	0.0003	39,916.635	0.0000	11.1080	0.0900	11.0847
100	5529.041	0.0002	61,422.675	0.0000	11.1091	0.0900	11.0930

10% Interest Factors for Annual Compounding

n	Single Payment		Equal Payment Series				Uniform Gradient-Series Factor
	Compound-Amount Factor	Present-Worth Factor	Compound-Amount Factor	Sinking-Fund Factor	Present-Worth Factor	Capital-Recovery Factor	
	To find *F* given *P*, *F/P, i, n*	To find *P* given *F*, *F/P, i, n*	To find *F* given *A*, *F/A, i, n*	To find *A* given *F*, *A/F, i, n*	To find *P* given *A*, *P/A, i, n*	To find *A* given *P*, *A/P, i, n*	To find *A* given *G*, *A/G, i, n*
1	1.100	0.9091	1.000	1.0000	0.9091	1.1000	0.0000
2	1.210	0.8264	2.100	0.4762	1.7355	0.5762	0.4762
3	1.331	0.7513	3.310	0.3021	2.4869	0.4021	0.9366
4	1.464	0.6830	4.641	0.2155	3.1699	0.3155	1.3812
5	1.611	0.6209	6.105	0.1638	3.7908	0.2638	1.8101
6	1.772	0.5645	7.716	0.1296	4.3553	0.2296	2.2236
7	1.949	0.5132	9.487	0.1054	4.8684	0.2054	2.6216
8	2.144	0.4665	11.436	0.0874	5.3349	0.1874	3.0045
9	2.358	0.4241	13.579	0.0736	5.7590	0.1736	3.3724
10	2.594	0.3855	15.937	0.0627	6.1446	0.1627	3.7255
11	2.853	0.3505	18.531	0.0540	6.4951	0.1540	4.0641
12	3.138	0.3186	21.384	0.0468	6.8137	0.1468	4.3884
13	3.452	0.2897	24.523	0.0408	7.1034	0.1408	4.6988
14	3.797	0.2633	27.975	0.0357	7.3667	0.1357	4.9955
15	4.177	0.2394	31.772	0.0315	7.6061	0.1315	5.2789
16	4.595	0.2176	35.950	0.0278	7.8237	0.1278	5.5493
17	5.054	0.1978	40.545	0.0247	8.0216	0.1247	5.8071
18	5.560	0.1799	45.599	0.0219	8.2014	0.1219	6.0526
19	6.116	0.1635	51.159	0.0195	8.3649	0.1195	6.2861
20	6.727	0.1486	57.275	0.0175	8.5136	0.1175	6.5081
21	7.400	0.1351	64.002	0.0156	8.6487	0.1156	6.7189
22	8.140	0.1228	71.403	0.0140	8.7715	0.1140	6.9189
23	8.954	0.1117	79.543	0.0126	8.8832	0.1126	7.1085
24	9.850	0.1015	88.497	0.0113	8.9847	0.1113	7.2881
25	10.835	0.0923	98.347	0.0102	9.0770	0.1102	7.4580
26	11.918	0.0839	109.182	0.0092	9.1609	0.1092	7.6186
27	13.110	0.0763	121.100	0.0083	9.2372	0.1083	7.7704
28	14.421	0.0693	134.210	0.0075	9.3066	0.1075	7.9137
29	15.863	0.0630	148.631	0.0067	9.3696	0.1067	8.0489
30	17.449	0.0573	164.494	0.0061	9.4269	0.1061	8.1762
31	19.194	0.0521	181.943	0.0055	9.4790	0.1055	8.2962
32	21.114	0.0474	201.138	0.0050	9.5264	0.1050	8.4091
33	23.225	0.0431	222.252	0.0045	9.5694	0.1045	8.5152
34	25.548	0.0391	245.477	0.0041	9.6086	0.1041	8.6149
35	28.102	0.0356	271.024	0.0037	9.6442	0.1037	8.7086
36	30.913	0.0323	299.127	0.0033	9.6765	0.1033	8.7965
37	34.004	0.0294	330.039	0.0030	9.7059	0.1030	8.8789
38	37.404	0.0267	364.043	0.0027	9.7327	0.1027	8.9562
39	41.145	0.0243	401.448	0.0025	9.7570	0.1025	9.0285
40	45.259	0.0221	442.593	0.0023	9.7791	0.1023	9.0962
45	72.890	0.0137	718.905	0.0014	9.8628	0.1014	9.3740
50	117.391	0.0085	1,163.909	0.0009	9.9148	0.1009	9.5704
55	189.059	0.0053	1,880.591	0.0005	9.9471	0.1005	9.7075
60	304.482	0.0033	3,034.816	0.0003	9.9672	0.1003	9.8023
65	490.371	0.0020	4,893.707	0.0002	9.9796	0.1002	9.8672
70	789.747	0.0013	7,887.470	0.0001	9.9873	0.1001	9.9113
75	1271.895	0.0008	12,708.954	0.0001	9.9921	0.1001	9.9410
80	2048.400	0.0005	20,474.002	0.0000	9.9951	0.1000	9.9609
85	3298.969	0.0003	32,979.690	0.0000	9.9970	0.1000	9.9742
90	5313.023	0.0002	53,120.226	0.0000	9.9981	0.1000	9.9831
95	8556.676	0.0001	85,556.760	0.0000	9.9988	0.1000	9.9889
100	13780.612	0.0001	137,796.123	0.0000	9.9993	0.1000	9.9927

11% Interest Factors for Annual Compounding

n	Single Payment		Equal Payment Series				Uniform Gradient-Series Factor
	Compound-Amount Factor	Present-Worth Factor	Compound-Amount Factor	Sinking-Fund Factor	Present-Worth Factor	Capital-Recovery Factor	
	To find *F* given *P*, *F/P*, *i*, *n*	To find *P* given *F*, *F/P*, *i*, *n*	To find *F* given *A*, *F/A*, *i*, *n*	To find *A* given *F*, *A/F*, *i*, *n*	To find *P* given *A*, *P/A*, *i*, *n*	To find *A* given *P*, *A/P*, *i*, *n*	To find *A* given *G*, *A/G*, *i*, *n*
1	1.110	0.9009	1.000	1.0000	0.9009	1.1100	0.0000
2	1.232	0.8116	2.110	0.4739	1.7125	0.5839	0.4739
3	1.368	0.7312	3.342	0.2992	2.4437	0.4092	0.9306
4	1.518	0.6587	4.710	0.2123	3.1024	0.3223	1.3700
5	1.685	0.5935	6.228	0.1606	3.6959	0.2706	1.7923
6	1.870	0.5346	7.913	0.1264	4.2305	0.2364	2.1976
7	2.076	0.4817	9.783	0.1022	4.7122	0.2122	2.5863
8	2.305	0.4339	11.859	0.0843	5.1461	0.1943	2.9585
9	2.558	0.3909	14.164	0.0706	5.5370	0.1806	3.3144
10	2.839	0.3522	16.722	0.0598	5.8892	0.1698	3.6544
11	3.152	0.3173	19.561	0.0511	6.2065	0.1611	3.9788
12	3.498	0.2858	22.713	0.0440	6.4924	0.1540	4.2879
13	3.883	0.2575	26.212	0.0382	6.7499	0.1482	4.5822
14	4.310	0.2320	30.095	0.0332	6.9819	0.1432	4.8619
15	4.785	0.2090	34.405	0.0291	7.1909	0.1391	5.1275
16	5.311	0.1883	39.190	0.0255	7.3792	0.1355	5.3794
17	5.895	0.1696	44.501	0.0225	7.5488	0.1325	5.6180
18	6.544	0.1528	50.396	0.0198	7.7016	0.1298	5.8439
19	7.263	0.1377	56.939	0.0176	7.8393	0.1276	6.0574
20	8.062	0.1240	64.203	0.0156	7.9633	0.1256	6.2590
21	8.949	0.1117	72.265	0.0138	8.0751	0.1238	6.4491
22	9.934	0.1007	81.214	0.0123	8.1757	0.1223	6.6283
23	11.026	0.0907	91.148	0.0110	8.2664	0.1210	6.7969
24	12.239	0.0817	102.174	0.0098	8.3481	0.1198	6.9555
25	13.585	0.0736	114.413	0.0087	8.4217	0.1187	7.1045
26	15.080	0.0663	127.999	0.0078	8.4881	0.1178	7.2443
27	16.739	0.0597	143.079	0.0070	8.5478	0.1170	7.3754
28	18.580	0.0538	159.817	0.0063	8.6016	0.1163	7.4982
29	20.624	0.0485	178.397	0.0056	8.6501	0.1156	7.6131
30	22.892	0.0437	199.021	0.0050	8.6938	0.1150	7.7206
31	25.410	0.0394	221.913	0.0045	8.7331	0.1145	7.8210
32	28.206	0.0355	247.324	0.0040	8.7686	0.1140	7.9147
33	31.308	0.0319	275.529	0.0036	8.8005	0.1136	8.0021
34	34.752	0.0288	306.837	0.0033	8.8293	0.1133	8.0836
35	38.575	0.0259	341.590	0.0029	8.8552	0.1129	8.1594
36	42.818	0.0234	380.164	0.0026	8.8786	0.1126	8.2300
37	47.528	0.0210	422.982	0.0024	8.8996	0.1124	8.2957
38	52.756	0.0190	470.511	0.0021	8.9186	0.1121	8.3567
39	58.559	0.0171	523.267	0.0019	8.9357	0.1119	8.4133
40	65.001	0.0154	581.826	0.0017	8.9511	0.1117	8.4659
45	109.530	0.0091	986.639	0.0010	9.0079	0.1110	8.6763
50	184.565	0.0054	1,668.771	0.0006	9.0417	0.1106	8.8185
55	311.002	0.0032	2,818.204	0.0004	9.0617	0.1104	8.9135
60	524.057	0.0019	4,755.066	0.0002	9.0736	0.1102	8.9762
65	883.067	0.0011	8,018.790	0.0001	9.0806	0.1101	9.0172
70	1488.019	0.0007	13,518.356	0.0001	9.0848	0.1101	9.0438
75	2507.399	0.0004	22,785.443	0.0000	9.0873	0.1100	9.0610
80	4225.113	0.0002	38,401.025	0.0000	9.0888	0.1100	9.0720
85	7119.561	0.0001	64,714.188	0.0000	9.0896	0.1100	9.0790
90	11996.874	0.0001	109,053.398	0.0000	9.0902	0.1100	9.0834
95	20215.430	0.0000	183,767.546	0.0000	9.0905	0.1100	9.0862
100	34064.175	0.0000	309,665.230	0.0000	9.0906	0.1100	9.0880

12% Interest Factors for Annual Compounding

n	Single Payment		Equal Payment Series				Uniform Gradient-Series Factor
	Compound-Amount Factor	Present-Worth Factor	Compound-Amount Factor	Sinking-Fund Factor	Present-Worth Factor	Capital-Recovery Factor	
	To find *F* given *P*, *F/P*, *i*, *n*	To find *P* given *F*, *F/P*, *i*, *n*	To find *F* given *A*, *F/A*, *i*, *n*	To find *A* given *F*, *A/F*, *i*, *n*	To find *P* given *A*, *P/A*, *i*, *n*	To find *A* given *P*, *A/P*, *i*, *n*	To find *A* given *G*, *A/G*, *i*, *n*
1	1.120	0.8929	1.000	1.0000	0.8929	1.1200	0.0000
2	1.254	0.7972	2.120	0.4717	1.6901	0.5917	0.4717
3	1.405	0.7118	3.374	0.2963	2.4018	0.4163	0.9246
4	1.574	0.6355	4.779	0.2092	3.0373	0.3292	1.3589
5	1.762	0.5674	6.353	0.1574	3.6048	0.2774	1.7746
6	1.974	0.5066	8.115	0.1232	4.1114	0.2432	2.1720
7	2.211	0.4523	10.089	0.0991	4.5638	0.2191	2.5515
8	2.476	0.4039	12.300	0.0813	4.9676	0.2013	2.9131
9	2.773	0.3606	14.776	0.0677	5.3282	0.1877	3.2574
10	3.106	0.3220	17.549	0.0570	5.6502	0.1770	3.5847
11	3.479	0.2875	20.655	0.0484	5.9377	0.1684	3.8953
12	3.896	0.2567	24.133	0.0414	6.1944	0.1614	4.1897
13	4.363	0.2292	28.029	0.0357	6.4235	0.1557	4.4683
14	4.887	0.2046	32.393	0.0309	6.6282	0.1509	4.7317
15	5.474	0.1827	37.280	0.0268	6.8109	0.1468	4.9803
16	6.130	0.1631	42.753	0.0234	6.9740	0.1434	5.2147
17	6.866	0.1456	48.884	0.0205	7.1196	0.1405	5.4353
18	7.690	0.1300	55.750	0.0179	7.2497	0.1379	5.6427
19	8.613	0.1161	63.440	0.0158	7.3658	0.1358	5.8375
20	9.646	0.1037	72.052	0.0139	7.4694	0.1339	6.0202
21	10.804	0.0926	81.699	0.0122	7.5620	0.1322	6.1913
22	12.100	0.0826	92.503	0.0108	7.6446	0.1308	6.3514
23	13.552	0.0738	104.603	0.0096	7.7184	0.1296	6.5010
24	15.179	0.0659	118.155	0.0085	7.7843	0.1285	6.6406
25	17.000	0.0588	133.334	0.0075	7.8431	0.1275	6.7708
26	19.040	0.0525	150.334	0.0067	7.8957	0.1267	6.8921
27	21.325	0.0469	169.374	0.0059	7.9426	0.1259	7.0049
28	23.884	0.0419	190.699	0.0052	7.9844	0.1252	7.1098
29	26.750	0.0374	214.583	0.0047	8.0218	0.1247	7.2071
30	29.960	0.0334	241.333	0.0041	8.0552	0.1241	7.2974
31	33.555	0.0298	271.293	0.0037	8.0850	0.1237	7.3811
32	37.582	0.0266	304.848	0.0033	8.1116	0.1233	7.4586
33	42.092	0.0238	342.429	0.0029	8.1354	0.1229	7.5302
34	47.143	0.0212	384.521	0.0026	8.1566	0.1226	7.5965
35	52.800	0.0189	431.663	0.0023	8.1755	0.1223	7.6577
36	59.136	0.0169	484.463	0.0021	8.1924	0.1221	7.7141
37	66.232	0.0151	543.599	0.0018	8.2075	0.1218	7.7661
38	74.180	0.0135	609.831	0.0016	8.2210	0.1216	7.8141
39	83.081	0.0120	684.010	0.0015	8.2330	0.1215	7.8582
40	93.051	0.0107	767.091	0.0013	8.2438	0.1213	7.8988
45	163.988	0.0061	1,358.230	0.0007	8.2825	0.1207	8.0572
50	289.002	0.0035	2,400.018	0.0004	8.3045	0.1204	8.1597
55	509.321	0.0020	4,236.005	0.0002	8.3170	0.1202	8.2251
60	897.597	0.0011	7,471.641	0.0001	8.3240	0.1201	8.2664
65	1581.872	0.0006	13,173.937	0.0001	8.3281	0.1201	8.2922
70	2787.800	0.0004	23,223.332	0.0000	8.3303	0.1200	8.3082
75	4913.056	0.0002	40,933.799	0.0000	8.3316	0.1200	8.3181
80	8658.483	0.0001	72,145.693	0.0000	8.3324	0.1200	8.3241
85	15259.206	0.0001	127,151.714	0.0000	8.3328	0.1200	8.3278
90	26891.934	0.0000	224,091.119	0.0000	8.3330	0.1200	8.3300
95	47392.777	0.0000	394,931.472	0.0000	8.3332	0.1200	8.3313
100	83522.266	0.0000	696,010.548	0.0000	8.3332	0.1200	8.3321

This page intentionally left blank

Appendix E
Normal Distribution Tables

This page intentionally left blank

Values of the Standard Normal Distribution Function

$$\text{Where } \Phi(z) = \int_{-\infty}^{z} \frac{1}{\sqrt{2\Pi}} e^{-u^2/2} (du = P(Z \leq z))$$

Each value represents the proportion of the total area under the normal curve that is under the portion of the curve from $-\infty$ to z, where:

$$z = \frac{X_i - \bar{X}}{\sigma'}$$

and X_i represents any desired value of the variable *X*.

Table E-1. Standard Normal Distribution Table

z	0	1	2	3	4	5	6	7	8	9
-3.0	0.0013	0.0010	0.0007	0.0005	0.0003	0.0002	0.0002	0.0001	0.0001	0.0000
-2.9	0.0019	0.0018	0.0018	0.0017	0.0016	0.0016	0.0015	0.0015	0.0014	0.0014
-2.8	0.0026	0.0025	0.0024	0.0023	0.0023	0.0022	0.0021	0.0021	0.0020	0.0019
-2.7	0.0035	0.0034	0.0033	0.0032	0.0031	0.0030	0.0029	0.0028	0.0027	0.0026
-2.6	0.0047	0.0045	0.0044	0.0043	0.0041	0.0040	0.0039	0.0038	0.0037	0.0036
-2.5	0.0062	0.0060	0.0059	0.0057	0.0055	0.0054	0.0052	0.0051	0.0049	0.0048
-2.4	0.0082	0.0080	0.0078	0.0075	0.0073	0.0071	0.0069	0.0068	0.0066	0.0064
-2.3	0.0107	0.0104	0.0102	0.0099	0.0096	0.0094	0.0091	0.0089	0.0087	0.0084
-2.2	0.0139	0.0136	0.0132	0.0129	0.0125	0.0122	0.0119	0.0116	0.0113	0.0110
-2.1	0.0179	0.0174	0.0170	0.0166	0.0162	0.0158	0.0154	0.0150	0.0146	0.0143
-2.0	0.0228	0.0222	0.0217	0.0212	0.0207	0.0202	0.0197	0.0192	0.0188	0.0183
-1.9	0.0287	0.0281	0.0274	0.0268	0.0262	0.0256	0.0250	0.0244	0.0239	0.0233
-1.8	0.0359	0.0351	0.0344	0.0336	0.0329	0.0322	0.0314	0.0307	0.0301	0.0294
-1.7	0.0446	0.0436	0.0427	0.0418	0.0409	0.0401	0.0392	0.0384	0.0375	0.0367
-1.6	0.0548	0.0537	0.0526	0.0516	0.0505	0.0495	0.0485	0.0475	0.0465	0.0455
-1.5	0.0668	0.0655	0.0643	0.0630	0.0618	0.0606	0.0594	0.0582	0.0571	0.0559
-1.4	0.0808	0.0793	0.0778	0.0764	0.0749	0.0735	0.0721	0.0708	0.0694	0.0681
-1.3	0.0968	0.0951	0.0934	0.0918	0.0901	0.0885	0.0869	0.0853	0.0838	0.0823
-1.2	0.1151	0.1131	0.1112	0.1093	0.1075	0.1056	0.1038	0.1020	0.1003	0.0985
-1.1	0.1357	0.1335	0.1314	0.1292	0.1271	0.1251	0.1230	0.1210	0.1190	0.1170
-1.0	0.1587	0.1562	0.1539	0.1515	0.1492	0.1469	0.1446	0.1423	0.1401	0.1379
-0.9	0.1841	0.1814	0.1788	0.1762	0.1736	0.1711	0.1685	0.1660	0.1635	0.1611
-0.8	0.2119	0.2090	0.2061	0.2033	0.2005	0.1977	0.1949	0.1922	0.1894	0.1867
-0.7	0.2420	0.2389	0.2358	0.2327	0.2296	0.2266	0.2236	0.2206	0.2177	0.2148
-0.6	0.2743	0.2709	0.2676	0.2643	0.2611	0.2578	0.2546	0.2514	0.2483	0.2451
-0.5	0.3085	0.3050	0.3015	0.2981	0.2946	0.2912	0.2877	0.2843	0.2810	0.2776
-0.4	0.3446	0.3409	0.3372	0.3336	0.3300	0.3264	0.3228	0.3192	0.3156	0.3121
-0.3	0.3821	0.3783	0.3745	0.3707	0.3669	0.3632	0.3594	0.3557	0.3520	0.3483
-0.2	0.4207	0.4168	0.4129	0.4090	0.4052	0.4013	0.3974	0.3936	0.3897	0.3859
-0.1	0.4602	0.4562	0.4522	0.4483	0.4443	0.4404	0.4364	0.4325	0.4286	0.4247
0.0	0.5000	0.4960	0.4920	0.4880	0.4840	0.4801	0.4761	0.4721	0.4681	0.4641

Table E-1. Standard Normal Distribution Table (Cont'd)

z	0	1	2	3	4	5	6	7	8	9
0.0	0.5000	0.5040	0.5080	0.5120	0.5160	0.5199	0.5239	0.5279	0.5319	0.5359
0.1	0.5398	0.5438	0.5478	0.5517	0.5557	0.5596	0.5636	0.5675	0.5714	0.5753
0.2	0.5793	0.5832	0.5871	0.5910	0.5948	0.5987	0.6026	0.6064	0.6103	0.6141
0.3	0.6179	0.6217	0.6255	0.6293	0.6331	0.6368	0.6406	0.6443	0.6480	0.6517
0.4	0.6554	0.6591	0.6628	0.6664	0.6700	0.6736	0.6772	0.6808	0.6844	0.6879
0.5	0.6915	0.6950	0.6985	0.7019	0.7054	0.7088	0.7123	0.7157	0.7190	0.7224
0.6	0.7257	0.7291	0.7324	0.7357	0.7389	0.7422	0.7454	0.7486	0.7517	0.7549
0.7	0.7580	0.7611	0.7642	0.7673	0.7704	0.7734	0.7764	0.7794	0.7823	0.7852
0.8	0.7881	0.7910	0.7939	0.7967	0.7995	0.8023	0.8051	0.8078	0.8106	0.8133
0.9	0.8159	0.8186	0.8212	0.8238	0.8264	0.8289	0.8315	0.8340	0.8365	0.8389
1.0	0.8413	0.8438	0.8461	0.8485	0.8508	0.8531	0.8554	0.8577	0.8599	0.8621
1.1	0.8643	0.8665	0.8686	0.8708	0.8729	0.8749	0.8770	0.8790	0.8810	0.8830
1.2	0.8849	0.8869	0.8888	0.8907	0.8925	0.8944	0.8962	0.8980	0.8997	0.9015
1.3	0.9032	0.9049	0.9066	0.9082	0.9099	0.9115	0.9131	0.9147	0.9162	0.9177
1.4	0.9192	0.9207	0.9222	0.9236	0.9251	0.9265	0.9279	0.9292	0.9306	0.9319
1.5	0.9332	0.9345	0.9357	0.9370	0.9382	0.9394	0.9406	0.9418	0.9429	0.9441
1.6	0.9452	0.9463	0.9474	0.9484	0.9495	0.9505	0.9515	0.9525	0.9535	0.9545
1.7	0.9554	0.9564	0.9573	0.9582	0.9591	0.9599	0.9608	0.9616	0.9625	0.9633
1.8	0.9641	0.9649	0.9656	0.9664	0.9671	0.9678	0.9686	0.9693	0.9699	0.9706
1.9	0.9713	0.9719	0.9726	0.9732	0.9738	0.9744	0.9750	0.9756	0.9761	0.9767
2.0	0.9772	0.9778	0.9783	0.9788	0.9793	0.9798	0.9803	0.9808	0.9812	0.9817
2.1	0.9821	0.9826	0.9830	0.9834	0.9838	0.9842	0.9846	0.9850	0.9854	0.9857
2.2	0.9861	0.9864	0.9868	0.9871	0.9875	0.9878	0.9881	0.9884	0.9887	0.9890
2.3	0.9893	0.9896	0.9898	0.9901	0.9904	0.9906	0.9909	0.9911	0.9913	0.9916
2.4	0.9918	0.9920	0.9922	0.9925	0.9927	0.9929	0.9931	0.9932	0.9934	0.9936
2.5	0.9938	0.9940	0.9941	0.9943	0.9945	0.9946	0.9948	0.9949	0.9951	0.9952
2.6	0.9953	0.9955	0.9956	0.9957	0.9959	0.9960	0.9961	0.9962	0.9963	0.9964
2.7	0.9965	0.9966	0.9967	0.9968	0.9969	0.9970	0.9971	0.9972	0.9973	0.9974
2.8	0.9974	0.9975	0.9976	0.9977	0.9977	0.9978	0.9979	0.9979	0.9980	0.9981
2.9	0.9981	0.9982	0.9982	0.9983	0.9984	0.9984	0.9985	0.9985	0.9986	0.9986
3.0	0.9987	0.9990	0.9993	0.9995	0.9997	0.9998	0.9998	0.9999	0.9999	1.0000

Appendix F
Acronyms and Abbreviations

This page intentionally left blank

Acronyms and Abbreviations - The following list includes terms used in this Toolkit and additional terms related to acquisition, reliability, and maintainability.

A_a	Achieved Availability
ADT	Administrative Delay Time
A_i	Inherent Availability
A_o	Operational Availability
BIT	Built-In Test
CAD	Computer-Aided Design
CAM	Computer-Aided Manufacturing
CAS	Computer-Aided Support
CBS	Cost Breakdown Structure
CDR	Conceptual Design Review
CE	Cost Effectiveness
CER	Cost Estimating Relationship
CHI	Comfort Health Index
CI	Configuration Item
CLM	Council of Logistics Management
CLS	Contractor Logistics Support
CM	Configuration Management
CMMI	Capability Maturity Model Integration
COTS	Commercial Off-The Shelf
CPPI	Continuous Product/Process Improvement
CSA	Configuration Status Accounting
CSCI	Computer Software Configuration Item
CSCMP	Council of Supply Chain Management Professionals
CSU	Computer Software Unit
DAU	Defense Acquisition University
DOD	Department of Defense
DRR	Design Readiness Review
DSMC	Defense Systems Management College
DT&E	Development Test And Evaluation
EC	Electronic Commerce
ECO	Engineering Change Order
ECP	Engineering Change Proposal
EDI	Electronic Data Interchange
EOQ	Economic Order Quantity
FFBD	Functional Flow Block Diagram
FMECA	Failure Mode, Effects And Criticality Analysis
FOM	Figure of Merit
f_{pt}	Preventive Maintenance Frequency
FTA	Fault Tree Analysis
GPS	Global Positioning System
HF	Human Factors
HFE	Human Factors Engineering
HOQ	House of Quality
IFB	Invitation For Bid
ILS	Integrated Logistic Support

ILSP	Integrated Logistic Support Plan
INCOSE	International Council on Systems Engineering
LCC	Life Cycle Cost
LCCA	Life Cycle Cost Analysis
LDT	Logistics Delay Time
LE	Logistics Effectiveness
LMI	logistics management information
LMI	Logistics Management Institute
LORA	Level-Of-Repair Analysis
LSA	Logistic Support Analysis
M	Maintainability
$\overline{M}$	Mean Active Maintenance Time
MA	Maintenance Action
MADS	Maintenance Analysis Data System
MAIS	Major Automation Information System
$\overline{M}_{ct}$	Mean Corrective Maintenance Time
$\tilde{M}_{ct}$	Median Corrective Maintenance Time
MDT	Maintenance Downtime
MDAPS	Mandatory Procedures For Major Defense Acquisition Programs
MAM	Maintenance Assist Module
MEA	Maintenance Engineering Analysis
MEAR	Maintenance Engineering Analysis Record
MI	Maintainability Improvement
MLA	Maintenance Level Analysis
MLH	Maintenance Labor-Hours
MLH/MA	Maintenance Labor-Hours per Maintenance Action
MLH/OH	Maintenance Labor-Hours per Operating Hour
M_{max}	Maximum Maintenance Time
MOE	Measure of Effectiveness
MP	Maintenance Prevention
$\overline{M}_{pt}$	Mean Preventive Maintenance Time
MTA	Maintenance Task Analysis
MTBF	Mean Time Between Failure
MTBM	Mean Time Between Maintenance
MTBR	Mean Time Between Replacement
MTTF	Mean Time To Failure
MTTR	Mean Time To Repair
OEE	Overall Equipment Effectiveness
O&M	Operation And Maintenance
OEM	Original Equipment Manufacturer
OSD	Office of the Secretary of Defense
OSD	Operational Sequence Diagram
OTA	Operator Task Analysis
OT&E	Operational Test And Evaluation
PBL	Performance Based Logistics
PDF	Probability Density Function
PDR	Preliminary Design Review

PH&T	Packaging, Handling And Transportation
PM	Preventive Maintenance
PV	Present Value
QFD	Quality Function Deployment
R	Reliability
RAC	Reliability Analysis Center
RCM	Reliability-Centered Maintenance
RFID	Radio Frequency Identification
RFP	Request For Proposal
RPN	Risk Priority Number
SA	Supportability Analysis
SAP	Supportability Analysis Plan
SC	Supply Chain
SCM	Supply Chain Management
SE	System Effectiveness
SECAM	System Engineering Capability Assessment Model
SE-CMM	Systems Engineering Capability Maturity Model
SEMP	System Engineering Management Plan
SHA	Safety Hazard Analysis
SOE	System Operational Effectiveness
SOW	Statement Of Work
SPC	Statistical Process Control
SRR	System Requirements Review
TAAF	Test, Analyze, And Fix
TAT	Turnaround Time
TD&I	Technical Data And Information
TOC	Total Ownership Cost
TPM	Technical Performance Measure
TPM	Total Productive Maintenance
TQM	Total Quality Management
VOC	Voice Of The Customer

This page intentionally left blank

Index

Index (Cont'd)

Index (Cont'd)

Index (Cont'd)

Index (Cont'd)

Index (Cont'd)